QUANTUM MECHANICS WITH APPLICATIONS

by

Iraj R. Afnan

School of Chemical and Physical Sciences,
Flinders University

To Trudy

Contents

Foreword

This eBook contains the lecture notes that Iraj Afnan produced for students who were taking his two-semester course on quantum mechanics. In 1990 I had the privilege of taking the class that covered the second half of this material. Those lectures were given to Flinders University Physics students who were in the last year of their undergraduate training. It is a testament to the sophistication of Iraj's presentation that the content of that course covers the curriculum of many first-year graduate classes on quantum mechanics at US institutions.

As Iraj points out in his preface, the way he taught the subject emphasized on computation of the observable consequences of the theory over derivations, theorems and philosophical issues. This was of great benefit to the students who took his course, since we learnt how quantum calculations are really *done*. These notes show how to start from first principles and use the formalism of quantum mechanics to elucidate the consequences of the theory for various systems. The symmetries of nature also play a prominent role in Iraj's treatment, and respect for those symmetries was something that he inculcated in all his students. The powerful way in which the quantum theory describes the subatomic world is thus brought home to the reader throughout the text, especially as Iraj draws his examples and problems from many subfields of physics: condensed-matter, atomic, molecular, and nuclear.

But, Iraj is, as usual, too modest about his own achievements. Although his writing does not lay excessive emphasis on the formal structure of the theory, his deep understanding of quantum mechanics means that its beauty still shines through. Iraj always said "Ugly physics is bad physics", and it was a maxim he lived out in his carefully crafted, finely honed, fifty-minute expositions of a topic. Indeed, Iraj would amaze his class by barely glancing at these notes (which he had given us typed copies of at the start of the semester), while presenting an elegant derivation that stretched over the entire lecture period. He would carefully cover several blackboards with systematic calculation in his very neat handwriting, without having to erase anything or correct himself. Iraj was able to do this because he knows quantum theory inside and out. That knowledge, gained as a practitioner and outstanding researcher, makes his insights into the subject very valuable.

I did my Ph.D. under Iraj's supervision between 1992 and 1995. When I got stuck in my research I would often turn to these notes from his quantum-mechanics course and so force myself back to basics. A few years later, one of the research papers I wrote while a post-doc had its genesis in the derivation of the effective-range expansion presented in Chapter 14. Indeed, my research on effective field theories for nuclear physics has been significantly shaped by the excellent presentation of scattering theory and separable potentials that is provided here. I have found that if my own Ph.D. students can master the material presented in this eBook then they have a thorough grounding which they can use to read many key research papers on the theory of few-body quantum systems.

I am delighted that the new medium of the eBook is allowing Iraj to share his careful and clear presentation, as well as his unique insights into the subject, with a broader audience. I am sure that readers of this text will benefit greatly from his wonderful exposition of quantum mechanics.

Daniel Phillips
Athens, Ohio
August 2010

Preface

The present set of lecture notes are a byproduct of lectures given by the author to undergraduate students at Flinders University over the period 1970 to 2004. The material in the lecture notes (eBook) was to cover a two semester course. The first semester, at second year level, consisted of two lectures plus one tutorial per week for a period of thirteen weeks, while the second semester, at third year level, was given at the rate of three lectures plus one tutorial per week. The aim of the lectures was to bring the student to a level that would allow them to read the literature in atomic, molecular and nuclear physics and be able to perform some simple calculations based on non-relativistic quantum mechanics for one- and two-body systems.

With the limited knowledge of calculus, the first semester concentrates on systems in one dimension with extensive applications to atomics, molecules, solids and nuclei. In Chapter 2 we review classical mechanics and the role of space symmetry in conservation laws. Although this chapter is not a prerequisite to later chapters, it gives the student the insight into the relation between classical and quantum physics. The emphasis in the following chapters is to illustrate the mathematical structure of quantum mechanics with examples rather than mathematical proofs. The first semester brings the students to the level of writing the Schrödinger equations for central potentials, *e.g.* the Coulomb potential.

The second semester of the course commences by considering the solution of the Schrödinger equation as a second order differential equation for simple potentials that admit analytic solutions. This allows us to study the properties of many of the special functions encountered in problems in atomic and nuclear physics. These functions are then used in the application of quantum mechanics to problems in atomic and nuclear physics. The Dirac notation is introduced more as a shorthand notation than as a treatise on Hilbert spaces to understand the importance of the different representations of the Schrödinger equation. Since symmetry plays a fundamental role in the simplification of problems encountered in quantum physics, a chapter is devoted to space symmetry and the corresponding conservation laws with emphasis on rotational symmetry. As most problems in physics require the introduction of approximation methods, a chapter is devoted to time independent perturbation and variational methods. The last chapter concentrates on scattering theory based on the Lippmann-Schwinger equation. Here, the solution of the Lippmann-Schwinger equation is illustrated by considering a rank one separable potential for which one can derive an analytic solution for the scattering amplitude bound state and scattering wave function.

CHAPTER 1

Introduction

Abstract: In this chapter we briefly review the failure of classical physics and the experiments and ideas that led to the formulation of quantum mechanics in 1925.

The first quarter of the 20th century saw the introduction of two new developments in physics that have revolutionized our understanding of the subject. These two developments are:

1. Relativity
2. Quantum Mechanics

Although the first of these two developments has had minor influence on our daily activity until recently, the second has dominated much of the technological development of the 20th century. It is quantum mechanics that has set the framework for our theoretical understanding of the basic properties of atoms, molecules and solids. Having this theoretical understanding, it is hoped that we can predict the nature of chemical reactions, the behavior of semiconductors, and the properties of superconductors at low and high temperatures. Thus quantum mechanics forms the cornerstone of both the chemical and electronics industry.

However, the influence of this theory goes beyond science and technology. In fact its influence extends to all aspects of human thinking from Art to Philosophy. When Heisenberg, in 1927 [1], came out with the idea of quantum theory based on the uncertainty principle, he found that his greatest opposition came from the German philosophers of the time. This was basically because the uncertainty principle violated not only the micro-causality of Newton and Lagrange, but also the basic philosophical ideas of Kant [2].

In the present eBook, we want to concentrate on the scientific influence of quantum mechanics, and in particular the tools it gives us to understand nature at the microscopic level. In fact, over the next chapters in this eBook, we will develop the ideas and the

Iraj R. Afnan

mathematical tools required to understand the properties of atoms, molecules, nuclei, solids and finally the subatomic world of particle physics. We will see how the symmetries of nature play a vital role in unfolding the underlying dynamics of each of the systems we will consider.

However, before we set out on our exploration into the world of quanta, I would like to give you a glimpse of the experiments and the models that led to this most exciting theory. To appreciate its novel ideas, we should go back and very briefly summarize the theories that existed some one hundred years ago.

1.1 Classical Physics

At the turn of the 20th century, most physical phenomena were considered to be governed by one of two theories;

1. Mechanics as originally formulated by Newton and developed by Euler, Lagrange, Hamilton and others.

2. Electrodynamics which is a unification of electricity and magnetism as observed by Faraday and formulated by Maxwell.

1.1.1 Newtonian Mechanics

In any introductory course in physics the student is introduced to the famous equation $\vec{F} = m\vec{a}$. This in fact is a statement of Newton's second law and should be written, in the case of a single particle, as

$$\frac{d\vec{p}}{dt} = \vec{F} \ , \tag{1.1}$$

where $\vec{p}$ is the momentum of the particle, $\vec{F}$ is the force on that particle and t stands for time. Thus the motion of the particle is governed by Eq. (1.1), which states that the rate of change in momentum of a body is equal to the force exerted on it. A more practical form of this equation is

$$m\frac{d^2\vec{r}}{dt^2} = \vec{F} \ , \tag{1.2}$$

where we have taken the momentum $\vec{p}$ to be given in terms of the velocity $\vec{v}$ or the position of the particle $\vec{r}$ by the relation

$$\vec{p} = m\vec{v} = m\frac{d\vec{r}}{dt} \ . \tag{1.3}$$

In Eq. (1.2), we have a second order differential equation which for certain simple forces can be integrated to give the position of the particle as a function of time. In general, the integrated equation has two arbitrary constants, which as we will see, are determined by the initial condition, e.g., the position and velocity of the particle at some initial time t_0.

Example: For a particle with mass m in a gravitational force $\vec{F} = -mg\hat{y}$, the motion in the vertical direction (i.e along the y-axis) is governed by the second order differential equation

$$m\frac{d^2y}{dt^2} = -mg \; .$$

Upon integration, this equation gives the familiar result

$$y = y_0 + v_0 t - \frac{1}{2}gt^2$$

where y_0 and v_0 are the position and velocity of the particle at some initial time $t_0 = 0$. Here we observe that given y_0 and v_0, we can determine the position of the particle at any time $t \geq t_0$.

This result is in general true and its implications are enormous. The result basically implies that if we know the position and velocity of all the particles in the universe at a given time, we could predict the behavior of the universe from that instant of time onward, i.e., we can tell what the future has in store for us. In practice that is not quite true, for we can never state the initial condition with sufficient precision to uniquely determine the behavior of the system for all times in the future.[1]

1.1.2 Maxwell's Electrodynamics

In a brilliant first unification, Maxwell [4], with the help of the experimental results of Faraday, was able to write four equations for the electric and magnetic field of any distribution of charges and currents. These equations, known as Maxwell's equations, were not only able to describe electricity and magnetism, but also predicted that an oscillating electric and magnetic field will generate radiation, presently known as electromagnetic radiation, light being a special example. Hertz in 1887 put Maxwell's prediction to the test by observing the radiation from an oscillating electric and magnetic field, and showing that this radiation has the same behavior as light. The electric and magnetic fields for this radiation are governed by the equations

$$\left(\frac{1}{c^2}\frac{\partial^2}{\partial t^2} - \nabla^2\right)\vec{E}(\vec{r},t) = 0 \qquad (1.4)$$

$$\left(\frac{1}{c^2}\frac{\partial^2}{\partial t^2} - \nabla^2\right)\vec{B}(\vec{r},t) = 0 \; . \qquad (1.5)$$

These equations are a special form of Maxwell's equation known as the wave equations. With this result it was finally established that light is a wave as Huygens predicted, and not a particle as Newton thought.

[1]This phenomena is often referred to as *The Butterfly Effect*. For a popular description of this and other related phenomena see [3].

1.2 The Particle Nature of Radiation

At the turn of the 20th century, many physicists thought it was merely a question of time before the minor problems still unresolved would be overcome, to render physics to the status of a dead subject. One of these minor problems was the spectrum of blackbody radiation.

1.2.1 Blackbody radiation

A blackbody is a closed box for which there is thermodynamic equilibrium between the radiation in the cavity and the walls of the box. If we now make a hole in the wall, with the hole being small enough not to disturb this equilibrium, then the spectrum of the radiation emitted through the hole is known as *blackbody radiation.* The name 'blackbody' is a result of the fact that the radiation from the box is emitted and not reflected radiation, a property common with any black surface that does not reflect light. The spectrum of this radiation is shown in Figure 1.1 on the left, where we have the energy density de in the frequency range between ω and $\omega+d\omega$ plotted versus the frequency ω. The solid curve in Figure 1.1 on the right represents the experimental result, while the dashed curve was the best that the classical theory of Rayleigh-Jeans could do in predicting the spectrum of the radiation to be

$$de = \frac{kT\omega^2}{\pi^2 c^3}\, d\omega \ , \tag{1.6}$$

where T is the temperature, $k = 8.617 \times 10^5$ eV K^{-1} is Boltzmann's constant, c is the velocity of light, and ω is the frequency of the radiation emitted by the blackbody.

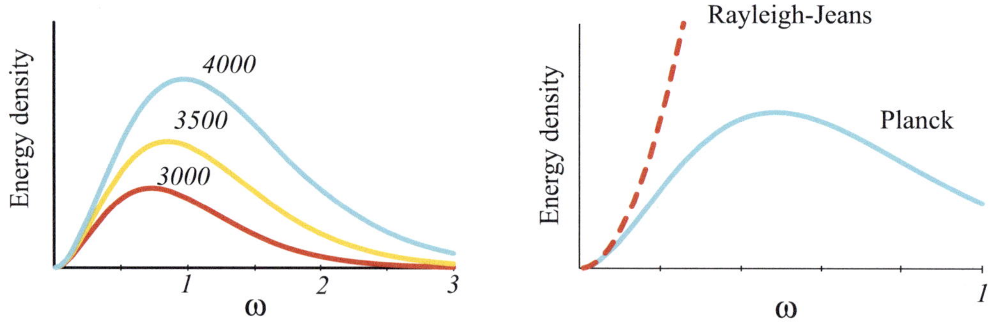

Figure 1.1: The Planck distribution for the energy density. The figure on the left gives the distribution for the energy at a temperature of 3000, 3500, and 4000°K, while the figure on the right is a comparison of the Planck distribution and the Rayleigh-Jeans result as defined in Eq. 1.6.

In this classical theory it was assumed that the energy in the cavity was the electromagnetic waves of Maxwell, in equilibrium with the walls. For this classical result, the total energy, $\int de$, was infinite, and therefore unacceptable. To overcome this discrepancy between theory and experiment, Max Planck, in 1901 [5], suggested that the energy of the radiation emitted or absorbed by the walls of the cavity is proportional to its frequency i.e.,

$$E = h\nu = \hbar w \qquad \hbar = \frac{h}{2\pi} , \tag{1.7}$$

where h is Planck's constant, $\hbar = 0.6582 \times 10^{-15}$ eV sec.[2] Planck then derived the energy density to be

$$de = \frac{\hbar}{\pi^2 c^3} \frac{\omega^3}{e^{\hbar\omega/kT} - 1} d\omega , \tag{1.8}$$

which gives a perfect fit to the experimental data, and is known as the *Planck's distribution law.*

This quantization of the energy that is emitted and/or absorbed by the walls of the cavity, removed the discrepancy between theory and experiment at the *cost* of introducing, for the first time, the concept that energy is transferred only in discreet quantities ($\Delta E = \hbar\omega$). Because of the small value of $\hbar$, the amount of energy transferred is too small to be observed in our day to day activities, but is important at the microscopic level of atoms and molecules.

1.2.2 Photoelectric effect

A second minor problem with classical physics involved the properties of the electrons emitted when radiation was incident on a metallic surface, see Figure 1.2. Here the electromagnetic radiation incident on a metallic surface results in the emission of electrons. In particular, it was observed that:

1. The energy of the individual electrons emitted was independent of the intensity of the radiation, but was linearly proportional to the frequency of the radiation, i.e.,

   $$E_e \propto \hbar\omega ,$$

   where E_e is the energy of the electron and ω the frequency of the radiation.

2. The number of electrons emitted was independent of the frequency of the radiation ω, provided the frequency was greater than some critical frequency. However, the number of electrons emitted was dependent on the intensity of the radiation, i.e.,

   $$N_e \propto N_\gamma ,$$

[2] In units of energy-length, we have

$$\hbar c = 197.327 \text{ MeV-fm} = 197.327 \text{ eV-nm} .$$

where N_e is the number of electrons emitted while N_γ is the intensity of the radiation or the number of photons.

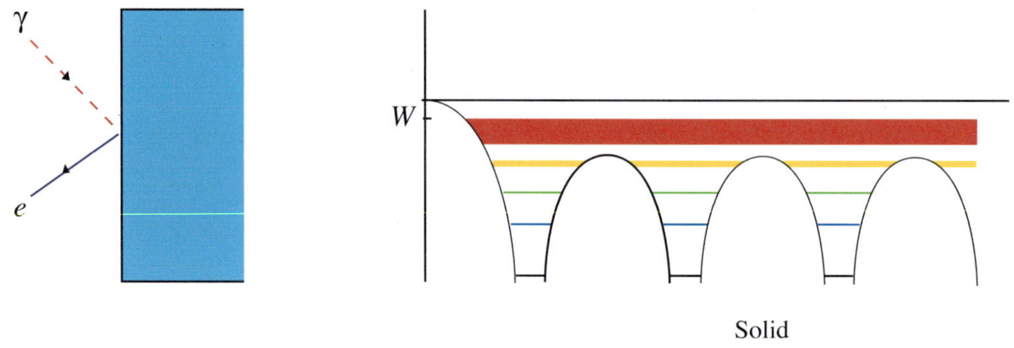

Figure 1.2: The photoelectric effect, schematic diagram on the right, while on the left we illustrate the atomic potential in a solid that gives rise to the work function W.

In 1905, Einstein [6] took Planck's postulate one step further by assuming that the electrons emitted in the photoelectric effect are the result of the fact that the radiation of frequency ω, consisted of photons of energy $E = \hbar\omega$, and that the electron absorbs the full energy of the photon. In other words, the energy of the electron E_e is given by

$$E_e = \frac{1}{2} m_e v^2 = \hbar\omega - W \ , \tag{1.9}$$

where v is the velocity of the electron emitted. In Eq. (1.9), W is the work function, and depends on the type of surface being radiated. W was introduced in order to explain the presence of a critical frequency below which no electrons are emitted from the surface.

Here again, we have an experimental observation whose explanation relies on the fact that radiation, which is electromagnetic in nature, comes in quanta with fixed energy $E = \hbar\omega$, i.e., electromagnetic radiation consists of bundles of energy like particles, and this energy in total is transferred to a single electron. This result was in contradiction with Maxwell's equations which was, and is still, considered one of the major achievements of nineteenth century science.

1.2.3 Compton scattering

Some twenty years later Compton [7] was scattering radiation from the electrons in an atom. Here, we find that the energy of the scattered radiation as a function of the scattering angle, can only be explained if we take the collision between the radiation and the electron to satisfy energy and momentum conservation. For the radiation, Compton took Planck's postulate, i.e., $E = \hbar\omega$, while for the momentum of the radiation he postulated

$$p = \frac{\hbar\omega}{c} = \frac{h}{\lambda} \ , \tag{1.10}$$

where λ is the wave length of this radiation. This postulate is equivalent to assuming that the radiation can be represented by particles called photons. These photons travel with the velocity of light and according to relativity, are therefore massless. Their energy can then be written as [3]

$$\begin{aligned} E &= \sqrt{p^2c^2 + m^2c^4} \\ &= pc \qquad \text{for} \qquad m = 0 \ . \end{aligned} \tag{1.11}$$

which is consistent with the definition of the momentum as given in Eq. (1.10). The kinematics for this reaction can be treated as if the photons in the initial and final state are a particle with energy $\hbar\omega$ and momentum $\hbar\omega/c$, i.e., the conservation of energy and momentum are given by

$$\begin{aligned} \hbar\omega + mc^2 &= \hbar\omega' + \sqrt{p^2c^2 + m^2c^4} \\ \frac{\hbar\omega}{c} &= \frac{\hbar\omega'}{c} \cos\theta + p \cos\phi \\ 0 &= \frac{\hbar\omega'}{c} \sin\theta - p \sin\phi \ . \end{aligned} \tag{1.12}$$

Thus Compton scattering established the fact that radiation can transfer momentum to an electron just like a particle does.

The above three experiments established the fact that electromagnetic radiation in some instances behaves like massless particles with energy $E = \hbar\omega = pc$, while in other instances it is known to behaves like a wave. This result, which violates classical physics i.e., Maxwell's equations, is an illustration of *particle-wave duality* of radiation.

1.3 The Wave Nature of Particles

Having established that electromagnetic radiation satisfies particle-wave duality, we can ask the question: Can particles that are governed by Newton's Laws sometime behave like waves? i.e., can particles satisfy the same duality that radiation seems to have? In 1923, this hypothesis was made by the young French nobleman Louis de Broglie. In fact, in a Ph.D. thesis submitted in 1924, de Broglie worked out the consequence of his hypothesis. The basis of this hypothesis [8] is that a particle with momentum p has a wave length λ given by

$$\lambda = \frac{h}{p} \quad \text{or} \quad p = \frac{h}{\lambda} = \frac{h\nu}{c} = \frac{\hbar\omega}{c} \ . \tag{1.13}$$

Note, this is identical to the relation between the momentum and the wave length of the radiation as given in Eq. (1.10). The main consequence of this hypothesis is that a beam of particles incident on a diffraction grating will produce an interference pattern.

[3]Here, we take the relativistic relation between the energy E and the momentum p.

The problem is that the wave length λ is very short and thus to observe the interference pattern requires a very fine grating.

While de Broglie was working out the consequences of his postulate in France, on the other side of the Atlantic Ocean, two American scientists, Davisson and Germer, while carrying out an experiment on electron scattering from a metal target, discovered the diffraction pattern resulting from the scattering of electrons from a crystal [9].

We thus have established that matter, e.g., electrons and neutrons can behave like waves under the right conditions. This implies that their motion cannot be governed by Newton's Second Law $\vec{F} = m\vec{a}$, since this equation does not admit solutions that describe the behavior of a wave.

We now have a complete complementarity, to the extent that both matter and radiation can behave like particles or waves depending on the circumstance. In Chapter 3 we will make use of this particle-wave duality to develop quantum mechanics as formulated by Schrödinger.

1.4 Atomic Spectra

Another of the minor problems to be resolved at the turn of the 20th century was the question of the radiation emitted by matter. Here it was found that the frequencies of this form of radiation suggested that the energy emitted comes in certain discreet quantities. In fact, for hydrogen, the energy of the radiation emitted could be given by

$$E_{nm} = R_\infty \left(\frac{1}{n^2} - \frac{1}{m^2}\right) \qquad n < m \ , \tag{1.14}$$

where n and m are integers and $R_\infty = 13.6058$ eV is the Rydberg. For $n = 1$ and $m = 2, 3, \ldots$ we have a series of possible frequencies or energies for the radiation. This series is known as the Lyman series. On the other hand the series resulting from taking $n = 2$ and $m = 3, 4, \ldots$ is known as the Balmer [10] series. These results were based on detailed experimental observations, and there was no theoretical justification for them at the time.

Since radiation is described by Maxwell's equations, one may be tempted to use these equations to understand the radiation. However, Maxwell's equations state that radiation results from the acceleration of charged particles have a continuous spectrum, i.e., all frequencies are possible. Clearly, one needs to go beyond the classical theories to understand the fundamental structure of the atom that is emitting the radiation.

In 1911 Rutherford [11] suggested a model for explaining the results of the scattering of α-particles from a thin gold foil. The experimental observation was that the number of particles scattered at large angles was consistent with the hypothesis that most of the mass of the atom is concentrated at its centre and that the centre is positively charged. At the suggestion of Rutherford, Bohr [12] constructed a planetary model of the atom in which the electrons are in orbits around the nucleus, and he allowed only certain orbits. The radiation then was the result of the electron jumping from one orbit to the next. In this way only certain frequencies were allowed. This model, which was based on Newtonian mechanics with an additional quantization rule, gave the observed spectrum of hydrogen. However, the extension of Bohr's ideas to more than one electron atom faced problems that could not be overcome, and by the mid 1920's it was getting clear that a new theory was required for the description of microscopic systems which seem to exhibit the effects of quantization of energy and particle-wave duality.

1.5 Problems

1. Given the Planck energy density distribution

$$\frac{de}{d\omega} = \frac{\hbar}{\pi^2 c^3} \frac{\omega^3}{e^{\hbar\omega/kT} - 1} ,$$

   where $k = 8.617 \times 10^{-5}$ eV K^{-1} is the Boltzmann constant and T the temperature in degrees Kelvin. Use MAPLE or Mathematica to plot this energy distribution for the following temperatures: 5, 300, 1000, and 5000 degrees Kelvin. From the plots, find the wave length at which the distribution is maximum.

2. Use energy and momentum conservation to show that in Compton scattering, the final energy of the photon E_2, is expressed as a function of the initial energy of the photon E_1 and the angle of deflection θ by

$$E_2 = \frac{E_1}{1 + (E_1/mc^2)(1 - \cos\theta)} ,$$

   where c is the velocity of light and m is the mass of the electron.

   Use either MAPLE or Mathematica to plot the energy of the final photon as a function of the angle θ, given the incident photon has a wave number of 0.07 Å. Note 1 Å=10^{-8} cm.

   Hint: Use relativistic kinematics for the electron, i.e., the energy of the electron is given by $E = \sqrt{p^2c^2 + m^2c^4}$, where p is the momentum and m the mass of the electron.

3. We have a mono-energetic source of X-rays (photons) of energy 50 KeV. However, for a specific experiment we need a beam of X-rays of energy 48 KeV.

   (a) Can we use Compton scattering to generate the beam of X-rays with the required energy? How?

(b) At what angle to the original beam do we extract the final beam to get the energy of 48 KeV?

(c) Plot, using MAPLE or Mathematica the energy of the final beam as a function of the extraction angle.

4. The energy released by the hydrogen atom in $2p \rightarrow 1s$ transition is given by

$$\Delta E = R_\infty \left(\frac{1}{n^2} - \frac{1}{m^2} \right)$$

where $n = 1$, $m = 2$ and $R_\infty = 13.6$ eV.

(a) What is the wave length of the light emitted as a result of this transition? Assume the atom does not recoil.

(b) To what accuracy do you have to measure the wave length of the emitted radiation to observe the effect of the recoil of the atom? Take the mass of the atom to be m, with $mc^2 = 940$ MeV. Here c is the velocity of light. (Use energy momentum conservation.)

(c) In a nuclear transition the typical energy release is a factor of 10^6 larger than the atomic transition. To what accuracy do you have to measure the energy of the photon in a nuclear transition, to detect the effect of the recoil of the atom?

Hint: We have that $\hbar c = 197.3$ eV-nm.

CHAPTER 2

Classical Mechanics - A Review

Abstract: This chapter is devoted to a short review of classical mechanics with the aim of defining the classical Hamiltonian, and establishing the relation between space symmetry and conservation laws.

In the present chapter, I would like to review classical mechanics with the aim of introducing the Euler-Lagrange formulation of mechanics based on Hamilton's principle of least action, and the Hamiltonian formulation of mechanics. The motivation is twofold:

1. The Euler-Lagrange formulation of mechanics allows us to examine the relationship between the symmetry of space-time and the conservation laws in mechanics. The same correspondence between symmetry and conservation can be utilized extensively in quantum mechanics to reduce the complexity of the problems at hand.

2. In the Hamiltonian formulation of mechanics, the independent variables are the momentum and the position or coordinate of the particles, rather than the velocity and coordinate. In quantizing a classical system we find that the momentum and the coordinate are the canonical variables needed for quantization. This allows us to introduce quantization rules that are general enough to be applied not only to classical mechanics, but also to electromagnetic theory.

2.1 Hamilton's Principle

In the Newtonian formulation of mechanics, the equations of motion or the Laws of mechanics are presented in terms of two vectors, the momentum $\vec{p}$, and the force $\vec{F}$. On the other hand, the Euler-Lagrange formulation of mechanics is based on two scalar quantities – the kinetic energy, T, and the potential energy, V. The fact that we have replaced two vector quantities by two scalar quantities is balanced by the introduction of a principle called *Hamilton's principle* or *the principle of least action*. The basic idea of this

Iraj R. Afnan

principle is as follows. Consider two points in space-time, P_1 and P_2, which correspond to the position of the particle at times t_1 and t_2 with $t_1 < t_2$, see Figure 2.1. The motion of the particle from P_1 to P_2 can be along any one of an infinity of possible paths. In Figure 2.1 we have three such paths. The integral of the difference between the kinetic energy, T, and potential energy, V, along any one of these paths is called the *action*, S, i.e.,

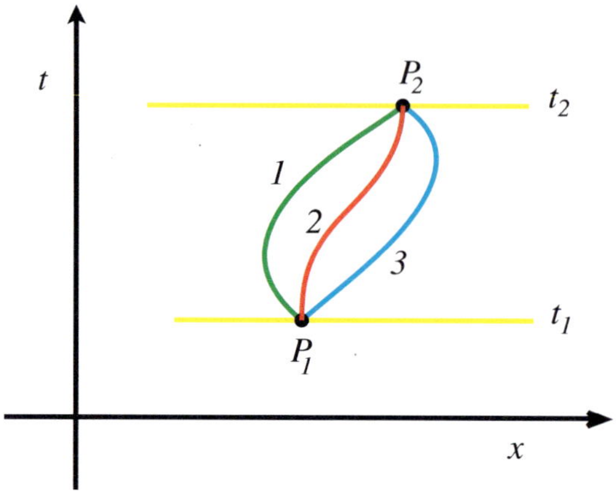

Figure 2.1: Possible trajectories for the motion of a particle starting at the point P_1 at time t_1 and ending at the point P_2 at time t_2 with $t_1 < t_2$.

$$S[q, \dot{q}] \equiv \int_{t_1}^{t_2} (T - V)\, dt \equiv \int_{t_1}^{t_2} L\,(q, \dot{q}; t)\, dt \ , \tag{2.1}$$

where $L\,(\dot{q}, q; t)$ is called the Lagrangian for the system. Here q is the position of the particle while the velocity of the particle is given by

$$\dot{q} \equiv \frac{dq}{dt} \ .$$

Note that both q and $\dot{q}$ are functions of time, t, i.e., $q(t)$ and $\dot{q}(t)$. Since the value of the action S depends on the path, we have indicated this by making the action a *functional* of q and $\dot{q}$.

Hamilton's principle states that the path which the particle actually takes is the one for which the action S, defined in Eq. (2.1), is a minimum. The problem now is how to determine this path for which S is a minimum.

In calculus, if we want to calculate the minimum of a function $y(t)$, we first take a time t_1 and calculate $y(t_1)$ and $y(t_1 + \delta t)$, where δt is a small change in the time at t_1. If $y(t_1) < y(t_1 + \delta t)$, then we calculate $y(t_1 - \delta t)$. If $y(t_1 - \delta t) > y(t_1)$, then we have established that $y(t)$ is a minimum at $t = t_1$. Otherwise we calculate $y(t_1 - 2\delta t)$ and compare it with $y(t_1 - \delta t)$, and so on. Alternatively, we can calculate

$$\delta y = y(t_1 + \delta t) - y(t_1) = \left(\frac{\partial y}{\partial t}\right)_{t=t_1} \delta t \ ,$$

where the final result is obtained by a Taylor series [1] expansion of $y(t_1 + \delta t)$ about $t = t_1$, keeping the first non-zero term in the final result. If $\delta y = 0$ and $\delta t \neq 0$, $\left(\frac{\partial y}{\partial t}\right)_{t=t_1} = 0$, then we have a minimum of the function $y(t)$.

In our case, the action S is a function of the path along which we are carrying out the integral, i.e., $[q(t), \dot{q}(t)]$. To change the path by an infinitesimal amount is equivalent to taking

$$q(t) \longrightarrow q(t) + \delta q(t) \tag{2.2}$$

for every time t, with $t_1 < t < t_2$. However, we need to keep the initial and final points P_1 and P_2 unchanged, i.e., $q(t_1) \longrightarrow q(t_1)$ and $q(t_2) \longrightarrow q(t_2)$ or $\delta q(t_1) = 0 = \delta q(t_2)$. A change in the path along which the particle moves, changes not only the position of the particle but also its velocity, i.e.,

$$\dot{q}(t) \longrightarrow \dot{q}(t) + \delta\dot{q}(t) \ . \tag{2.3}$$

We now can calculate the change in the action S, i.e.,

$$\begin{aligned}
\delta S &= S\left[q + \delta q, \dot{q} + \delta\dot{q}\right] - S\left[q, \dot{q}\right] \\
&= \int_{t_1}^{t_2} L\left(q + \delta q, \dot{q} + \delta\dot{q}\right) dt - \int_{t_1}^{t_1} L\left(q, \dot{q}\right) dt \\
&\equiv \int_{t_1}^{t_2} \delta L\left(q, \dot{q}\right) dt \ .
\end{aligned} \tag{2.4}$$

To calculate $\delta L(q, \dot{q})$, we need to expand $L(q + \delta q, \dot{q} + \delta\dot{q})$ about $L(q, \dot{q})$, and keep the lowest order terms in δq and $\delta\dot{q}$ since we are taking an arbitrarily small variation in the path, i.e.,

$$L(q + \delta q, \dot{q} + \delta\dot{q}) = L(q, \dot{q}) + \frac{\partial L}{\partial q}\delta q + \frac{\partial L}{\partial \dot{q}}\delta\dot{q} \ .$$

Here, we have dropped terms of the order $(\delta q)^2$ or higher, since δq is small. Therefore, we can write the variation in the action S, as

$$\delta S = \int_{t_1}^{t_2} \left(\frac{\partial L}{\partial q}\delta q + \frac{\partial L}{\partial \dot{q}}\delta\dot{q}\right) dt \ .$$

Making use of the fact that

$$\delta\dot{q} = \delta\left(\frac{dq}{dt}\right) = \frac{d}{dt}(\delta q) \ ,$$

[1] See Appendix B Sec. B.4 for the definition of a Taylor series.

we can rewrite the change in the action δS as

$$\delta S = \int_{t_1}^{t_2} \left(\frac{\partial L}{\partial q} \delta q + \frac{\partial L}{\partial \dot{q}} \frac{d}{dt}(\delta q) \right) dt \ .$$

Integrating the second term on the right hand side by parts, we get

$$\delta S = \frac{\partial L}{\partial \dot{q}} \delta q \Big]_{t_1}^{t_2} + \int_{t_1}^{t_2} \left[\frac{\partial L}{\partial q} - \frac{d}{dt} \left(\frac{\partial L}{\partial \dot{q}} \right) \right] \delta q \, dt \ .$$

Since we have taken the points P_1 and P_2 to be fixed, then $\delta q(t_1) = 0$ and $\delta q(t_2) = 0$. This renders the first term on the right hand side to be zero, and we have by Hamilton's Principle that

$$\delta S = \int_{t_1}^{t_2} \left[\frac{\partial L}{\partial q} - \frac{d}{dt} \left(\frac{\partial L}{\partial \dot{q}} \right) \right] \delta q \, dt = 0 \ . \tag{2.5}$$

This result is valid for any variation δq in the path taken between P_1 and P_2, and therefore the quantity in the square bracket should be zero if the variation in the action is to be zero, i.e.,

$$\frac{d}{dt} \left(\frac{\partial L}{\partial \dot{q}} \right) - \frac{\partial L}{\partial q} = 0 \ . \tag{2.6}$$

This equation is known as the *Euler-Lagrange* equation or the equation of motion for this one particle system.

2.2 The Euler-Lagrange Equation

In the above section, we found that Hamilton's principle or the principle of least action, when applied to the motion of a single particle in one dimension with kinetic energy T and potential energy V, gives a differential equation for the Lagrangian L, where $L \equiv T - V$. This result can be generalized to a system of n-particles in more than one dimension. The Lagrangian in this case is a function of all the coordinates and velocities, and is the difference between the kinetic energy of the system, T, and its potential energy V, i.e.,

$$L = L\left(q_1, \ldots, q_N, \dot{q}_1, \ldots, \dot{q}_N; t\right) \equiv T - V \ , \tag{2.7}$$

where $N = 3n$ for n particles in three dimensional space. The action S for this system is now given by

$$S[q_1, \ldots, q_N, \dot{q}_1, \ldots, \dot{q}_N] = \int_{t_1}^{t_2} L\left(q_1, \ldots, q_N, \dot{q}_1, \ldots, \dot{q}_N; t\right) dt \ , \tag{2.8}$$

and the Euler-Lagrange equations become

$$\frac{d}{dt}\left(\frac{\partial L}{\partial \dot{q}_i}\right) - \frac{\partial L}{\partial q_i} = 0 \qquad i = 1, \ldots, N \ . \tag{2.9}$$

Here, we observe that the number of equations is the same as was the case for Newton's equation, e.g., for one particle in three dimension, Newton's equation is a vector equation that can be written in terms of components of the vectors. This would give three equations for the three components. In the case of the Euler-Lagrange equation, we would have $N = 3$, and therefore three equations as well.

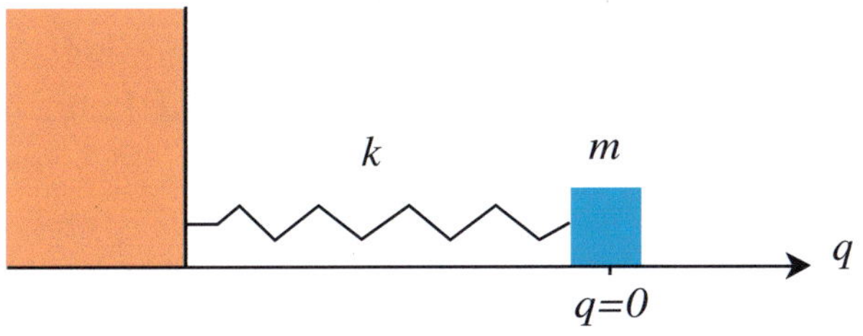

Figure 2.2: Mass m connected to a spring with spring constant k. The equilibrium position of the mass is $q = 0$.

Example: To illustrate the form of these equations for a familiar system, consider the problem of a mass at the end of a spring. This is a one dimensional system with the kinetic energy given by

$$T = \frac{1}{2}m\dot{q}^2 \ ,$$

while the potential energy is given by

$$V(q) = \frac{1}{2}mw^2q^2 = \frac{1}{2}kq^2 \ ,$$

with the frequency $w = \sqrt{\frac{k}{m}}$. Here k is the spring constant. The Lagrangian for this system can now be written as

$$L(q, \dot{q}) = T - V = \frac{1}{2}m\dot{q}^2 - \frac{1}{2}kq^2.$$

We now have that

$$\frac{\partial L}{\partial \dot{q}} = m\dot{q} \ ,$$

and

$$\frac{d}{dt}\left(\frac{\partial L}{\partial \dot{q}}\right) = m\frac{d\dot{q}}{dt} \equiv m\ddot{q} \ ,$$

while

$$\frac{\partial L}{\partial q} = -kq.$$

Thus the Euler-Lagrange equation, Eq. (2.6), is given by

$$m\ddot{q} + kq = 0 \ .$$

This is the equation of motion for a simple harmonic motion, and is identical to the result we get from Newton's equation in one dimension, i.e., $F = ma$.

Having established the fact that the Euler-Lagrange approach, based on Hamilton's principle of minimum action, is identical to Newton's second Law, the question is; what are the advantages and disadvantages of the two methods? The main advantage of the Euler-Lagrange method is that we can add any number of constraints to the motion without changing the formalism. All we need to do is minimize the action subject to the constraint. In addition, the approach can be extended to electrodynamics providing us with a classical theory of fields. In particular, Maxwell's equations turn out to be the Euler-Lagrange equations for the electromagnetic field. Finally, one can quantize the theory based on the action using the Feynman path integral approach [13]. The main disadvantage of the Euler-Lagrange method is it's failure, in it's present form, when there are dissipative forces present, e.g., friction.

2.3 Hamiltonian Mechanics

In the above formulation of mechanics we have concentrated on the coordinate and velocity of the particle. In quantum mechanics it is advantageous to work with the coordinate and momentum. The Lagrangian, as we have seen above, is a function of the position, q, and velocity, $\dot{q}$. To construct a function that will replace the Lagrangian and is a function of position and momentum, we need to carry out a Legendre transformation on the Lagrangian $L(q, \dot{q})$. We first define the *canonical momentum* p as

$$p \equiv \frac{\partial L}{\partial \dot{q}} \ . \tag{2.10}$$

Then the Hamiltonian $H(p, q)$, which is a function of the position q and the canonical momentum p, is defined in terms of the Lagrangian by the *Legendre transformation*

$$\begin{aligned} H(p, q) &= \dot{q}\frac{\partial L}{\partial \dot{q}} - L(q, \dot{q}) \\ &= \dot{q}p - L(q, \dot{q}) \ . \end{aligned} \tag{2.11}$$

Assuming the velocity $\dot{q}$ is a function of the position q and the momentum p, we can write

$$\begin{aligned} dH(p,q) &= \dot{q}dp + p\left(\frac{\partial \dot{q}}{\partial q}dq + \frac{\partial \dot{q}}{\partial p}dp\right) - \frac{\partial L}{\partial q}dq - \frac{\partial L}{\partial \dot{q}}\left(\frac{\partial \dot{q}}{\partial q}dq + \frac{\partial \dot{q}}{\partial p}dp\right) \\ &= \dot{q}dp - \frac{\partial L}{\partial q}dq + \left(p - \frac{\partial L}{\partial \dot{q}}\right)\left(\frac{\partial \dot{q}}{\partial q}dq + \frac{\partial \dot{q}}{\partial p}dp\right) \end{aligned}$$

Making use of the definition of the canonical momentum, Eq. (2.10), we can write

$$\begin{aligned} dH(p,q) &= \dot{q}dp - \frac{\partial L}{\partial q}dq \\ &= \dot{q}dp - \dot{p}dq \ , \end{aligned} \tag{2.12}$$

since

$$\dot{p} = \frac{d}{dt}\left(\frac{\partial L}{\partial \dot{q}}\right) = \frac{\partial L}{\partial q} \ ,$$

where we have made use of the Euler-Lagrange equation, Eq. (2.6). Since the Hamiltonian H is a function of the position q and momentum p, we can write

$$dH(p,q) = \frac{\partial H}{\partial p}dp + \frac{\partial H}{\partial q}dq \ . \tag{2.13}$$

Comparing Eqs. (2.12) and Eq. (2.13), we can write the equation of motion in terms of the Hamiltonian H as,

$$\dot{q} = \frac{\partial H}{\partial p} \qquad \text{and} \qquad \dot{p} = -\frac{\partial H}{\partial q} \ . \tag{2.14}$$

These two equations are known as *Hamilton's equations* of motion.

For microscopic systems, the underlying forces are non-dissipative, and the potential energy is a function of the position only. In this case the Lagrangian is given by

$$L = \frac{1}{2}m\dot{q}^2 - V(q) \ , \tag{2.15}$$

and the canonical momentum p is given as

$$p = \frac{\partial L}{\partial \dot{q}} = m\dot{q} \ , \tag{2.16}$$

and the Hamiltonian is given by

$$H(p,q) = \frac{p^2}{2m} + V(q) \ . \tag{2.17}$$

In this case the Hamiltonian is the sum of the kinetic energy $p^2/2m$ and the potential energy $V(q)$. We will use this form for the Hamiltonian in conjunction with the quantization rules to discuss the corresponding quantum mechanical systems.

2.4 Symmetry and Conservation Laws

Newton, in his formulation of mechanics, introduced three laws. Only the second law gives the equation of motion for the system. The first and third laws are basically statements about the symmetry of space in which the system is placed. Having derived the equation of motion, i.e., the Euler-Lagrange equations, from Hamilton's principle, we will turn our attention to the conservation laws. In particular we will show that the standard conservation laws are a direct result of the symmetries of the system and the space in which it is placed. Here, we will consider only the simplest of these space-time symmetries, and the corresponding conservation laws.

2.4.1 Homogeneity of time

What do we mean by homogeneity in time? If we consider an experiment being performed today, and then again in a week's time, then for the results of the experiment to be meaningful, we should get the same results in both cases. This means that the results of the experiments do not depend on the time they were performed. If they did, it would make very little sense to perform the experiments. In making this statement, we should also make sure the conditions under which the experiment were performed have not changed - at least those conditions that can have influence on the final result of the experiments. These conditions can be stated mathematically, by making the statement that *the equations of motion do not change under translation in time.*

If our Lagrangian has no explicit time dependence, i.e., the potential energy or the forces in the system are time independent, then $\frac{\partial L}{\partial t} = 0$, and the time derivative of the Lagrangian is given by

$$\begin{aligned} \frac{dL}{dt} &= \sum_i \frac{\partial L}{\partial q_i}\frac{dq_i}{dt} + \sum_i \frac{\partial L}{\partial \dot{q}_i}\frac{d\dot{q}_i}{dt} \\ &= \sum_i \frac{\partial L}{\partial q_i}\dot{q}_i + \sum_i \frac{\partial L}{\partial \dot{q}_i}\ddot{q}_i \ . \end{aligned}$$

Making use of the Euler-Lagrange equations Eq. (2.9), we get

$$\begin{aligned} \frac{dL}{dt} &= \sum_i \frac{d}{dt}\left(\frac{\partial L}{\partial \dot{q}_i}\right)\dot{q}_i + \sum_i \frac{\partial L}{\partial \dot{q}_i}\ddot{q}_i \\ &= \frac{d}{dt}\left[\sum_i \left(\frac{\partial L}{\partial \dot{q}_i}\right)\dot{q}_i\right] \ , \end{aligned}$$

or

$$\frac{d}{dt}\left[\sum_i \left(\frac{\partial L}{\partial \dot{q}_i}\right)\dot{q}_i - L\right] = 0 \ . \tag{2.18}$$

The quantity in the square bracket is nothing but the Hamiltonian which is the sum of the kinetic and potential energy of the system. Making use of Eq. (2.11) we have

$$\frac{dH(p,q)}{dt} = 0 \ ,$$

i.e., the total energy of the particles does not change with time. Thus, we have found that for systems where the Lagrangian has no explicit time dependence, the total energy of the system does not change with time. We thus can conclude that: *Homogeneity of time, which is a symmetry of space-time, leads to the conservation of energy.*

2.4.2 Homogeneity of space

The homogeneity of space means that if we move our experiment from one corner of a room to the next corner, the properties of the space in which we are doing the experiment do not change. Mathematically this means that if we displace a system at $\vec{q}$ by an amount $\vec{\epsilon}$, i.e.,

$$\vec{q} \rightarrow \vec{q} + \vec{\epsilon} \ ,$$

then the Lagrangian does not change, i.e.,

$$\delta L = 0 \ .$$

For a system consisting of one particle in three-dimensions, we have in rectangular coordinates

$$\delta L = \sum_{i=1}^{3} \frac{\partial L}{\partial q_i} \delta q_i = \sum_{i=1}^{3} \frac{\partial L}{\partial q_i} \epsilon_i \ ,$$

where $\epsilon_i, (i = 1, \ 2, \ 3)$ are the components of the vector $\vec{\epsilon}$. Making use of the Euler-Lagrange equations and the fact that $\delta q_i = \epsilon_i$, we can write the above change in the Lagrangian as,

$$\begin{aligned} \delta L &= \sum_{i=1}^{3} \frac{d}{dt}\left(\frac{\partial L}{\partial \dot{q}_i}\right) \delta q_i \\ &= \frac{d}{dt}\left(\sum_{i=1}^{3} p_i \, \epsilon_i\right) = 0 \ . \end{aligned}$$

In writing the second line we have made use of the fact that the ϵ_i's $i = 1, 2, 3$ are not a function of time t. Since the ϵ_i are arbitrary, the above result means that the components of the particles momenta are constant, i.e.,

$$\vec{p} = \ \text{constant} \ .$$

For space to be homogeneous, we should have no external forces acting on the system. For a system consisting of one particle, this homogeneity of space gives us the result that the momentum of the particle is constant, i.e., does not change with time. This is nothing but Newton's first law. Thus Newton's first law is a statement that space is homogeneous and has nothing to do with the properties of the system itself.

Let us now consider a system of two particles. For simplicity, let us first consider the problem in one dimension. Then q_i , $i = 1, 2$, labels the position of the two particles. The transformation of translation of the total system by a distance ϵ is now given by

$$q_i \rightarrow q_i + \epsilon \quad (i = 1, 2)$$

where both particles are displaced by the same amount ϵ, i.e., the whole apparatus for the experiment has been moved by ϵ. The change in the Lagrangian is then,

$$\delta L = \epsilon \sum_{i=1}^{2} \frac{\partial L}{\partial q_i} .$$

Making use of the equation of motion, i.e., the Euler-Lagrange equation, we get

$$\begin{aligned} \delta L &= \epsilon \sum_{i=1}^{2} \frac{d}{dt} \left(\frac{\partial L}{\partial \dot{q}_i} \right) \\ &= \epsilon \frac{d}{dt} \left(\sum_{i=1}^{2} \frac{\partial L}{\partial \dot{q}_i} \right) = 0 . \end{aligned} \tag{2.19}$$

Making use of the definition of momentum Eq. (2.10), and the fact that $\epsilon \neq 0$ and is a constant, we can write this equation as,

$$\frac{d}{dt} (p_1 + p_2) = 0$$

and therefore,

$$p_1 + p_2 = \text{ constant } .$$

In general, for n-particles in three dimensions, we have

$$\sum_{i=1}^{n} \vec{p}_i = \text{ constant } , \tag{2.20}$$

i.e., the total momentum of the system is a constant of the motion if the space in which the system is placed is homogeneous. An alternative way of examining this result is to make use of the fact that the kinetic energy is a function of the velocity $\dot{q}_i$ only. In this case

$$\frac{\partial L}{\partial q_i} = -\frac{\partial V}{\partial q_i} = F_i ,$$

where F_i is the force on the ith particle. Using this result, we can write Eq. (2.19) for the case when $\delta L = 0$ as

$$\vec{F}_1 + \vec{F}_2 = 0 .$$

This is a statement of Newton's third Law, i.e., *for every action there is an equal and opposite reaction.*

In the above analysis we have examined two properties of space-time, and the corresponding conservation laws. In fact within the framework of the special theory of relativity, time and space combine to form a four dimensional space where the homogeneity of space-time gives the conservation of the total four-momentum of the system. This is identical to the conservation of energy and momentum.

In addition to the homogeneity of space, we can also accept the isotropy of space, i.e., the orientation of the apparatus in the room does not effect the final result of the experiment. This isotropy of space gives rise to the conservation of the total angular momentum of the system.

We will come back to a more detailed discussion of the relation between symmetries and the corresponding conservation laws in quantum systems Chapter 12, when we consider symmetry and conservation in quantum mechanics.

2.5 Problems

1. Consider a particle of mass m moving along the x-axis under the influence of an external force
$$F = F_0 \sin \omega t \ .$$
   (a) Integrate the equation of motion to determine the position of the particle at any time t, given the initial position is given to be $x = 0$ and the initial velocity to be v_0.
   (b) Use *Mathematica* to plot the position of the particle as a function of time for $0 < t < 50$ sec. given $m = 0.05$ Kg. and $v_0 = 3$ m/sec. Take $\omega = \pi/2$ and $F_0 = 2$ N.

2. Consider a particle of mass m in an external potential that is proportional to the position, i.e., $V = Aq$, where A is a constant.
   (a) Write the Lagrangian for this system.
   (b) What is the equation of motion, i.e., the Euler-Lagrange equation for this system?
   (c) Integrate the equation of motion to determine the position of the particle as a function of time.
   (d) Write the Hamiltonian for this system.

Quantum Mechanics with Applications, **2011, 22-50**

CHAPTER 3

Wave Packet and the Uncertainty Principle

Abstract: We devote this chapter to the concept of the superposition of waves to construct wave packets that satisfy the uncertainty principle and the Schrödinger equation.

In Chapter 1, we discussed a number of experimental observations that raised questions regarding the validity of classical Newtonian mechanics in the realm of microscopic physics. The general conclusions we drew from these observations were that particles sometimes behave like waves, while radiation, that commonly exhibits a wave behavior, can transfer energy and momentum like a particle. In an attempt to understand this dual behavior of matter and radiation, let us attempt to form a localized object like a particle, out of a superposition of waves.

3.1 Superposition of Waves

To examine the wave nature of matter, let us consider a particle of momentum p and energy E. In accordance with de Broglie's postulate, the de Broglie wave length λ for this particle is related to its momentum p by the postulate

$$p = \frac{h}{\lambda} \equiv \hbar k \qquad \text{where} \qquad \hbar \equiv \frac{h}{2\pi} \ , \tag{3.1}$$

and the energy of the particle is given in terms of the frequency by

$$E = \hbar\omega \quad \text{with} \quad \omega = 2\pi\nu \ ,$$

Iraj R. Afnan

where ν is the frequency.[1] We now can write a wave in one dimension in one of three forms:

$$\sin(kx - \omega t) \quad , \quad \cos(kx - \omega t) \quad \text{or} \quad e^{i(kx-\omega t)} \ .$$

In all three cases the wave has a unit amplitude and is spread over all space. This can hardly be considered a particle, since we conceive a particle as something localized in space. Let us now take the sum of two waves, i.e.,

$$\Psi(x,t) = \Psi_1(x,t) + \Psi_2(x,t) \ ,$$

where Ψ_1 and Ψ_2 are two waves with different wave numbers and frequencies. Using the $\sin(kx - \omega t)$ representation for the waves in one dimension, we get,[2]

$$\begin{aligned} \Psi(x,t) &= \sin(k_1 x - \omega_2 t) + \sin(k_2 x - \omega_2 t) \\ &= 2\cos\left(\frac{k_1 - k_2}{2}x - \frac{\omega_1 - \omega_2}{2}t\right) \sin\left(\frac{k_1 + k_2}{2}x - \frac{\omega_1 + \omega_2}{2}t\right) \ . \end{aligned}$$

We now define the average wave number and average frequency as

$$\bar{k} = \frac{1}{2}(k_1 + k_2) \quad \text{and} \quad \bar{\omega} = \frac{1}{2}(\omega_1 + \omega_2) \ , \tag{3.2}$$

while the relative wave number and relative frequency are defined as

$$\Delta k = \frac{1}{2}(k_1 - k_2) \quad \text{and} \quad \Delta\omega = \frac{1}{2}(\omega_1 - \omega_2) \ . \tag{3.3}$$

This allows us to write the superposition of two waves as

$$\Psi(x,t) = 2\cos(\Delta k\, x - \Delta\omega\, t) \ \sin(\bar{k}x - \bar{\omega}t) \ . \tag{3.4}$$

Since in general, $\Delta k \ll \bar{k}$ and $\Delta\omega \ll \bar{\omega}$, we have in the above expression for the sum of two waves, a product of a wave with a short wave length $\lambda_S = 2\pi/\bar{k}$, and one with a long wave length $\lambda_L = 2\pi/\Delta k$. Furthermore, the wave with the longer wave length modulates (or envelopes) the wave with the shorter wave length. The velocity of the two waves are

$$v_p = \frac{\bar{\omega}}{\bar{k}} \qquad \text{(phase velocity)} \ , \tag{3.5}$$

and

$$v_g = \frac{\Delta\omega}{\Delta k} \qquad \text{(group velocity)} \ . \tag{3.6}$$

---

[1] In these lectures we always work with ω and refer to it as frequency.

[2] Note that

$$\sin A + \sin B = 2\cos\frac{A-B}{2}\sin\frac{A+B}{2} \ .$$

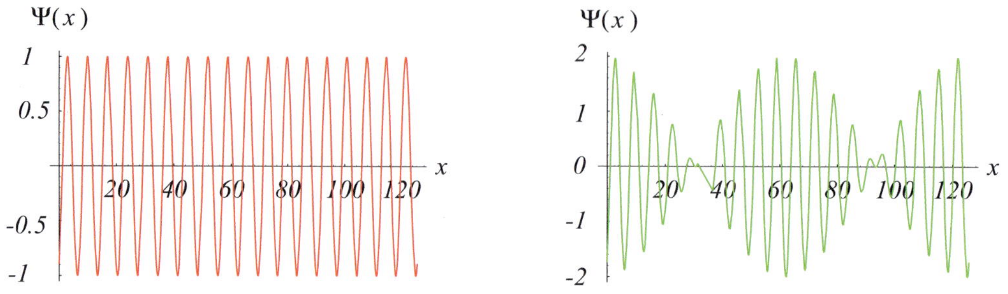

Figure 3.1: The plot on the left is of the wave $\sin(kx - \omega t)$ as a function of x for fixed t. On the right, we have the sum of two waves of equal amplitude but of a different wave number and frequency.

This superposition of the two waves allows us to construct a function with an amplitude that is suppressed at some points in space in comparison to other points (see plot on the right in Figure 3.1). This suggests that we might be able to localize our wave by the process of a superposition of a number of waves with different wave numbers. In other words, we can write a wave that is localized in space as

$$\Psi(x,t) = \int dk\; g(k)\, e^{i(kx-\omega t)} \; . \tag{3.7}$$

In writing Eq. (3.7), we have made use of the $e^{i(kx-\omega t)}$ representation of a wave rather than the $\sin(kx - \omega t)$ representation. This, we will see, allows us to make use of tools developed over a century ago by the French mathematician Fourier (1768-1830)[14] for the analysis of functions of the form presented in Eq. (3.7).

Before we proceed further with our discussion of how we are to choose the function $g(k)$ to get a description of a particle as a localized wave in space, let us re-examine the physical meaning of the phase and group velocity as defined in Eqs. (3.5) and (3.6) respectively. We know that for a wave, the velocity is given by the ratio of the frequency divided by the wave number, i.e.,

$$v = \frac{\omega}{k} \; .$$

Making use of this definition of velocity of a wave we can write the phase velocity as

$$v_p = \frac{\bar{\omega}}{\bar{k}} = \frac{\hbar\bar{\omega}}{\hbar\bar{k}} = \frac{\bar{E}}{\bar{p}} \; .$$

On the other hand, for a free particle of mass m, the energy and momentum of the particle are given by $E = \frac{1}{2}mv^2$ and $p = mv$. If we now assume that the energy and the momentum of the particle are $\bar{E}$ and and $\bar{p}$, then we have

$$v_p = \frac{\bar{E}}{\bar{p}} = \frac{\frac{1}{2}mv^2}{mv} = \frac{v}{2} \; .$$

In a similar manner we can write the group velocity as

$$v_g = \frac{\Delta\omega}{\Delta k} = \frac{\hbar\Delta\omega}{\hbar\Delta k} \approx \frac{dE}{dp} = v \ .$$

The above results suggest, very strongly, that the *group velocity* of the wave is in fact the velocity of the particle.

3.2 Fourier Integrals

To get a better understanding of the mathematics required to construct an object which, on the one hand is the sum of waves, and on the other hand a quantity that is localized in space, we need to study integrals of the form given in Eq. (3.7). This integral is often referred to as the integral representation of the function $\Psi(x,t)$. This subject comes under the heading of Fourier series and Fourier integrals in mathematics. Here I will introduce some of the basic ideas and results we need in order to examine the integral in Eq. (3.7), and in this way determine what functions we can take for $g(k)$ in order to make $\Psi(x,t)$ localized in space.

Consider the periodic function $f(x)$ with a period $2a$, i.e.,

$$f(x+2a) = f(x) \ . \tag{3.8}$$

Because of the periodic nature of this function, we can write it as a superposition of $\sin kx$ and $\cos kx$ since these functions are also periodic, i.e.,

$$f(x) = \sum_n A_n \cos k_n x + \sum_n B_n \sin k_n x \ . \tag{3.9}$$

To guarantee the periodicity of $f(x)$, i.e., $f(x) = f(x+2a)$ we should have

$$\cos k_n x = \cos k_n(x+2a) \quad \text{and} \quad \sin k_n x = \sin k_n(x+2a) \ .$$

To satisfy these two conditions, we need to take $2ak_n = 2\pi n$ or $k_n = \frac{n\pi}{a}$, with $(n = 0, 1, \ldots)$. As a result when $x \to x + 2a$, then $k_n x \to k_n x + 2k_n a = k_n x + 2n\pi$, and therefore the value of the function $f(x)$ remains unchanged.

Making use of the fact that

$$\cos k_n x = \frac{1}{2}\left(e^{ik_n x} + e^{-ik_n x}\right) \tag{3.10a}$$

and

$$\sin k_n x = \frac{1}{2i}\left(e^{ik_n x} - e^{-ik_n x}\right) \tag{3.10b}$$

we can rewrite the function $f(x)$ as a sum over exponentials of the form

$$\begin{aligned} f(x) &= \sum_{n=0}^{\infty} \frac{1}{2}\left(A_n - iB_n\right) e^{ik_n x} + \sum_{n=0}^{\infty} \frac{1}{2}\left(A_n + iB_n\right) e^{-ik_n x} \\ &= \sum_{n=-\infty}^{+\infty} C_n e^{in\pi x/a} \,. \end{aligned} \tag{3.11}$$

In this way we can write any periodic function of period $2a$ in terms of the sum over $e^{in\pi x/a}$ or $\cos(n\pi x/a)$ and $\sin(n\pi x/a)$ with n taking all integer values between $-\infty$ and $+\infty$. Making use of the orthonormality relation [3]

$$\frac{1}{2a}\int_{-a}^{+a} dx\; e^{\frac{i\pi x}{a}(n-m)} = \delta_{nm} = \begin{cases} 1 & n = m \\ 0 & n \neq m \end{cases} \,, \tag{3.12}$$

we can determine the coefficients of the expansion C_n, in Eq. (3.11). To achieve this we multiply both sides of Eq. (3.11) by $e^{-im\pi x/a}$, with m an integer, and integrate over x between $-a$ and $+a$. This gives, using the orthogonality in Eq. (3.12), the coefficient C_m to be

$$C_m = \frac{1}{2a}\int_{-a}^{+a} dx\; f(x)\, e^{-im\pi x/a} \,. \tag{3.13}$$

This result allows us to write any periodic function, $f(x)$, of period $2a$, as an infinite sum of exponentials of the form $e^{ik_n x}$, with $k_n = \frac{n\pi}{a}$ where n is an integer. The series in Eq. (3.11) is known as the *Fourier series* for the function $f(x)$, and the coefficients C_n are the *Fourier coefficients*.

In the problem we are considering, k_n will play the role of the wave number, or $\hbar k_n$ the momentum. Since the momentum of a particle is a continuous variable, we need to take the limit of the above results for the Fourier series, as the sum over n is replaced by an integral over k. Since

$$k_n = \frac{n\pi}{a} \quad \text{and} \quad \Delta k = k_{n+1} - k_n = \frac{\pi}{a}$$

k_n can be a continuous variable in the limit as $a \to \infty$. This in turn will make the result valid for any function $f(x)$, which is exactly what we want. This can be achieved by introducing $\Delta n = 1$ in Eq. (3.11), see Figure 3.2, so that

$$\begin{aligned} f(x) &= \sum_n C_n\, \Delta n\; e^{in\pi x/a} \\ &= \frac{a}{\pi}\sum_n C_n \frac{\pi \Delta n}{a}\; e^{in\pi x/a} \,. \end{aligned}$$

---

[3] We have made use of the fact that

$$\int dx\, e^{iax} = \int dx\, \cos ax + i\int dx\, \sin ax$$

to write Eq. (3.12).

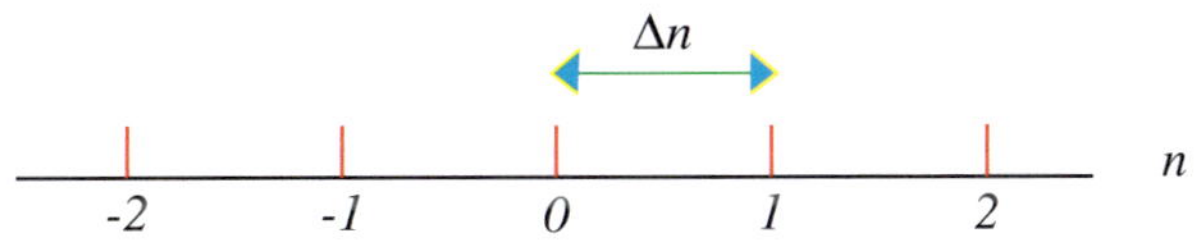

Figure 3.2: The n-axis which we convert, in the limit $a \to \infty$ to the k-axis.

Since $k_n = \frac{\pi n}{a}$, we can define $\Delta k = \frac{\pi \Delta n}{a}$ and the limit $a \to \infty$ gives $\Delta k \to dk$. We now rewrite $f(x)$ as:

$$f(x) = \sum_k \frac{g(k)}{\sqrt{2\pi}} \; e^{ikx} \; \Delta k$$

where the function $g(k)$ is defined by the relation,

$$\frac{g(k)}{\sqrt{2\pi}} = \frac{aC_n}{\pi} \; .$$

We should note that in writing the above expression we have converted $\sum_n$ to $\sum_k$, and this is the first step in converting the sum into an integral when $a \to \infty$. We are now in a position to take the limit as $a \to \infty$, in which case $\Delta k \to dk$ in the above sum. In this way we have converted the series expansion for the periodic function $f(x)$ to an integral for a non-periodic function in the form

$$f(x) = \frac{1}{\sqrt{2\pi}} \int_{-\infty}^{+\infty} dk \; g(k) \, e^{ikx} \; . \tag{3.14}$$

This is known as the *Fourier decomposition* of the non-periodic function $f(x)$. The function $g(k)$ can now be written using Eq. (3.13) as

$$\begin{aligned} g(k) \;\; &= \;\; \sqrt{2\pi} \left(\frac{aC_n}{\pi}\right) = \frac{1}{\sqrt{2\pi}} \int_{-a}^{+a} dx \; f(x) \; e^{-i(n\pi/a)x} \\ &\to \;\; \frac{1}{\sqrt{2\pi}} \int_{-\infty}^{+\infty} dx \; f(x) \; e^{-ikx} \quad \text{for} \quad a \to \infty \; . \end{aligned} \tag{3.15}$$

In Eq. (3.15) we have the Fourier decomposition of $g(k)$ in terms of the function $f(x)$. In other words, if we consider Eq. (3.14) as the Fourier transform, then Eq. (3.15) may be considered the inverse transform. By substituting the result of Eq. (3.15) into Eq. (3.14), we get

$$\begin{aligned} f(x) \;\; &= \;\; \frac{1}{2\pi} \int_{-\infty}^{+\infty} dk \; e^{ikx} \int_{-\infty}^{+\infty} dx' e^{-ikx'} f(x') \\ &= \;\; \int_{-\infty}^{+\infty} dx' f(x') \left\{ \frac{1}{2\pi} \int_{-\infty}^{+\infty} dk \; e^{ik(x-x')} \right\} \; . \end{aligned} \tag{3.16}$$

This result, if it is to be valid, suggests that the quantity in the curly bracket is zero except when $x = x'$ in which case it is one. We now define the *Dirac δ-function,* [15] in one dimension, as

$$\delta(x - x') \equiv \frac{1}{2\pi} \int_{-\infty}^{+\infty} dk\ e^{ik(x-x')} \ . \tag{3.17}$$

This function, often called a distribution rather than a function, plays a central role in Quantum Mechanics. It was first introduced by Dirac for this specific purpose, and has the property that

$$\int_{-\infty}^{+\infty} dx'\ \delta(x - x') f(x') = f(x) \ , \tag{3.18}$$

In other words $\delta(x - x') \neq 0$ for $x = x'$, otherwise it is zero. We can write our Dirac δ-function as a limit of the form

$$\begin{aligned} \delta(x) &= \lim_{L\to\infty} \frac{1}{2\pi} \int_{-L}^{+L} dk\ e^{ikx} \\ &= \lim_{L\to\infty} \frac{1}{2\pi} \frac{e^{ixL} - e^{-ixL}}{ix} \\ &= \lim_{L\to\infty} \frac{\sin Lx}{\pi x} \ . \end{aligned} \tag{3.19}$$

In fact, if we take any function that has a peak and normalize it so that the area under the peak equal to one, then the Dirac δ-function can be defined as the limit when the width of the peak tends to zero while the area under the peak stays fixed at one, e.g.,

$$\delta(x) = \lim_{\alpha\to 0} \frac{1}{\pi} \frac{\alpha}{x^2 + \alpha^2} \ , \tag{ 3.20a}$$

and

$$\delta(x) = \lim_{\alpha\to\infty} \frac{\alpha}{\sqrt{\pi}}\ e^{-\alpha^2 x^2} \ . \tag{ 3.20b}$$

A property of the Dirac δ-function that is very useful is

$$\int_{-\infty}^{+\infty} dx\ f(x)\ \delta(g(x)) = \sum_i \frac{f(x_i)}{|g'(x_i)|} \ , \tag{ 3.20c}$$

where x_i are the positions of the zeros of the function $g(x)$ and

$$g'(x_i) = \left.\frac{dg(x)}{dx}\right|_{x=x_i} \ . \tag{3.21}$$

Other properties of the Dirac δ-function will be introduced as we need them.

We thus have established the procedure for determining the function $g(k)$ given the fact that we want a function $f(x)$ that is localized in space. In particular we have introduced the Fourier integral representation for the function $f(x)$ given by

$$f(x) = \frac{1}{\sqrt{2\pi}} \int_{-\infty}^{+\infty} dk\ g(k)\ e^{ikx} \tag{3.22a}$$

with the inverse transform giving the function $g(k)$ in terms of $f(x)$ as

$$g(k) = \frac{1}{\sqrt{2\pi}} \int_{-\infty}^{+\infty} dx\ f(x)\ e^{-ikx}\ . \tag{3.22b}$$

We note here, the symmetry between the Fourier transform and its inverse. At this stage this symmetry was introduced by construction. However, when we proceed to use it in Quantum Mechanics, we will find that it is natural to have this symmetry between the Fourier transform and its inverse if we are going to give a physical meaning to this transform.

3.3 The Uncertainty Principle

With the above result in hand, let us consider the following integral

$$\psi(x) = \frac{1}{\sqrt{2\pi}} \int_{-\infty}^{+\infty} dk\ \phi(k)\, e^{ikx}$$

as a possible function that is localized in space, which could represent the wave nature of a particle. To get the localization in space, we need to carefully select our weighting function $\phi(k)$. We know from the results of Sec. 3.2 on Fourier transforms, that the function $\phi(k)$ is the inverse Fourier transform and is given by

$$\phi(k) = \frac{1}{\sqrt{2\pi}} \int_{-\infty}^{+\infty} dx\ \psi(x)\, e^{-ikx}\ .$$

If we want to consider $\psi(x)$ to be the amplitude of a 'wave' that represents a localized particle at $x = x_0$, then one possible choice for $\psi(x)$ is the Gaussian function (see Figure 3.3), i.e.,

$$\psi(x) = N\ e^{-(x-x_0)^2/4\sigma^2}\ , \tag{3.23}$$

where N is a normalization factor. As we will see, the width of the peak in the function is related to σ. Since we want $\psi(x)$ to be the amplitude of the wave, then the intensity of the wave at point x will be given by $|\psi(x)|^2$. Since the wave is going to represent a single particle, we need to choose our normalization N such that the integral of the intensity over all space is one, i.e.,

$$\int_{-\infty}^{+\infty} dx\ |\psi(x)|^2 = 1\ .$$

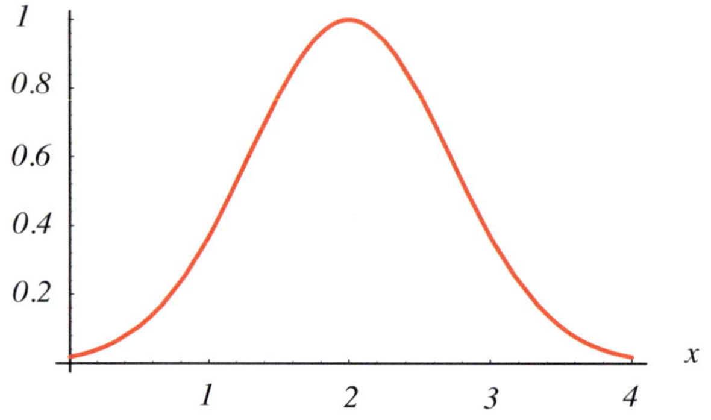

Figure 3.3: This is a plot of the function $e^{-(x-x_0)^2}$ for $x_0 = 2$, and corresponds to a function peaked at $x = 2$ with a maximum height of one.

This gives us a value for the normalization constant N of [4]

$$N^2 = \frac{1}{\sigma\sqrt{2\pi}}\ . \tag{3.24}$$

The intensity of the wave that is localized at $x = x_0$ is then given by

$$|\psi(x)|^2 = \frac{1}{\sigma\sqrt{2\pi}}\ e^{-(x-x_0)^2/2\sigma^2}\ . \tag{3.25}$$

We now define the width of the peak in $|\psi(x)|^2$ by first determining the value of x_1 for which

$$|\psi(x_1)|^2 = |\psi(x_0)|^2\ e^{-1}\ ,$$

where x_0 is the point at which $|\psi(x)|^2$ is maximum. In this case, this corresponds to

$$(x_1 - x_0)^2 = 2\sigma^2 \quad \text{or} \quad |x_1 - x_0| = \sqrt{2}\ \sigma\ .$$

[4]We have made use of the Gaussian integral

$$\int_{-\infty}^{+\infty} du\, e^{-u^2} = \sqrt{\pi}\ .$$

We now define the width to be [5]

$$\Delta x = 2|x_1 - x_0| = 2\sqrt{2}\,\sigma\ . \tag{3.26}$$

This can be considered as the uncertainty in the position of the particle about the value x_0.

We now make use of the inverse Fourier transform to determine $\phi(k)$ in terms of $\psi(x)$ to be

$$\begin{aligned}\phi(k) &= \frac{1}{\sqrt{2\pi}} \int_{-\infty}^{+\infty} dx\ \psi(x)\, e^{-ikx} \\ &= \frac{N}{\sqrt{2\pi}} \int_{-\infty}^{+\infty} dx\ e^{-(x-x_0)^2/4\sigma^2} e^{-ikx}\end{aligned}$$

To simplify this integral, we change the variable of integration to $\xi = x - x_0$ or $x = \xi + x_0$. We then have $d\xi = dx$ so that $\phi(k)$ can be written as

$$\phi(k) = \frac{N}{\sqrt{2\pi}}\, e^{-ikx_0} \int_{-\infty}^{+\infty} d\xi\ e^{-\left(\frac{\xi^2}{4\sigma^2}+ik\xi\right)}\ .$$

To evaluate this integral, we need to recast it into an integral over a Gaussian function. This can be achieved by completing the square in the exponent under the integral sign. This gives

$$\begin{aligned}\phi(k) &= \frac{N}{\sqrt{2\pi}}\, e^{-ikx_0}\, e^{-\sigma^2k^2} \int_{-\infty}^{+\infty} d\xi\ e^{-\left(\frac{\xi}{2\sigma}+i\sigma k\right)^2} \\ &= \frac{N}{\sqrt{2\pi}}\, e^{-ikx_0}\, e^{-\sigma^2k^2}\ 2\sigma \int_{-\infty}^{+\infty} du\ e^{-u^2} \\ &= \sqrt{2}\, N\, \sigma e^{-ikx_0}\ e^{-\sigma^2k^2}\ ,\end{aligned} \tag{3.27}$$

where in the second line we have again changed the variable of integration such that $u = \frac{\xi}{2\sigma} + i\sigma k$ and then $d\xi = 2\sigma du$. We now can write the magnitude of $\phi(k)$ as

$$\begin{aligned}|\phi(k)|^2 &= 2\,N^2\,\sigma^2\ e^{-2\sigma^2k^2} \\ &= \sqrt{\frac{2}{\pi}}\,\sigma\ e^{-2\sigma^2k^2}\ .\end{aligned} \tag{3.28}$$

---

[5]This definition of the width is chosen such that:

$$\int_{-\sqrt{2}\sigma}^{\sqrt{2}\sigma} dx\ e^{-x^2/2\sigma^2} = Erf(1) = 0.842701\ ,$$

and independent of σ.

In the next chapter we will come back to a discussion of the function $\phi(k)$ and we will be able to give it a physical meaning similar to that of $\psi(x)$. We now can determine the width of the Gaussian given by $|\phi(k)|^2$ using the same definition of width used to extract the width of $|\psi(x)|^2$ above. This is given by

$$\Delta k = \frac{2}{\sqrt{2}\,\sigma} \ . \tag{3.29}$$

This width can be considered as an uncertainty in the wave number k. If we combine the results of Eqs. (3.26) and (3.29) for Δx and Δk, we get

$$\Delta k\ \Delta x = 4 > 1$$

or after multiplication by $\hbar$ we have

$$\Delta p\ \Delta x > \hbar \ , \tag{3.30}$$

which is a statement of the *Heisenberg uncertainty relation.* We have thus found that if we want to localize the position of a particle to an uncertainty Δx, we have to take the sum of waves with different wave numbers (i.e., momenta), to get this localization in space. As a result of this superposition, we have an uncertainty in the momentum of the particle of magnitude Δp with $\Delta x \Delta p > \hbar$.

In Figure 3.4 we have the plot of $|\psi(x)|^2$ and $|\phi(k)|^2$ as given in Eqs. (3.25) and (3.28). We observe that both functions have a Gaussian form, and as we reduce the width of the peak in $\psi(x)$ we increase the width of the peak in $\phi(k)$. (For a more detailed examination of how $\psi(x)$ and $\phi(k)$ depend on σ, see the problems at the end of this chapter).

From the above analysis we come to the conclusion that if we treat the particle as the superposition of simple waves, then the momentum of the particle is uncertain. In fact, the harder we work to localize the position of the particle, the more uncertain its momentum becomes. This is not consistent with classical physics, where a particle by definition is localized in space and has a definite momentum at any instant of time. Thus the concept of associating a wave property with a particle has put us at odds with the classical concept of a particle being localized in space and having a definite momentum. This suggests that the experimental observation of the wave nature of matter is going to put us in contradiction with classical Newtonian mechanics.

So far we have considered the superposition of stationary (time independent) waves to form our localized wave packet. To see how such a wave packet propagates with time, we need to consider the superposition of traveling waves, e.g., for waves traveling along the positive x-axis, we use the basic wave

$$e^{i(kx-\omega t)} \ .$$

To study the propagation of such a wave we need to know the relationship between the angular frequency ω and the wave number k, e.g., for electromagnetic waves, we have

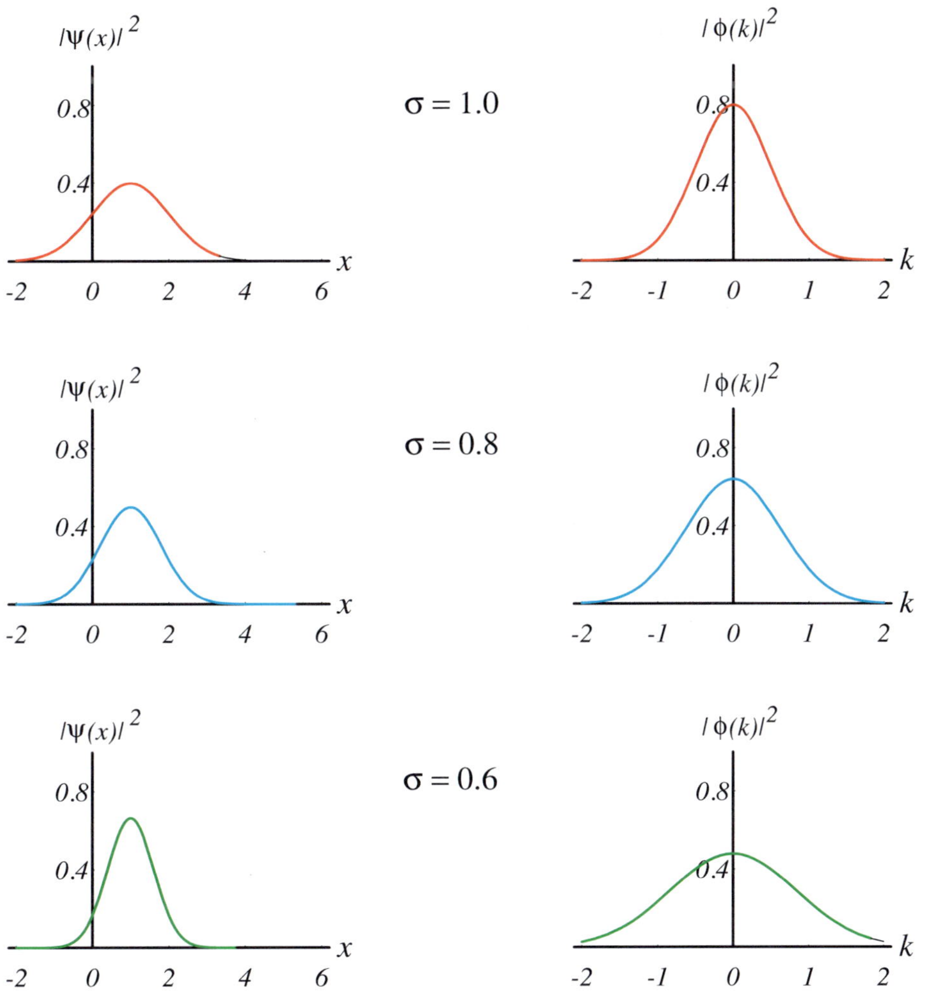

Figure 3.4: Here we have a plot of $|\psi(x)|^2$ as a function of x, and the corresponding $|\phi(k)|^2$ as a function of k. We have taken $x_0 = 1$ and $\sigma = 1.0, 0.8, 0.6$.

$E = \hbar\omega = pc = \hbar kc$ and thus $\omega = kc$. This allows us to describe an electromagnetic wave (e.g., light) as

$$e^{ik(x-ct)}$$

where c is the velocity of the electromagnetic wave, i.e., the velocity of light. In this case our wave packet takes the form

$$\begin{aligned} \Psi(x,t) &= \frac{1}{\sqrt{2\pi}} \int_{-\infty}^{+\infty} dk\ \phi(k) e^{ik(x-ct)} \\ &= \Psi(x-ct)\ . \end{aligned} \tag{3.31}$$

Because the wave packet is a function of $(x - ct)$, it travels along the x-axis with time without changing its form. To illustrate this, consider the case when the wave $\Psi(x,t)$ has its maximum, f_0, at $x = 0$ when $t = 0$, i.e., $\Psi(0,0) = f_0$. At a later time $t > 0$, the peak of the wave packet will have moved to x. Since $\Psi(x,t)$ is a function of $(x - ct)$, then $\Psi(x,t) = \Psi(0,0) = f_0$, if $x = ct$. This means that the peak of the wave travels with velocity of light c, and the shape of the wave packet has not changed with time.

For particles of mass m, the relation between ω and k is more complicated. For example,

$$E = \hbar\omega = \begin{cases} \sqrt{p^2c^2 + m^2c^4} & \text{for relativistic particles} \\ \dfrac{p^2}{2m} = \dfrac{\hbar^2k^2}{2m} & \text{for non-relativistic particles} \end{cases} ,$$

where c is the velocity of light. In general we can assume the frequency ω, to be a function of the wave number k, and we can write our wave packet as

$$\Psi(x,t) = \frac{1}{\sqrt{2\pi}} \int_{-\infty}^{+\infty} dk\ \phi(k)\, e^{i(kx-\omega(k)t)} . \tag{3.32}$$

Supposing $\phi(k)$ is peaked sharply about $k = k_0$, then the major contribution to the integral comes from values of k close to k_0. This allows us to expand $\omega(k)$ in the integrand in a Taylor series about $k = k_0$. If we retain all terms up to quadratic terms in $(k - k_0)$, we can write the frequency $\omega(k)$ as a quadratic in $(k - k_0)$ of the form

$$\begin{aligned} \omega(k) &\approx \omega(k_0) + (k-k_0)\left(\frac{d\omega}{dk}\right)_{k=k_0} + \frac{1}{2}(k-k_0)^2\left(\frac{d^2\omega}{dk^2}\right)_{k=k_0} \\ &\equiv \omega_0 + v_g(k-k_0) + \beta(k-k_0)^2 , \end{aligned} \tag{3.33}$$

where the group velocity, v_g, is given by

$$v_g = \left(\frac{d\omega}{dk}\right) .$$

If we now assume that $\phi(k)$ in Eq. (3.32) is a Gaussian, i.e.,

$$\phi(k) = N_k e^{-\sigma^2(k-k_0)^2} ,$$

and change the variables of integration in Eq. (3.32) to $\eta = k - k_0$, we get

$$\begin{aligned} \Psi(x,t) &= \frac{N_k}{\sqrt{2\pi}} e^{i(k_0x-\omega_0t)} \int_{-\infty}^{+\infty} d\eta\ e^{i\eta x}\, e^{-\sigma^2\eta^2}\, e^{-iv_g\eta t} e^{-i\beta\eta^2 t} \\ &= \frac{N_k}{\sqrt{2\pi}} e^{i(k_0x-\omega_0t)} \int_{-\infty}^{+\infty} d\eta\ e^{-(\sigma^2+i\beta t)\eta^2}\, e^{i(x-v_gt)\eta} . \end{aligned}$$

Completing the squares in the exponent of the integrand, we get

$$\Psi(x,t) = \frac{N_k}{\sqrt{2\pi}} \frac{e^{i(k_0 x - \omega_0 t)}}{\sqrt{\sigma^2 + i\beta t}} \exp\left\{-\frac{(x - v_g t)^2}{4(\sigma^2 + i\beta t)^2}\right\} \int_{-\infty}^{+\infty} du\, e^{-u^2} ,$$

where

$$u = \sqrt{\sigma^2 + i\beta t}\, \eta - \frac{i(x - v_g t)}{2\sqrt{\sigma^2 + i\beta t}} .$$

Since the above integral over u is $\sqrt{\pi}$, we can write the intensity of our wave packet as

$$|\Psi(x,t)|^2 = \frac{N_k^2}{2} \left(\frac{1}{\sigma^4 + \beta^2 t^2}\right)^{\frac{1}{2}} \exp\left\{-\frac{\sigma^2 (x - v_g t)^2}{2(\sigma^4 + \beta^2 t^2)}\right\} . \tag{3.34}$$

This is a wave packet whose peak travels with velocity v_g which is the velocity of the particle. However, the width of this wave increases with time. In fact at $t = 0$ the width is $2\sqrt{2}\sigma$, whereas at a time t later, it has increased to

$$2\sqrt{2}\sigma \left(1 + \frac{\beta^2 t^2}{\sigma^4}\right)^{\frac{1}{2}} .$$

Furthermore, if the wave packet is wide initially, the rate at which it spreads is smaller because the rate of increase in width is inversely proportional to σ, the width at time $t = 0$.

3.4 The Schrödinger Equation

In the last section, we showed how we can construct a function, $\Psi(x,t)$, which can describe the time evolution of the system (particle, e.g., an electron) that is consistent with wave-particle duality. This wave-particle duality was exhibited by the observation that the object or particle described by the function $\Psi(x,t)$ was localized in space like a particle and its development with time was also consistent with a particle having velocity v_g. On the other hand $\Psi(x,t)$ was nothing more than the superposition of waves. The two consequences of satisfying the above wave-particle duality are;

1. There is a limitation on the accuracy of our knowledge of the position and momentum of the particle described by $\Psi(x,t)$. This limitation is stated in the form of the uncertainty principle, Eq.(3.30), and illustrated in Figure 3.4.

2. The uncertainty in the position of the particle, i.e., the width of the wave packet, changes with time. The rate of change depends on the shape of the wave packet and the relation between the frequency ω and the wave number k.

In this section, we will show that this wave packet $\Psi(x,t)$ which describes the motion of a free particle in space, satisfies a differential equation first proposed by Schrödinger in 1926 [16]. The particular form of the differential equation is a consequence of the non-relativistic nature of the relation between the energy and the momentum of the particle. Thus for a free particle, i.e., for a particle that is not under the influence of any external force, we have

$$E = \hbar\omega(k) = \frac{p^2}{2m} = \frac{\hbar^2 k^2}{2m} \ .$$

Since our wave packet is given by

$$\Psi(x,t) = \frac{1}{\sqrt{2\pi}} \int_{-\infty}^{+\infty} dk\ \phi(k) e^{i(kx-\omega t)} \ ,$$

we can differentiate it with respect to time to get

$$i\hbar \frac{\partial \Psi(x,t)}{\partial t} = \frac{1}{\sqrt{2\pi}} \int_{-\infty}^{+\infty} dk\ \phi(k)\, \hbar\omega(k)\ e^{i(kx-\omega t)} \ . \tag{3.35}$$

On the other hand, differentiating the wave packet with respect to position x, twice, we get

$$\begin{aligned} -\frac{\hbar^2}{2m} \frac{\partial^2 \Psi(x,t)}{\partial x^2} &= \frac{1}{\sqrt{2\pi}} \int_{-\infty}^{+\infty} dk\ \phi(k)\, \frac{\hbar^2 k^2}{2m}\ e^{i(kx-\omega t)} \\ &= \frac{1}{\sqrt{2\pi}} \int_{-\infty}^{+\infty} dk\ \phi(k)\, \hbar\omega(k)\, e^{i(kx-\omega t)} \ . \end{aligned} \tag{3.36}$$

We therefore have, on comparing the results of Eqs. (3.35) and (3.36), that the wave packet amplitude $\Psi(x,t)$, satisfies the differential equation

$$i\hbar \frac{\partial \Psi(x,t)}{\partial t} = -\frac{\hbar^2}{2m} \frac{\partial^2 \Psi(x,t)}{\partial x^2} \ , \tag{3.37}$$

which describes the evolution of the wave packet that represents a free particle in space-time.

For a particle in an external potential $V(x)$, the total energy $E = \hbar\omega$ is the sum of the kinetic energy $\frac{p^2}{2m}$ and the potential energy $V(x)$, i.e.,

$$\begin{aligned} E = \hbar\omega &= \frac{p^2}{2m} + V(x) \\ &= \frac{\hbar^2 k^2}{2m} + V(x) \ . \end{aligned} \tag{3.38}$$

In this case we can follow the same procedure implemented above, for a free particle, to find that the function $\Psi(x,t)$ satisfies the relation

$$\begin{aligned}\left(-\frac{\hbar^2}{2m}\frac{\partial^2}{\partial x^2}+V(x)\right)\Psi(x,t) &= \frac{1}{\sqrt{2\pi}}\int_{-\infty}^{+\infty} dk\ \phi(k)\left(\frac{\hbar^2k^2}{2m}+V(x)\right)e^{i(kx-\omega t)} \\ &= \frac{1}{\sqrt{2\pi}}\int_{-\infty}^{+\infty} dk\ \phi(k)\hbar\omega\ e^{i(kx-\omega t)} \\ &= \frac{1}{\sqrt{2\pi}}\int_{-\infty}^{+\infty} dk\ \phi(k) i\hbar\frac{\partial}{\partial t}\ e^{i(kx-\omega t)} \\ &= i\hbar\frac{\partial}{\partial t}\Psi(x,t)\ .\end{aligned}$$

In writing the second line in the above equation we have made use of the relation between the energy and wave number as given in Eq. (3.38). We now can write the partial differential equation that determines the evolution of a wave packet describing a particle in an external potential $V(x)$. This equation is of the form

$$i\hbar\frac{\partial\Psi(x,t)}{\partial t} = \left(-\frac{\hbar^2}{2m}\frac{\partial^2}{\partial x^2}+V(x)\right)\Psi(x,t)\ , \tag{3.39}$$

and is known as the Schrödinger equation for the *wave amplitude* or *wave function* $\Psi(x,t)$.

In deriving this equation, our starting point was de Broglie's postulate which states that a particle behaves like a wave with momentum $p = \frac{h}{\lambda} = \hbar k$, with k being the wave number corresponding to the energy $E = \hbar\omega$. In an attempt to localize the particle in coordinate space, we found it necessary to take a linear superposition of waves with different wave numbers, each wave having a weighting determined by $\phi(k)$. This procedure led us to the uncertainty relationship between the position and the momentum of the particle, i.e.,

$$\Delta p\ \Delta x > \hbar\ .$$

We then made use of the relationship between the energy $E = \hbar\omega$ and the momentum $p = \hbar k$ (Eq. (3.38)) to derive an equation (a partial differential equation) that describes the properties of the wave amplitude $\Psi(x,t)$ in space-time. This equation is known as the Schrödinger equation.

From classical mechanics we know that the Hamiltonian for the system is given by (see Eq. (2.17))

$$H = \frac{p^2}{2m}+V(x)$$

and is equal to the total energy E, i.e.,

$$E = H(p,x)\ .$$

One procedure to quantize a system described by the above classical Hamiltonian is to carry out the following substitution for the energy and momentum

$$E \longrightarrow i\hbar\frac{\partial}{\partial t} , \qquad (3.40a)$$

$$p \longrightarrow -i\hbar\frac{\partial}{\partial x} . \qquad (3.40b)$$

In this way, the classical Hamiltonian is converted to a differential operator which we define as $\tilde{H}$, i.e.,

$$H(p,x) \longrightarrow -\frac{\hbar^2}{2m}\frac{\partial^2}{\partial x^2} + V(x) \equiv \tilde{H} \qquad (3.40c)$$

where the ˜ above the H indicates that the Hamiltonian is an operator.

The above substitutions may be considered the general rule for deriving the quantum mechanical equivalent to a classical system described by the Hamiltonian $H(p,x)$.[6] The Schrödinger equation can now be written as

$$i\hbar\frac{\partial \Psi}{\partial t} = \tilde{H}\,\Psi , \qquad (3.41)$$

where $\tilde{H}$ is referred to as the quantum mechanical Hamiltonian for the system.

So far we have considered one space dimension. In three dimensions, the same substitution is used to go from classical to quantum mechanics. However, now the momentum is a vector, and therefore

$$\vec{p} \rightarrow -i\hbar\,\vec{\nabla} ,$$

where $\vec{\nabla}$, the gradient operator, is defined in rectangular coordinates as

$$\vec{\nabla} = \hat{\imath}\frac{\partial}{\partial x} + \hat{\jmath}\frac{\partial}{\partial y} + \hat{k}\frac{\partial}{\partial z} .$$

Here $\hat{\imath}$, $\hat{\jmath}$, and $\hat{k}$ are unit vectors along the x, y and z axis. The quantum Hamiltonian for a particle in a potential $V(\vec{r})$ is then given by

$$\tilde{H} = -\frac{\hbar^2}{2m}\nabla^2 + V(\vec{r}) , \qquad (3.42)$$

where ∇^2, the Laplacian, is given in rectangular coordinates as

$$\nabla^2 = \vec{\nabla}\cdot\vec{\nabla} = \frac{\partial^2}{\partial x^2} + \frac{\partial^2}{\partial y^2} + \frac{\partial^2}{\partial z^2} .$$

The Schrödinger equation in this case is still given by Eq. (3.41) with the Hamiltonian defined by Eq. (3.42).

---

[6]For the general procedure for the quantization of classical Hamiltonian see Schiff [17] *Quantum Mechanics*, Sec. 24.

In the above discussion, we have shown that the amplitude of the wave packet, resulting from the need to describe a particle's motion in terms of a wave, satisfies the Schrödinger equation, and is consistent with Heisenberg's uncertainty relation. However, it should be pointed out that the procedure followed is by no means a derivation of the Schrödinger equation. On the other hand, if we postulate the quantization rules given in Eqs. (3.40a) - (3.40c), we can write the Schrödinger equation, Eq. (3.41), for any system, given the classical Hamiltonian for that system. To demonstrate the generality of the quantization rules we have introduced, let us consider the case of a free relativistic particle, so that the relation between the energy, i.e., the Hamiltonian, and the momentum is given by

$$E = \sqrt{p^2c^2 + m^2c^4} \ .$$

Using the quantization rules given in Eqs. (3.40), the corresponding quantum mechanical equation could be

$$i\hbar \frac{\partial \Psi}{\partial t} = \sqrt{-\hbar^2 c^2 \nabla^2 + m^2 c^4} \ \Psi(\vec{r}, t) \ .$$

This equation is highly non-local because the differential operator ∇^2 is under the square root sign. An alternative starting point would be to consider

$$E^2 = p^2c^2 + m^2c^4 \ . \tag{3.43}$$

Upon quantization, the relation between the energy and momentum gives the equation

$$-\hbar^2 \frac{\partial^2 \Psi(\vec{r}, t)}{\partial t^2} = \left(-\hbar^2 c^2 \nabla^2 + m^2 c^4\right) \Psi(\vec{r}, t) \ ,$$

or

$$\left(\frac{1}{c^2}\frac{\partial^2}{\partial t^2} - \nabla^2 + \frac{m^2c^2}{\hbar^2}\right) \Psi(\vec{r}, t) = 0 \ . \tag{3.44}$$

This equation, known as the Klein-Gordon equation[18, 19], reduces to the wave equation for a zero mass particle, and to the Schrödinger equation in the non-relativistic limit, i.e., $\frac{p}{m} \ll 1$. The main problem with this equation is the fact that it is second order in the time derivative which gives rise to possible problems in the interpretation of $\Psi(\vec{r}, t)$.[7]

3.5 Physical Interpretation of $\Psi(\vec{r}, t)$

In the above, we have referred to $\Psi(x, t)$ as the amplitude of the wave packet. This was the result of the fact that we had constructed this wave packet in terms of a superposition of waves with a wave number k and an amplitude $\phi(k)$. However, since our wave packet is to describe a particle, we need to interpret the function $\Psi(x, t)$ in terms of properties of

[7] To give a physical interpretation to the $\Psi(\vec{r}, t)$ resulting from the solution of the Klein-Gordan equation, we need to introduce the concept of antiparticle [23] see e.g., Schweber's book [24].

particles. In electromagnetic waves, and light in particular, the intensity is considered to be the magnitude squared of the amplitude, and is a measure of how much light is present at a given point in space and time. Since our particle, as a superposition of waves, is not localized in space any more, we could give $|\Psi(x,t)|^2$ an interpretation similar to the intensity of light. However, rather than using the word intensity, we make use of the word *probability density* to specify the probability of finding the particle at the point x at time t. This probability density, $\rho(x,t)$, is given in analogy with intensity as

$$\rho(x,t) = |\Psi(x,t)|^2 \ . \tag{3.45}$$

In this way, we can reconcile the fact that a particle is in fact a point object, while at the same time we can get interference behavior by using a beam of particles, e.g., electrons in an electron microscope. With the above interpretation for $\Psi(x,t)$, we require that

$$\int_{-\infty}^{+\infty} dx\ |\Psi(x,t)|^2 = 1 \ , \tag{3.46}$$

i.e., the probability of finding the particle anywhere between $x = -\infty$ and $x = +\infty$, at a given time t, be equal to one. For this integral to be finite, we require that in one dimension

$$\lim_{x\to\pm\infty} \Psi(x,t) \to x^{-\frac{1}{2}-\epsilon} \ ,$$

where $\epsilon > 0$.

Although we have given a meaning to the magnitude of the function $\Psi(x,t)$, the phase of this function is important. In fact, to get any form of interference we need to add two functions, such as

$$\Psi(x,t) = a_1\Psi_1(x,t) + a_2\Psi_2(x,t) \ ,$$

where a_1 and a_2 are in general complex constants. It is clear from this result that the relative phase of Ψ_1 and Ψ_2, and the value of the constants a_1 and a_2 determine the degree of interference. Finally, given the fact that $\Psi(x,t)$ satisfies the partial differential equation that is second order in x, the function $\Psi(x,t)$ has to be a continuous function of x.

To examine the properties of the probability density $\rho(x,t)$, let us consider the Schrödinger equation and its complex conjugate, i.e.,

$$i\hbar\,\frac{\partial\Psi(x,t)}{\partial t} = \left[-\frac{\hbar^2}{2m}\frac{\partial^2}{\partial x^2} + V(x)\right]\Psi(x,t) \ , \tag{3.47a}$$

and

$$-\,i\hbar\,\frac{\partial\Psi^*(x,t)}{\partial t} = \left[-\frac{\hbar^2}{2m}\frac{\partial^2}{\partial x^2} + V(x)\right]\Psi^*(x,t) \ . \tag{3.47b}$$

In writing Eq. (3.47b) we have assumed the potential $V(x)$ to be a real function. If we now multiply Eq. (3.47a) from the left by $\Psi^*(x,t)$ and Eq. (3.47b) from the right by $\Psi(x,t)$ and subtract the second equation from the first equation, we get

$$i\hbar \frac{\partial}{\partial t}(\Psi^*\Psi) = i\hbar \frac{\partial \rho}{\partial t} = -\frac{\partial}{\partial x}\left\{\frac{\hbar^2}{2m}\left(\Psi^*\frac{\partial \Psi}{\partial x} - \frac{\partial \Psi^*}{\partial x}\Psi\right)\right\} . \qquad (3.48)$$

The result in Eq. (3.48) can be written as an equation for the conservation of probability by recasting it into the form of a continuity equation, i.e.,

$$\frac{\partial \rho}{\partial t} + \frac{\partial j}{\partial x} = 0 , \qquad (3.49)$$

where the probability current $j(x,t)$ is given by

$$j(x,t) = \frac{\hbar}{2im}\left(\Psi^*\frac{\partial \Psi}{\partial x} - \frac{\partial \Psi^*}{\partial x}\Psi\right) . \qquad (3.50)$$

To see how the continuity equation, Eq. (3.49), is a statement of conservation of probability, let us integrate Eq. (3.49) over all space, to get

$$\int_{-\infty}^{+\infty} dx\, \frac{\partial \rho}{\partial t} = -\int_{-\infty}^{+\infty} dx\, \frac{\partial j}{\partial x}$$

or

$$\frac{d}{dt}\int_{-\infty}^{+\infty} dx\, \rho(x,t) = -j(x,t)\Big|_{x=-\infty}^{x=+\infty} . \qquad (3.51)$$

Since $\Psi(x,t) \to 0$ for $x \to \pm\infty$, then $j(x,t) \to 0$ for $x \to \pm\infty$, and the right hand side of Eq. (3.51) is zero. We therefore have

$$\frac{d}{dt}\int_{-\infty}^{+\infty} dx\, \rho(x,t) = 0 . \qquad (3.52)$$

Since $\rho(x,t)$ is the probability density, then $\int \rho(x,t)\, dx$ is the probability of having a particle at time t. This integrated probability being time independent is a statement of the fact that the number of particles in the system does not change with time, i.e., particles are not created or destroyed, and we have a conservation of particle number or conservation of probability.

Having established that we can take $\Psi(x,t)$ to be the probability amplitude of finding the particle at position x at time t, we can now calculate the average position of the particle as

$$\begin{aligned}\langle x\rangle &= \int_{-\infty}^{+\infty} dx\, x\, \rho(x,t) \\ &= \int_{-\infty}^{+\infty} dx\, \Psi^*(x,t)\, x\, \Psi(x,t)\ . \end{aligned} \tag{3.53}$$

In fact, making use of the fact that the average of x^n is given by

$$\langle x^n\rangle = \int_{-\infty}^{+\infty} dx\, \Psi^*(x,t)\, x^n\, \Psi(x,t)\ , \tag{3.54}$$

we can calculate the average of any function $f(x)$ that can be written as a power series in x as

$$\langle f(x)\rangle = \int_{-\infty}^{+\infty} dx\, \Psi^*(x,t)\, f(x)\, \Psi(x,t)\ . \tag{3.55}$$

What if we want to calculate the average momentum of the particle? Since we do not know the momentum as a function of position we can not use the above procedure. However, we could try the classical definition, i.e.,

$$\langle p\rangle = m\frac{d}{dt}\langle x\rangle = m\frac{d}{dt}\int_{\infty}^{\infty} dx\, \rho(x,t)\, x = m\int_{\infty}^{\infty} dx\, x\, \frac{\partial\rho}{\partial t}\ . \tag{3.56}$$

Making use of the continuity equation, Eq.(3.49), we can write

$$\langle p\rangle = m\int_{\infty}^{\infty} dx\, x\, \frac{\partial\rho}{\partial t} = -m\int_{\infty}^{\infty} dx\, x\, \frac{\partial}{\partial x} j(x,t)\ . \tag{3.57}$$

Integrating by parts, we get

$$\langle p\rangle = -mxj(x,t)\Big|_{-\infty}^{+\infty} + m\int_{-\infty}^{+\infty} dx\ j(x,t)\ .$$

Since the amplitude $\Psi(x,t)$ goes to zero faster than x^{-1} as $x\to\pm\infty$, then $x\,j(x,t)\to 0$ as $x\to\pm\infty$ and we have

$$\begin{aligned}\langle p\rangle &= m\int_{-\infty}^{+\infty} dx\ j(x,t) = \frac{\hbar}{2i}\int_{-\infty}^{+\infty} dx\left(\Psi^*\frac{\partial\Psi}{\partial x} - \frac{\partial\Psi^*}{\partial x}\Psi\right) \\ &= \int_{-\infty}^{+\infty} dx\ \Psi^*(x,t)\left(-i\hbar\frac{\partial}{\partial x}\right)\Psi(x,t)\ . \end{aligned} \tag{3.58}$$

To get the last line of Eq. (3.58), we have integrated by parts and used the fact that $\Psi(x,t) \rightarrow 0$ as $x \rightarrow \pm\infty$. Thus, to calculate the average momentum using the wave function $\Psi(x,t)$ we have to write the momentum p in terms of the coordinate x as

$$\tilde{p} = -i\hbar\frac{\partial}{\partial x} , \tag{3.59}$$

the ˜ on top of the p indicates that it represents the momentum as a differential operator. In a similar manner, we can show that the average of $\tilde{p}^n$ is given by

$$\langle \tilde{p}^n \rangle = \int_{-\infty}^{+\infty} dx \; \Psi^*(x,t) \left(-i\hbar\frac{\partial}{\partial x}\right)^n \Psi(x,t) , \tag{3.60}$$

and the average of any function of momentum, $f(\tilde{p})$, that can be written as a power series in $\tilde{p}$ is given by

$$\langle f(\tilde{p}) \rangle = \int_{-\infty}^{+\infty} dx \; \Psi^*(x,t)\, f\left(-i\hbar\frac{\partial}{\partial x}\right) \Psi(x,t) . \tag{3.61}$$

Here, we can raise the question of what is meant by the function $f\left(-i\hbar\frac{\partial}{\partial x}\right)$. In general, any analytic function $f(p)$ can be written as

$$f(p) = a_0 + a_1 p + a_2 p^2 + \cdots$$

with the a_i, $i = 0, 1, \cdots$ as complex constants. This allows us to define any function of an operator in terms of a power series in the operator. e.g., a function of the momentum operator is given by the power series

$$f\left(-i\hbar\frac{\partial}{\partial x}\right) = a_0 + a_1\left(-i\hbar\frac{\partial}{\partial x}\right) + a_2\left(-i\hbar\frac{\partial}{\partial x}\right)^2 + \cdots . \tag{3.62}$$

This is a well defined operator for which we can calculate the above integral with the expectation that the resultant series for $\langle f(\tilde{p})\rangle$ can be summed to give a finite result.

An alternative way of writing the average momentum makes use of the fact that we can write a Fourier expansion for $\Psi(x,t)$, i.e

$$\Psi(x,t) = \frac{1}{\sqrt{2\pi}} \int_{-\infty}^{+\infty} dk \; \Phi(k,t)\, e^{ikx} ,$$

where

$$\Phi(k,t) = \phi(k)\, e^{-i\omega t} .$$

We now can write the average momentum in terms of the function $\Phi(k,t)$ as

$$\begin{aligned}
\langle p\rangle &= \frac{1}{2\pi}\int_{-\infty}^{+\infty} dx\,dk\,dk'\; e^{-ikx}\Phi^*(k,t)\left(-i\hbar\frac{\partial}{\partial x}\right)\Phi(k',t)\,e^{ik'x} \\
&= \frac{1}{2\pi}\int_{-\infty}^{+\infty} dk\,dk'\;\Phi^*(k,t)\;\hbar k'\;\Phi(k',t)\int_{-\infty}^{+\infty} dx\,e^{i(k'-k)x} \\
&= \int_{-\infty}^{+\infty} dk\,dk'\;\Phi^*(k,t)\;\hbar k'\;\Phi(k',t)\,\delta(k-k') \\
&= \int_{-\infty}^{+\infty} dk\;\Phi^*(k,t)\;\hbar k\;\Phi(k,t) \\
&= \int_{-\infty}^{+\infty} dk\;|\Phi(k,t)|^2\,\hbar k \\
&\equiv \int_{-\infty}^{+\infty} dk\;\rho(k,t)\,\hbar k\;. \qquad (3.63)
\end{aligned}$$

This suggests that we can consider $\Phi(k,t)$ to be the probability amplitude of finding the particle with momentum $p = \hbar k$ at time t, and $\rho(k,t) = |\Phi(k,t)|^2$ as the corresponding probability of finding the particle with momentum $p = \hbar k$ at time t. In other words, while $\Psi(x,t)$ describes the particle in coordinate space, $\Phi(k,t)$ describes the same particle in momentum space. We will see later, when we consider problems in Atomic and Solid State Physics, that experiments often give a direct measurement of $\Phi(k,t)$ and not $\Psi(x,t)$, and to get a physical image of what the experimental measurement gives, we will have to Fourier transform the experimental results (or data) from momentum space to coordinate space. From this point on we will refer to $\Psi(x,t)$ as the *wave function in coordinate space*, and $\Phi(k,t)$ as the *wave function in momentum space*.[8]

So far we have established that in coordinate space, the average position and the average momentum of the particle are given by integrals involving the function $\Psi(x,t)$, and by the operators corresponding to the position and momentum of the particle, i.e.,

$$\tilde{x} = x \qquad \text{and} \qquad \tilde{p} = -i\hbar\frac{\partial}{\partial x}\;, \qquad (3.64)$$

[8]Since the 1930's there have been objections to the interpretation of $|\Psi(x,t)|^2$ as a probability by no lesser persons than Schrödinger (Schrödinger Cat [20]) and Einstein (The EPR experiment [21]). This has raised many questions regarding the interpretation of the wave function in quantum mechanics and the measurement process. For a recent review of the subject see G. Greenstein and A.G. Zajonc [22]. At this stage there is no experimental evidence suggesting that the probabilistic interpretation presented in this eBook is not valid.

On the other hand, in momentum space, the average momentum of the particle is given by an integral involving the wave function $\Phi(k,t)$ and the momentum operator $\tilde{p} = p = \hbar k$. To complete the symmetry between the two spaces, we need to determine the average position of the particle in terms of the momentum space wave function $\Phi(k,t)$. To establish this relation, we commence with the definition of the average position as given in Eq. (3.53), i.e.,

$$\langle x \rangle = \int_{-\infty}^{+\infty} dx \; \Psi^*(x,t)\, x\, \Psi(x,t) \; ,$$

and write the coordinate space wave function $\Psi(x,t)$ in terms of the momentum space wave function $\Phi(k,t)$. This allows us to write the average position as

$$\begin{aligned}
\langle x \rangle &= \frac{1}{2\pi} \int_{-\infty}^{+\infty} dx\, dk\, dk' \;\; \Phi^*(k,t)\, e^{-ikx}\; x\, \Phi(k',t)\, e^{ik'x} \\
&= \frac{1}{2\pi} \int_{-\infty}^{+\infty} dx\, dk\, dk' \;\; \Phi^*(k,t)\, \Phi(k',t) \left(-i\frac{\partial}{\partial k'}\right) e^{i(k'-k)x} \\
&= \int_{-\infty}^{+\infty} dk\, dk' \; \Phi^*(k,t)\, \Phi(k',t) \left(-i\frac{\partial}{\partial k'}\right) \delta(k'-k) \; .
\end{aligned}$$

We now integrate by parts over k' with the result that

$$\begin{aligned}
\langle x \rangle &= \int_{-\infty}^{+\infty} dk\, dk' \;\; \Phi^*(k,t)\, \delta(k'-k) \left(i\frac{\partial}{\partial k'}\right) \Phi(k',t) \\
&= \int_{-\infty}^{+\infty} dk \;\; \Phi^*(k,t) \left(i\hbar\frac{\partial}{\partial p}\right) \Phi(k,t) \; . \qquad (3.65)
\end{aligned}$$

Thus in momentum space we have, for the position and momentum operators,

$$\tilde{x} = i\hbar\frac{\partial}{\partial p} \qquad \text{and} \qquad \tilde{p} = p \; . \qquad (3.66)$$

The results of these quantization rules for the energy, position, and the momentum of a particle are summarized in Table 3.1. These quantization rules will allow us to quantize any system for which we know the classical Hamiltonian. Furthermore, the resultant equations can be written in coordinate space or momentum space. To get a better feeling for the behavior of quantum systems, we will initially consider all examples in coordinate space.

If we now consider the momentum and position operators as $\tilde{p}$ and $\tilde{x}$, then it is clear from the above definitions of these operators that

$$\tilde{p}\,\tilde{x}\;\Psi(x,t) \neq \tilde{x}\,\tilde{p}\;\Psi(x,t) \; .$$

Table 3.1: The quantization rules for the energy, position and momentum in coordinate and momentum space.

Classical variable	Quantum Mechanical operator	
	Coordinate space	Momentum space
E	$i\hbar\frac{\partial}{\partial t}$	$i\hbar\frac{\partial}{\partial t}$
x	$\tilde{x} = x$	$\tilde{x} = i\hbar\frac{\partial}{\partial p}$
p	$\tilde{p} = -i\hbar\frac{\partial}{\partial x}$	$\tilde{p} = p$

In fact, in coordinate space we have that

$$\begin{aligned} \tilde{p}\,\tilde{x}\;\Psi(x,t) &= -i\hbar\frac{\partial}{\partial x}\left(x\,\Psi(x,t)\right) \\ &= -i\hbar\Psi(x,t) - i\hbar x\,\frac{\partial\Psi}{\partial x} \\ &= -i\hbar\,\Psi(x,t) + \tilde{x}\,\tilde{p}\;\Psi(x,t)\;. \end{aligned} \tag{3.67}$$

or

$$(\tilde{x}\tilde{p} - \tilde{p}\tilde{x})\,\Psi(x,t) = i\hbar\Psi(x,t)\;.$$

Since this result is valid for any function $\Psi(x,t)$, we have that

$$\tilde{x}\tilde{p} - \tilde{p}\tilde{x} = i\hbar\;. \tag{3.68}$$

If we now define for any two operators $\tilde{A}$ and $\tilde{B}$, the commutator

$$\left[\tilde{A},\tilde{B}\right] \equiv \tilde{A}\tilde{B} - \tilde{B}\tilde{A}\;,$$

we can write

$$[\tilde{x},\tilde{p}] = i\hbar\;. \tag{3.69}$$

This is known as the commutation relation between the coordinate $\tilde{x}$ and the canonical momentum $\tilde{p}$.

We are now in a position to outline the procedure for quantizing any classical theory. To illustrate this, let us consider a Lagrangian $L(q,\dot{q})$ for a single particle in one-dimension. For the position q we can define the canonical momentum by the standard procedure, i.e.,

$$p = \frac{\partial L}{\partial \dot{q}}\;. \tag{3.70}$$

The corresponding classical Hamiltonian is now given by

$$H(p,q) = \dot{q}p - L(q,\dot{q})\;. \tag{3.71}$$

The quantization procedure involves replacing the position and momentum variable by the corresponding position and momentum operators, i.e.,

$$q \rightarrow \tilde{q} \qquad \text{and} \qquad p \rightarrow \tilde{p} \,. \tag{3.72}$$

These operators then satisfy the commutation relation

$$[\tilde{q}, \tilde{p}] = i\hbar \,. \tag{3.73}$$

If we now make this substitution for the coordinate and corresponding momentum in the Hamiltonian, i.e.,

$$H(p,q) \rightarrow \tilde{H}\left(\tilde{p},\tilde{q}\right) \,, \tag{3.74}$$

then the states of the system are described by the Schrödinger equation

$$i\hbar\frac{\partial\Psi}{\partial t} = \tilde{H}\left(\tilde{p},\tilde{q}\right)\Psi \,, \tag{3.75}$$

where Ψ can be the wave function in either coordinate or momentum space. Although the above procedure for quantization has been carried out for one particle in one-dimension, the generalization to more than one particle in more than one dimension is simply achieved by giving the position and momentum operators as vector operators with each set of position and momentum operators having a particle label, i.e.,

$$\begin{aligned} & \tilde{p} \rightarrow \tilde{\vec{p}}_i \quad \text{and} \quad \tilde{x} \rightarrow \tilde{\vec{r}}_i \quad \text{with} \quad i = 1, \cdots, n \\ \text{or} \quad & \tilde{p} \rightarrow \tilde{p}_j \quad \text{and} \quad \tilde{x} \rightarrow \tilde{x}_j \quad \text{with} \quad j = 1, \cdots, 3n, \end{aligned} \tag{3.76}$$

where n is the number of particles in three dimensional space. We now can write the general commutation relations between the position and momentum operators as

$$[\tilde{x}_i, \tilde{p}_j] = i\hbar\,\delta_{i.j} \,, \tag{3.77}$$

where $i, j = 1, \cdots, 3n$.

In the event that we have a Lagrangian that is Lorentz invariant, the corresponding equation to Eq. (3.75) will be the Klein-Gordon equation given in Eq. (3.44). Finally, we note that this procedure for quantizing a classical theory can be applied to fields, and thus we are able to derive the quantum theory of radiation by quantizing Maxwell's equation for the electromagnetic fields from a Lagrangian for the electromagnetic fields. In this way, we have a unified theory for both particles and waves, and a procedure for quantizing the corresponding classical theory.[9]

[9]The above quantization is a simplification of the more general canonical quantization procedure in which the Poisson brackets for canonical variables are replaced by the corresponding commutation relation in the classical equation of motion (see Dirac [15] Sec. 21). For an illustration and limitation of this general canonical quantization see e.g. Schiff [17] Sec.24.

The above quantization program is known as canonical quantization since the commutation relation is written between canonical variables, in this case, the position and momentum. There is an alternative quantization program based on path integrals. Here, the quantization is achieved by summing over all possible paths each weighted by a factor of [13]

$$\exp\{-\frac{i}{\hbar}S\}$$

where S is the classical action defined in the last chapter.

3.6 Problems

1. A free electron bounces elastically back and forth in one dimension between two walls that are $L = 0.50$ nm apart.

   (a) Assuming that the electron is represented by a de Broglie standing wave with a node at each wall, show that the permitted de Broglie wave lengths are $\lambda = 2L/n,\ (n = 1, 2, \ldots)$.

   (b) Find the values of the kinetic energy of the electron for $n = 1, 2,$ and 3.

   (c) Use *Mathematica* to plot the wave function for $n = 1, 2, 3, 4$. Can you see any difference between the odd n and even n wave functions if the walls are at $x = \pm\frac{L}{2}$?

2. Use *Mathematica* to plot two waves with wave numbers 0.9 and 1.1, and frequencies 0.85 and 1.0. Show that the sum of these waves has an amplitude that is not uniform in space. What happens if you add a third wave with wave number 1.0 and frequency 1.1?

3. Given the function $f(x)$ that is localized in space between $\{-1, +1\}$, and defined as:

$$f(x) = \begin{cases} 1 - |x| & \text{for} \quad |x| < 1 \\ 0 & \text{for} \quad |x| \geq 1 \end{cases} .$$

   (a) Plot $f(x)$ as function of x.

   (b) Find the Fourier transform $g(k)$ using Eq. (3.22b).

   (c) Plot $g(k)$ as function of k for $-6 < k < +6$.

4. The momentum distribution for a wave packet is given by

$$g(k) = \frac{N}{k^2 + \beta^2} .$$

where β is a constant and N is the normalization of the wave packet.

(a) Calculate the normalization of the packet.

(b) Calculate the width of the packet in coordinate space by calculating $f(x)$.

(c) Taking the width of the wave packet to be the width of the distribution when the height is e^{-1} of the maximum, show that

$$\Delta k \, \Delta x > 1$$

is independent of β.

Hint: You may use the following integrals. See Appendix B for the method of evaluation of such integrals.

$$\int_{-\infty}^{+\infty} dk \, \frac{1}{(k^2+\beta^2)^2} = \frac{\pi}{2\beta^3} \quad \text{and} \quad \int_{-\infty}^{+\infty} dk \, \frac{e^{ikx}}{k^2+\beta^2} = \frac{\pi}{\beta} e^{-\beta|x|}$$

5. A beam of electrons is to be fired over a distance of 10^4 km. If the size of the initial packet is 1 mm, what will be its size upon arrival, if its kinetic energy is:

(a) 13.6 eV,

(b) 100 MeV.

(c) Write a *Mathematica* program that calculates the spreading in the width of a wave packet with energy over the distance of 10^4 km. (This program should cover both of the energies considered in the previous parts of this question.)

Note, the relation between kinetic energy and momentum is not always K.E. $= p^2/2m$. When the momentum of the particle becomes comparable to its mass, we need to use the relativistic relation between the energy and momentum.

6. Suppose that $V(x)$ is complex, i.e., $V(x) = V_R(x) + iV_I(x)$.

(a) Derive an expression for the rate of change of the density $\frac{\partial \rho}{\partial t}$, for the Hamiltonian

$$H = -\frac{\hbar^2}{2m}\frac{\partial^2}{\partial x^2} + V(x)$$

(b) Calculate

$$\frac{d}{dt}\int dx \, \rho(x,t) \, .$$

(c) Show that for absorption,

$$\frac{d}{dt}\int dx\,\rho(x,t)$$

must be negative. What does this tell us about $V(x)$?

7. The coordinate space wave function for a particle is given by

$$\psi(x) = \frac{N}{x^2 + a^2} \ .$$

(a) Calculate N, the normalization of the wave function $\psi(x)$.

(b) For what values of n is the integral $\langle x^n \rangle$ defined?

(c) Use the above coordinate space wave function to calculate $\langle x^n \rangle$.

(d) Calculate the momentum space wave function $\phi(k)$.

(e) Calculate $\langle p^2 \rangle$ directly in coordinate and momentum space.

(f) Use the definitions

$$\begin{aligned} \Delta x &= \sqrt{\langle x^2 \rangle - \langle x \rangle^2} \\ \Delta p &= \sqrt{\langle p^2 \rangle - \langle p \rangle^2} \end{aligned}$$

to calculate $\Delta x \Delta p$. Is this consistent with the Heisenberg uncertainty relation?

8. Show that the operator relation

$$e^{ipa/\hbar}\, x\, e^{-ipa/\hbar} = x + a$$

is valid. The operator e^A is defined to be

$$e^A = \sum_{n=0}^{\infty} \frac{A^n}{n!} \ .$$

[*Hint:* Calculate $e^{ipa/\hbar}\, x\, e^{-ipa/\hbar} f(p)$ where $f(p)$ is any function of p, and use the representation $x = i\hbar\frac{d}{dp}$.]

CHAPTER 4

The Schrödinger Equation

Abstract: In this chapter we turn to a solution of the Schrödinger equation for a simple one dimensional potential as a tool for understanding the meaning of the wave function with an introduction to symmetry and conservation laws.

In Chapter 2 we showed that a particle of mass m moving in one dimension in an external potential $V(x)$, is described by the classical Hamiltonian

$$H(p,x) = \frac{p^2}{2m} + V(x) \ . \tag{4.1}$$

If this system is then quantized, the wave function for the particle $\Psi(x,t)$ satisfies the Schrödinger equation

$$\begin{aligned} i\hbar\frac{\partial\Psi(x,t)}{\partial t} &= \tilde{H}(\tilde{p},\tilde{x})\Psi(x,t) \\ &= \left(-\frac{\hbar^2}{2m}\frac{\partial^2}{\partial x^2} + V(x)\right)\Psi(x,t) \ . \end{aligned} \tag{4.2}$$

In writing the second line of Eq. (4.2), we made use of the fact that in quantizing the system, the momentum operator $\tilde{p}$ is replaced by $-i\hbar\frac{\partial}{\partial x}$, while the coordinate operator $\tilde{x}$ is replaced by x.

In the present chapter we would like to discuss some of the general properties of the wave function $\Psi(x,t)$. To extract these general properties we will make use of the simple example of a particle in a box. As a first step in that direction, we will show how Eq. (4.2), a partial differential equation in two variables, can be reduced to a set of two ordinary differential equations. We then proceed to show that the solution of the Schrödinger equation can form a basis for the expansion of any function. This is a generalization of the results of Fourier where the sin and cos functions formed the basis for the expansion of a periodic function. We also demonstrate that the symmetries of the

Iraj R. Afnan

system are reflected in the wave function $\Psi(x,t)$ that describes the quantum mechanical behavior of the system. Finally, we consider the condition under which two or more quantities can be measured at the same time to any desired accuracy.

Although all the results demonstrated in this chapter are based on the simple example of a particle in a box, the results hold true in general for any potential $V(x)$, and form the basis for the mathematical structure of quantum mechanics. In Chapter 11 we will return to this mathematical structure and prove these results in their more general form.

4.1 Method of Separation of Variables

Before we can proceed to a general discussion of the solutions of the Schrödinger equation, we need to solve the above partial differential equation. This can be achieved in one of two ways such that the time and space dependence of the solution are separated. The first method is that of separation of variables, while the second approach involves the Fourier decomposition of the time dependence.

To implement the method of separation of variables, we rewrite the Schrödinger equation as

$$\left(i\hbar\frac{\partial}{\partial t} - \tilde{H}\right)\Psi(x,t) = 0 \tag{4.3}$$

or

$$\tilde{L}\Psi(x,t) = 0\ . \tag{4.4}$$

The operator $\tilde{L}$ is the sum of two parts. The first part depends on time, while the second part depends on the coordinate x.[1] Under these circumstances, the solution to Eq. (4.4), $\Psi(x,t)$, is the product of two functions - one depending on time only, the other depending on the coordinate x, i.e.,

$$\Psi(x,t) = F(t)\psi(x)\ . \tag{4.5}$$

With this form for the wave function we can write Eq. (4.3) as

$$i\hbar\frac{\partial F(t)}{\partial t}\,\psi(x) - F(t)\tilde{H}\psi(x) = 0\ .$$

If we now divide this equation by $F(t)\psi(x)$, we can rewrite it as

$$\frac{i\hbar}{F(t)}\,\frac{\partial F(t)}{\partial t} = \frac{1}{\psi(x)}\,\tilde{H}\psi(x)\ . \tag{4.6}$$

The left hand side of this equation depends on time only, while the right hand side depends on the position x only. Since time and position are independent variables, each side of Eq. (4.6) must be a constant for this equality to be valid, i.e.,

$$\frac{1}{\psi(x)}\,\tilde{H}\psi(x) = E \qquad \text{and} \qquad \frac{i\hbar}{F(t)}\,\frac{dF(t)}{dt} = E\ .$$

[1] Here we have assumed that the Hamiltonian $\tilde{H}$ is time independent.

Note, we have replaced the partial derivative by the total derivative since the function F depends on time only. At this stage, E is a constant with the same dimension as the Hamiltonian, i.e., it has the dimension of energy. The above procedure allows us to reduce the one partial differential equation in two variables, Eq. (4.2), to two ordinary differential equations in one variable. These two equations are

$$\tilde{H}\,\psi(x) = E\,\psi(x) \tag{4.7a}$$

known as the time independent Schrödinger equation, and

$$i\hbar\,\frac{dF(t)}{dt} = E\,F(t)\ . \tag{4.7b}$$

In Eq. (4.7b) we have a first order linear differential equation with the solution given by [2]

$$F(t) = F_0\ e^{-iEt/\hbar}\ . \tag{4.8}$$

This means that our total wave function is of the form

$$\Psi(x,t) = e^{-iEt/\hbar}\ \psi(x)\ . \tag{4.9}$$

In writing Eq. (4.9), we have incorporated, without any loss of generality, the constant F_0 into the function $\psi(x)$. This is justified on the grounds that the total wave function will be normalized. Here we observe that the solution given in Eq. (4.9) is general enough, provided that the potential $V(x)$, and therefore the Hamiltonian $\tilde{H}$ is time independent. To complete our solution of the Schrödinger equation we need to solve Eq. (4.7a) and for that we have to specify Hamiltonian $\tilde{H}$, and therefore the potential $V(x)$.

4.2 Method of Fourier Analysis

An alternative procedure for reducing partial differential equations to ordinary differential equations, involves the application of Fourier transforms. In this case we can reduce the dimensionality of the partial differential equation by the Fourier decomposition of the solution in one of the variables. For the Schrödinger equation, with a Hamiltonian that has no explicit time dependence, it is most convenient to Fourier decompose the time variable in the wave function, i.e.,

$$\Psi(x,t) = \frac{1}{\sqrt{2\pi}}\int_{-\infty}^{+\infty} dw\,\Psi(x,w)\,e^{-iwt}\ . \tag{4.10}$$

[2]It is straight forward to show, by substitution, that the function $F(t)$, given in Eq. (4.8), is a solution to the first order differential equation given in Eq. (4.7b).

Then by straight forward differentiation we have that:

$$i\hbar\,\frac{\partial\Psi}{\partial t} = \frac{1}{\sqrt{2\pi}}\int_{-\infty}^{+\infty} dw\;\hbar w\,\Psi(x,w)\,e^{-iwt}\ , \tag{4.11}$$

and

$$\tilde{H}\,\Psi(x,t) = \frac{1}{\sqrt{2\pi}}\int_{-\infty}^{+\infty} dw\;\tilde{H}\,\Psi(x,w)\,e^{-iwt}\ . \tag{4.12}$$

Therefore, the Schrödinger equation can be written as

$$\frac{1}{\sqrt{2\pi}}\int_{-\infty}^{+\infty} dw\,\left[\hbar w - \tilde{H}\right]\Psi(x,w)\,e^{-iwt} = 0\ . \tag{4.13}$$

This in general is only valid if

$$\left[\hbar w - \tilde{H}\right]\Psi(x,w) = 0\ , \tag{4.14}$$

where w is a parameter of the differential equation since it only appears in $\hbar w$. Equation (4.14), which now is an ordinary differential equation in x, is identical to Eq. (4.7a), if we take

$$E = \hbar w \qquad \text{and} \qquad \psi_E(x) = \Psi(x,w)\ . \tag{4.15}$$

In writing Eq. (4.15) we had to label the function $\psi(x)$ with the subscript E since the function Ψ depends on the variable ω and therefore $E = \hbar\omega$. Our solution for the time dependent Schrödinger equation is now of the form

$$\Psi(x,t) = \frac{1}{\sqrt{2\pi\hbar}}\int_{-\infty}^{+\infty} dE\;\psi_E(x)\,e^{-iEt/\hbar}\ . \tag{4.16}$$

In fact, this way of writing the time dependent wave function is a special case of a more general form which is given by

$$\Psi(x,t) = \frac{1}{\sqrt{2\pi\hbar}}\int dE\;C(E)\,\psi_E(x)\,e^{-iEt/\hbar} \tag{4.17}$$

which satisfies the Schrödinger equation for any function $C(E)$. This result is a consequence of the fact that the operator $\tilde{L} = i\hbar\frac{\partial}{\partial t} - \tilde{H}$ is linear. The question then is, what is a *linear operator*? An operator $\tilde{L}$ is said to be linear if

$$\tilde{L}\left(\Psi_1 + \Psi_2\right) = \tilde{L}\Psi_1 + \tilde{L}\Psi_2\ , \tag{4.18}$$

where in our case, Ψ_1 and Ψ_2 are functions of (x,t). A more general definition of a linear operator is [3]

$$\tilde{L}\left(a\Psi_1 + b\Psi_2\right) = a\tilde{L}\Psi_1 + b\tilde{L}\Psi_2 \ , \tag{4.19}$$

where a and b are any two complex numbers.

At this stage, we should point out that the integral over E (4.17) consists of a sum over a set of discrete energies, and an integral with a finite lower limit, i.e.,

$$\int dE \cdots = \sum_n \cdots + \int_{E_0}^{\infty} dE \cdots \ . \tag{4.20}$$

The fact that we have a sum over discrete energies is a result of the fact that Eq. (4.7a) will have solutions only for certain values of the energy $E_n < E_0$.

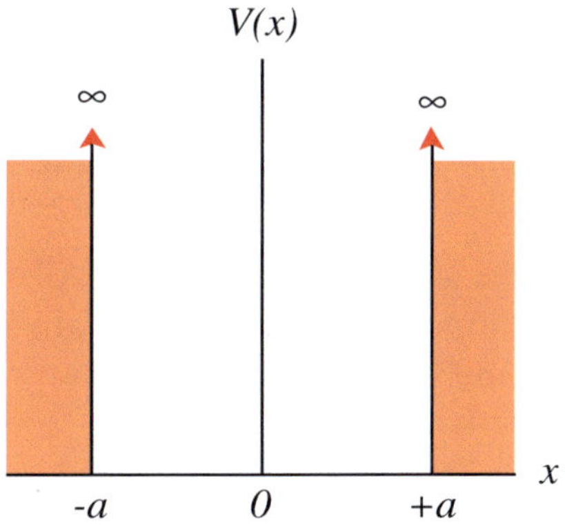

Figure 4.1: The potential for a particle in a one dimensional box with sides at $x = \pm a$.

4.3 Particle in a Box

To illustrate some of the general properties of the wave function $\psi(x)$,[4] we need to solve Eq. (4.7a) for a specific potential $V(x)$. The simplest such potential that also exhibits the properties of the wave function $\psi(x)$, is the potential that corresponds to a particle in a one dimensional box of length $2a$, see Figure 4.1 i.e.,

$$V(x) = \begin{cases} +\infty & x < -a \\ 0 & -a < x < a \\ +\infty & x > a \end{cases} \ . \tag{4.21}$$

[3]In Chapter 12 we will come across an anti-linear operator for which

$$\tilde{L}a\Psi = a^*\tilde{L}\Psi$$

where a is a complex number, and a^* is the complex conjugate of a.

[4]To simplify the notation, we have for the moment dropped the E subscript on $\psi(x)$.

Since the potential is infinite for $|x| > a$, the particle is constrained to move in the region $|x| < a$. In other words,

$$\psi(x) = 0 \qquad \text{for} \qquad |x| \geq a \ . \tag{4.22}$$

We therefore have to solve the time independent Schrödinger equation,

$$\tilde{H}\,\psi(x) = E\,\psi(x)$$

in the domain $|x| < a$. In this region since the potential is zero, the time independent Schrödinger equation takes the form

$$-\frac{\hbar^2}{2m}\,\frac{d^2\psi(x)}{dx^2} = E\,\psi(x) \qquad |x| < a$$

or, upon multiplication by $-2m/\hbar^2$, we have

$$\frac{d^2\psi(x)}{dx^2} = -k^2\,\psi(x) \quad \text{for} \quad |x| < a \ , \tag{4.23}$$

where $k^2 = 2mE/\hbar^2$. This equation is a second order linear differential equation, and therefore has two independent solutions. These could be $\sin kx$ and $\cos kx$. The general solution to this second ordered differential equation can be written as a linear combination of the two solutions, i.e.,

$$\psi(x) = B\sin kx + C\cos kx \ . \tag{4.24}$$

The constants B and C, and the parameter k can now be determined by the *boundary conditions* on the general solution given in Eq. (4.24).

The first set of boundary conditions are that the wave function $\psi(x)$ is zero at $x = \pm a$. This follows from the fact that the particle is constrained to the region $-a < x < +a$, i.e., $\psi(x) = 0$ for $|x| > a$. The application of these two boundary conditions gives us two equations for the constants B, C and k, which are

$$B\sin ka + C\cos ka = 0 \qquad \text{for} \qquad x = a \tag{4.25a}$$

$$-B\sin ka + C\cos ka = 0 \qquad \text{for} \qquad x = -a \ . \tag{4.25b}$$

The second set of boundary conditions for a second order differential equation are that the derivative of the wave function should be zero for $x = \pm a$. These boundary conditions result in Eqs. (4.25a) and (4.25b) with B and C interchanged. Since the final wave function is to be normalized (see Eq. (4.36)), this second set of boundary conditions give no additional constraints on the wave function and are not considered any further for this problem.

If we now add the two Eqs. (4.25a) and (4.25b), we get the condition

$$2\,C\cos ka = 0 \ . \tag{4.26}$$

There are two possible solutions to this equation. These are: (i) $C = 0$ which gives a zero wave function and therefore is of no interest to us. (ii) $C \neq 0$, in which case $\cos ka = 0$. For this condition to be satisfied, ka can take only certain values, i.e.,

$$ka = \frac{n\pi}{2} \quad \text{or} \quad k = \frac{n\pi}{2a} \quad \text{with} \quad n = 1, 3, 5, \cdots \ . \tag{4.27}$$

On the other hand, if we subtract Eq. (4.25b) from Eq. (4.25a), we get

$$2\,B \sin ka = 0 \ . \tag{4.28}$$

Here again, there are two possible solutions with the non-trivial solution corresponding to $B \neq 0$, in which case $\sin ka = 0$ and ka can take on only certain values, i.e.,

$$ka = \frac{n\pi}{2} \quad \text{or} \quad k = \frac{n\pi}{2a} \quad \text{with} \quad n = 2, 4, \ldots \ . \tag{4.29}$$

We can combine the results in Eqs. (4.27) and (4.29) by noting that when $C \neq 0$, k is given by Eq. (4.27) with n odd, while for $B \neq 0$, k is given by the same equation with n even. Furthermore, the parameter k is related to E by

$$k = \sqrt{\frac{2mE}{\hbar^2}} \ .$$

Thus, in general, we can write k as [5]

$$k_n = \sqrt{\frac{2mE_n}{\hbar^2}} = \frac{\pi}{2a}\,(n+1) \qquad \text{for} \qquad n = 0, 1, 2, \ldots \ , \tag{4.30}$$

where we have introduced a subscript for both k and E to indicate that these quantities depend on n. This result proves that the energy E_n can have only certain discrete values, and these are given by

$$E_n = \frac{\hbar^2 k_n^2}{2m} = \frac{\hbar^2 \pi^2}{8ma^2}\,(n+1)^2 \quad \text{with} \quad n = 0, 1, 2, \ldots \ , \tag{4.31}$$

while the corresponding wave functions are given by

$$\psi_n(x) = \begin{cases} B_n \cos k_n x & \text{for} \quad n = 0, 2, \ldots \\ \\ B_n \sin k_n x & \text{for} \quad n = 1, 3, \ldots \end{cases} \ . \tag{4.32}$$

In writing Eq. (4.32), we have: (i) Introduced a subscript for the wave function n. This subscript replaces the energy, given the fact that the energy is a function of n. (ii) Replaced the constants B and C by the constant B_n, which corresponds to taking $B \to B_n$ for n an even integer, or $C \to B_n$ for n an odd integer. In this way we have emphasized the fact

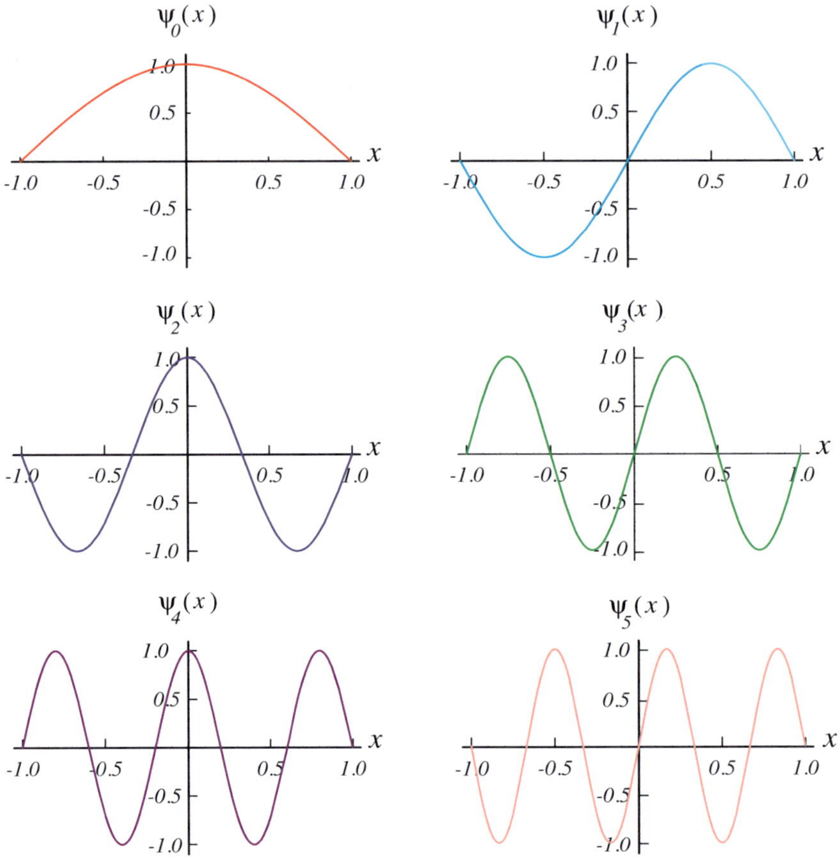

Figure 4.2: Here we have a plot of the wave function for a particle in a box with $a = 1$, $\hbar = m = 1$, and $n = 0, 1, \ldots, 5$..

that we have one arbitrary constant, with the overall magnitude of the wave function, B_n, to be determined.

As stated earlier, the third constant in the wave function B_n, is determined by the normalization of the wave function, i.e., the overall magnitude of the wave function B_n, is determined by the condition that,

$$\int_{-\infty}^{+\infty} dx \ \psi_n^*(x)\, \psi_n(x) = \int_{-a}^{+a} dx \ \psi_n^*(x)\, \psi_n(x) = 1 \ , \tag{4.33}$$

or

$$B_n^2 \int_{-a}^{+a} dx \ \sin^2 k_n x = 1 \qquad \text{for} \qquad n = \text{odd} \ , \tag{4.34a}$$

[5]We have replaced $n \to (n+1)$ in order to have $n = 0, 1, 2, \ldots$. In this way the lowest energy, i.e., the ground state, corresponds to $n = 0$. This is a convention used in most books on the subject.

and

$$B_n^2 \int_{-a}^{+a} dx \cos^2 k_n x = 1 \qquad \text{for} \qquad n = \text{even} \; . \tag{4.34b}$$

Evaluating these integrals gives us [6]

$$B_n = \frac{1}{\sqrt{a}} \; . \tag{4.35}$$

Therefore the normalized wave function for a particle in a one dimensional box is given for $|x| < a$ by

$$\psi_n(x) = \frac{1}{\sqrt{a}} \begin{cases} \cos k_n x & \text{for} \quad n = \text{even} \\ \\ \sin k_n x & \text{for} \quad n = \text{odd} \end{cases} , \tag{4.36}$$

while $\psi_n(x) = 0$ for $|x| \geq a$. The wave functions $\psi_n(x)$ for a particle in a box are illustrated in Figure 4.2 for $n = 0, \cdots, 5$ for the case when $a = 1$ and $\hbar = m = 1$.

4.4 General Properties of the Wave Function $\psi_n(x)$

Having determined the general solution of the Schrödinger equation for the simple case of a particle in a box, we proceed in this section to examine some of the mathematical properties of this solution that hold true in general for all solutions of the Schrödinger equation for any potential. In this way we can illustrate the mathematical structure of Quantum Mechanics without getting involved in the mathematical foundation of the theory.

With the solution of the Schrödinger equation for a particle in a box, we can now calculate the following integral

$$\begin{aligned} \int_{-\infty}^{+\infty} dx \; \psi_n^*(x)\, \psi_m(x) &= \frac{1}{a} \int_{-a}^{+a} dx \;\; \cos k_n x \, \cos k_m x \quad \text{for } n \text{ and } m \text{ even} \\ &= \frac{1}{a} \int_{-a}^{+a} dx \;\; \sin k_n x \, \sin k_m x \quad \text{for } n \text{ and } m \text{ odd} \\ &= \frac{1}{a} \int_{-a}^{+a} dx \;\; \sin k_n x \, \cos k_m x \quad \text{for } n \text{ odd and } m \text{ even} \; . \end{aligned}$$

---

[6] We make use of the following integrals,

$$\int_{-\pi/2}^{+\pi/2} dx \, \sin^2 nx = \int_{-\pi/2}^{+\pi/2} dx \, \cos^2 nx = \frac{\pi}{2} \quad \text{for } n \text{ integer} \; .$$

Here we find that for $n \neq m$, the integral is zero,[7] i.e.,

$$\int_{-\infty}^{+\infty} dx\; \psi_n^*(x)\,\psi_m(x) = \begin{cases} 0 & n \neq m \\ 1 & n = m \end{cases}$$
$$\equiv \delta_{nm} \,. \tag{4.37}$$

In writing the last line in the above equation, we have defined a new function δ_{mn} (The *Kronecker delta* function) that takes the value of one if $n = m$, and zero otherwise. The motivation for introducing this new function is that we will encounter integrals of the above form repeatedly in Quantum Mechanics.

To give a physical meaning to the result in Eq. (4.37), let us assume we want to evaluate the integral on the left-hand-side of Eq. (4.37) numerically. This involves replacing the integral by a sum, i.e.,

$$\int_{-\infty}^{+\infty} dx\; \psi_n^*(x)\,\psi_m(x) = \sum_{i=1}^{N} \psi_n^*(x_i)\,\psi_m(x_i) \,.$$

This sum looks like the product of a row matrix of length N and a column matrix of length N with the result that for $n \neq m$ the product is zero, while for $n = m$ the product is one. In linear algebra this property is known as orthogonality and normalization. In other words, the functions $\psi_n(x)$ are normalized and orthogonal, i.e., they are *orthonormal.* Note, the process of replacing the integral in Eq. (4.37) by a sum, replaces the function $\psi_n(x)$ by a column matrix $\psi_n(x_i)$ in which the elements of the matrix are labeled by x_i, and these cover the domain over which the function $\psi_n(x)$ is defined.

A second property the functions $\psi_n(x)$ have is that any function $f(x)$ that satisfies the boundary condition $f(a) = 0 = f(-a)$, can be written as

$$f(x) = \sum_{n=0}^{\infty} A_n\; \psi_n(x) \,. \tag{4.38}$$

In this case, because the wave function $\psi_n(x)$ is $\sin k_n x$ or $\cos k_n x$, the above expansion reduces to the Fourier series for the function $f(x)$ since Eq. (4.38) can be written as

$$f(x) = \sum_{n=\text{ even}} \frac{A_n}{\sqrt{a}} \cos k_n x + \sum_{n=\text{ odd}} \frac{A_n}{\sqrt{a}} \sin k_n x \,.$$

---

[7]We have made use of the fact that

$$\int_{-\pi/2}^{+\pi/2} dx\; \sin nx\; \sin mx = \int_{-\pi/2}^{+\pi/2} dx\; \cos nx\; \cos mx = 0 \quad \text{for } n \neq m, n \text{ and } m \text{ integer} \,.$$

However, this result is true for any set of functions $\psi_n(x)$ that are the solution of an equation (e.g., the time independent Schrödinger equation) of the form

$$\tilde{H}\psi_n = E_n\psi_n \tag{4.39}$$

provided $\tilde{H}$ is a linear Hermitian operator.[8] Here again we can recast Eq. (4.39) in the language of linear algebra. In particular, we consider the problem of eigenstates and eigenvalues of a matrix. If we think of the Hamiltonian $\tilde{H}$ as a matrix of dimension $N \times N$, then Eq. (4.39) suggests that we refer to the wave function $\psi_n(x)$ as the eigenstate of the matrix $\tilde{H}$, with E_n as the corresponding eigenvalue, i.e., the eigenstate $\psi_n(x)$ is taken to be a column matrix of length N.

This equivalence between *Quantum Mechanics* (i.e., the solution of the Schrödinger equation) and *linear algebra* allows us to make use of many of the results derived in linear algebra when examining quantum mechanical systems. To illustrate this analogy, consider the eigenstates of a real symmetric matrix where the eigenstates belonging to different eigenvalues are orthogonal. In the case of matrices, this orthogonality takes the form of a multiplication of a row matrix by a column matrix, while in quantum mechanics this orthogonality takes the form given in Eq. (4.37). These two forms become identical if we replace the integral in Eq. (4.37) by a sum over quadratures, which is how we perform integration on a computer.

To illustrate the matrix nature of the Hamiltonian $\tilde{H}$, we first consider the integral

$$\int dx\; \psi_m^*(x)\, \tilde{H}\, \psi_n(x) \;.$$

Making use of the fact that $\psi_n(x)$ is an eigenstate of $\tilde{H}$ with eigenvalue E_n, and the eigenstates form an orthonormal basis, we can write

$$\begin{aligned} \int_{-a}^{+a} dx\; \psi_m^*(x)\, \tilde{H}\, \psi_n(x) &= \int_{-a}^{+a} dx\; \psi_m^*(x)\, E_n\, \psi_n(x) \\ &= E_n \int_{-a}^{+a} dx\; \psi_m^*(x)\, \psi_n(x) = E_n\, \delta_{mn} \;. \end{aligned} \tag{4.40}$$

[8]The matrix H is Hermitian if $(H^*)^T \equiv H^\dagger = H$, where $(H^*)^T$ in the complex conjugate transpose, i.e., if we define the (n, m) elements of the matrix H as h_{nm}, i.e.,

$$[H]_{nm} = h_{nm} \quad \text{then} \quad [(H^*)^T]_{nm} = h_{mn}^* \;.$$

Therefore, for a Hermitian matrix we have

$$h_{nm} = h_{mn}^* \;.$$

The quantity we have calculated in Eq. (4.40) we refer to as the *matrix element* of the Hamiltonian $\tilde{H}$ between wave functions which, in this case, are the eigenstates of this same Hamiltonian. This gives us the energy of the system in a given state n. In other words, the Hamiltonian matrix is diagonal and the elements of this diagonal matrix are the energies of the system. In Quantum Mechanics terminology this can be stated as: *The Hamiltonian $\tilde{H}$ is an operator corresponding to the energy observable E_n.* On the other hand in the language of *linear algebra* we say that ψ_n is an eigenstate of the Hermitian matrix $\tilde{H}$ with eigenvalue E_n.

Let us now consider a system described by the wave function $f(x)$. This wave function can be written as in Eq. (4.38) in terms of the functions $\psi_n(x)$, which in this case are solutions of the Schrödinger equation for a given potential, e.g., the potential given in Eq. (4.21). To give a physical meaning to the coefficients A_n in Eq. (4.38), we first multiply Eq. (4.38) from the left by $\psi^*_m(x)$ and integrate over the coordinate x. Because of the orthonormality (see Eq. (4.37)) of the wave function $\psi_n(x)$, we can write the coefficients A_n, as

$$A_n = \int_{-\infty}^{+\infty} dx\,\psi^*_n(x) f(x) = \int_{-a}^{+a} dx\;\psi^*_n(x)\; f(x)\;. \tag{4.41}$$

We now consider the integral

$$\int dx\;f^*(x)\;\tilde{H}\;f(x)\;.$$

Using first the expansion for the function $f(x)$ in Eq. (4.38), we can then write the matrix element of the Hamiltonian $\tilde{H}$ with respect to the state $f(x)$ as

$$\int dx\;f^*(x)\,\tilde{H}\,f(x) = \sum_n\sum_m A^*_n A_m \int dx\,\psi^*_n(x)\,\tilde{H}\,\psi_m(x)\;.$$

Making use of the orthonormality of the eigenstates of the Hamiltonian $\tilde{H}$, we can write this integral as

$$\begin{aligned}\int dx\;f^*(x)\,\tilde{H}\,f(x) &= \sum_{nm} A^*_n A_m E_m \int dx\,\psi^*_n(x)\,\psi_m(x)\\ &= \sum_n |A_n|^2 E_n\;.\end{aligned} \tag{4.42}$$

On the other hand, the normalization of the wave function $f(x)$ implies that

$$\int dx\;f^*(x) f(x) = \sum_N |A_n|^2 = 1\;. \tag{4.43}$$

From the above results given in Eqs. (3.42) and (3.43), we may conclude that:

1. From the structure of Eq. (4.43) we may consider $|A_n|^2$ as a probability. In particular, this $|A_n|^2$ is the probability that the particle described by the wave function $f(x)$ is in a state with energy E_n. In this case the normalization is another way of stating that the probability of finding the particle in any state ψ_n is one.

2. If $|A_n|^2$ is a probability of finding the particle with energy E_n, then $\sum\limits_n |A_n|^2 E_n$ is the average energy of the particle described by the function $f(x)$.

Since $|A_n|^2$ is a probability, then A_n, defined in Eq. (4.40), is the *probability amplitude.* In particular, for a particle described by the wave function $f(x)$, A_n is the probability amplitude of finding the particle in a state with energy E_n.

4.5 Symmetry Under Inversion - Parity

Before we can commence our discussion on symmetry we should make sure we all agree as to what is meant by symmetry.

Consider a chain with beads spaced at intervals of a cm along the chain. If you move the chain by a distance a along its length while we blink our eyes, then we will not be able to tell if you have done anything to the chain, provided of course, that we can't see the ends of the chain. Under these conditions we would say that the chain has a symmetry corresponding to translation of the chain by a distance a cm.

Our problem of a particle in a box, discussed in Sec. 4.3, has a symmetry under inversion, i.e., when $x \to -x$, the Hamiltonian for the system does not change. This is clear from Figure 4.1. This symmetry is known as parity, and the operator for inversion, $\tilde{P}$, has the property that

$$\tilde{P}\,\psi(x) = \psi(-x)\ . \tag{4.44}$$

The wave functions for the particle in a box as given in Eq. (4.36) can be divided into two classes - those that change sign under the parity operation, and those that do not change sign, i.e.,

$$\psi_n(-x) = \begin{cases} +\,\psi_n(x) & \text{for} \quad n \text{ even} \\ \\ -\,\psi_n(x) & \text{for} \quad n \text{ odd} \end{cases}\ . \tag{4.45}$$

In other words

$$\tilde{P}\,\psi_n(x) = \psi_n(-x) = (-1)^n\,\psi_n(x)\ . \tag{4.46}$$

Thus, the wave function or eigenstate ψ_n, which is an eigenstate of the Hamiltonian, is also an eigenstate of the parity operator $\tilde{P}$. This is established by the fact that when the operator $\tilde{P}$ acts on the eigenstate $\psi_n(x)$ we get the corresponding eigenvalue $(-1)^n$ times the same eigenstate. This means we can label the states of the particle $\psi_n(x)$, by both the energy and the parity. In this case it turns out that n can label both the energy and the parity of the state, but this is a special case. We could have written the eigenstate of $\tilde{H}$ and $\tilde{P}$ as $\psi_{E,\pi}$ where E labels the energy, and $\pi = \pm 1$ labels the parity of the state.

The reason we can construct one eigenstate which is both an eigenstate of the Hamiltonian $\tilde{H}$ and an eigenstate of parity $\tilde{P}$ is because of the symmetry of the Hamiltonian $\tilde{H}$ under the parity operation. In this case the Hamiltonian does not change under the transformation $x \rightarrow -x$ in $\tilde{H}$. This leads us to the concept of invariance of the theory under that symmetry. To illustrate this, consider the time dependent Schrödinger equation for the Hamiltonian $\tilde{H}$, i.e.,

$$i\hbar \frac{\partial \Psi(x,t)}{\partial t} = \tilde{H}\Psi(x,t) \ . \tag{4.47}$$

If we operate on this equation with the parity operator, $\tilde{P}$, we get

$$i\hbar \tilde{P}\frac{\partial \Psi}{\partial t} = i\hbar \frac{\partial(\tilde{P}\Psi)}{\partial t} = \tilde{P}\tilde{H}\Psi \ , \tag{4.48}$$

where we have assumed that the parity operator does not depend on time and can be taken inside the differentiation. But in general the action of the operator $\tilde{P}$ on a state Ψ gives a new state, i.e.,

$$\tilde{P}\,\Psi = \Psi' \qquad \text{and} \qquad \Psi = \tilde{P}^{-1}\Psi' \ , \tag{4.49}$$

where in writing the second statement we have assumed that the operator $\tilde{P}$ has an inverse, which in the case of the parity operator is true. This is also true for the operation of moving the chain with beads. In this case if the operator corresponds to moving the chain to the right by a cm, then the inverse would correspond to moving the chain to the left by a distance a cm. With the definition of Ψ' and the inverse operator for parity, we now can write Eq. (4.48) as

$$i\hbar \frac{\partial \Psi'}{\partial t} = \tilde{P}\tilde{H}\,\Psi = \tilde{P}\tilde{H}\tilde{P}^{-1}\Psi' \ . \tag{4.50}$$

But for the theory to be invariant under a given symmetry, the equations of motion have to keep their form (i.e., no change) under the transformation of the symmetry. For our equations to be the same under parity, we require that

$$\tilde{P}\tilde{H}\tilde{P}^{-1} = \tilde{H} \ . \tag{4.51}$$

Then both Ψ and Ψ' satisfy the same identical equation, and therefore the solution is the same and the physics has not changed as a result of the transformation. Thus the condition for the theory to be invariant under the symmetry transformation is

$$\tilde{P}\tilde{H} - \tilde{H}\tilde{P} \equiv [\tilde{P},\, \tilde{H}] = 0 \ . \tag{4.52}$$

The fact that $\tilde{P}$ commutes with the Hamiltonian implies that $\tilde{H}$ does not change under the transformation $x \rightarrow -x$. The fact that our equation does not change under parity means that the parity of a given state is a constant of the motion, and this is emphasized by the fact that the wave function is labeled by both the energy and the parity.

4.6 Eigenstates of the Momentum

The time independent Schrödinger equation, $\tilde{H}\psi_n = E_n\psi_n$, is an eigenvalue problem in which the eigenvalues, E_n, are the energies of the system described by the Hamiltonian $\tilde{H}$. In actual fact, to every observable, e.g., position and momentum, we have a Hermitian operator, and given that this operator is both linear and Hermitian, we can set up an eigenvalue problem with the eigenvalues being the observables corresponding to that operator. As an example let us consider the momentum. The corresponding operator in coordinate space is

$$\tilde{p} = -i\hbar\frac{d}{dx} \tag{4.53}$$

The eigenvalue problem we can now set up is

$$\tilde{p}\,\chi_p = p\,\chi_p \ ,$$

where χ_p is the eigenstate, and p is the corresponding eigenvalue of the operator $\tilde{p}$. Since the momentum operator in coordinate space is the differential operator given in Eq. (4.53), we can write this eigenvalue problem as a first order linear differential equation of the form

$$-i\hbar\frac{d\chi_p(x)}{dx} = p\,\chi_p(x) \ .$$

The solution to this first order differential equation is obtained by multiplying both sides of the equation by dx, dividing by χ_p and integrating both sides, i.e.,

$$-i\hbar\int\frac{d\chi_p}{\chi_p} = \int p\,dx \ .$$

Since p is a constant, the result of the integration is

$$\log\chi_p = \frac{i}{\hbar}\,xp + C,$$

where C is a constant of integration of the differential equation. This result can be rewritten as

$$\chi_p(x) = N\ e^{ipx/\hbar} = N\ e^{ikx} \ ,$$

where the constant of integration is given by N.[9] To normalize this wave function we have to calculate the integral

$$\begin{aligned} \int_{-\infty}^{+\infty} dx\,\chi_{p'}^*(x)\ \chi_p(x) &= N^2\int_{-\infty}^{+\infty} dx\,e^{i(p-p')x/\hbar} \\ &= N^2\,(2\pi\hbar)\,\delta(p-p') \ , \end{aligned}$$

[9]The normalization constant can be written in terms of the constant C as

$$C = \log N \ .$$

Note: We have used log for the natural logarithm.

where we have made use of the definition of the Dirac δ-function, Eq. (3.17), to write the second line in the above expression. The normalization of the wave function N, is now given as,

$$N = \frac{1}{\sqrt{2\pi\hbar}} \ .$$

We are now in a position to write the normalized eigenstates of the momentum operator as

$$\chi_p(x) = \frac{1}{\sqrt{2\pi\hbar}} e^{ipx/\hbar} \ . \tag{4.54}$$

We should note that this function is, up to normalization, the space part of the wave $e^{i(kx-wt)}$ if we observe that $p = \hbar k$.

To understand the relation between the eigenstate of the momentum operator and a wave with wave number k and frequency w, we next consider the time independent Schrödinger equation for a free particle, i.e., $V = 0$. This Schrödinger equation in one dimension takes the form

$$-\frac{\hbar^2}{2m}\frac{d^2\psi(x)}{dx^2} = E\psi(x) \tag{4.55}$$

or

$$\frac{d^2\psi(x)}{dx^2} = -k^2\psi(x) \quad \text{with} \quad k^2 = \frac{2mE}{\hbar^2} \ . \tag{4.56}$$

This second order differential equation is the space part of the wave equation and has a solution of the form

$$\psi_k(x) = N \ e^{ikx} \ . \tag{4.57}$$

Taking into consideration the fact that $p = \hbar k$, we have established that the eigenstates of the momentum operator are identical to the eigenstates of the Hamiltonian for a free particle. This result is a consequence of the fact that the Hamiltonian for a free particle $\tilde{H} = \frac{\tilde{p}^2}{2m}$ commutes with the momentum operator $\tilde{p}$, i.e.,

$$[\tilde{H}, \tilde{p}] = 0 \ . \tag{4.58}$$

When this condition is satisfied, we can find one eigenstate that is an eigenstate of both operators.

The physical implication of this result is that a measurement of the momentum of the particle does not in any way effect the energy of the particle, and both the energy and momentum can be measured to any desired accuracy. Thus when two operators commute, there is no uncertainty principle between the corresponding observables and we can construct one wave function that is an eigenstate of both operators. Note that if the particle is in an external force, i.e., $V \neq 0$, then the Hamiltonian will not commute with the momentum operator and in that case we can not measure both the energy and momentum to any desired accuracy.

At this stage let us go back to our description of a wave packet in which we wrote

$$\psi(x) = \frac{1}{\sqrt{2\pi}} \int_{-\infty}^{+\infty} dk\, \phi(k)\, e^{ikx} \ . \tag{4.59}$$

If we now change variables from k to $p = \hbar k$ we have

$$\begin{aligned} \psi(x) &= \frac{1}{\sqrt{2\pi\hbar}} \int_{-\infty}^{+\infty} dp\, \phi(p)\, e^{ipx/\hbar} \\ &= \int_{-\infty}^{+\infty} dp\ \phi(p)\, \chi_p(x) \ . \end{aligned} \tag{4.60}$$

We now can understand this result as an expansion of the wave packet in terms of the complete set of eigenstates of the momentum operator. In this case, $\phi(p)$ is the probability amplitude of finding the particle in a state with momentum p, and $\chi_p(x)$ is the wave function for this momentum eigenstate. Thus, what we considered previously as a Fourier decomposition of the wave function $\psi(x)$, we now can interpret as nothing more than an expansion of the wave function $\psi(x)$ in terms of the eigenstates of the momentum operator, $\chi_p(x)$.[10]

4.7 Problems

1. Consider the two operators $\tilde{A}$ and $\tilde{B}$ defined by the relations

$$\tilde{A}\,\psi(x) = x^2\,\psi(x) \quad \text{and} \quad \tilde{B}\,\psi(x) = x\frac{d}{dx}\,\psi(x) \ .$$

   (a) Are these operators linear?

   (b) Determine the commutation relation $\left[\tilde{A}, \tilde{B}\right]$.
   Note: The procedure to calculate $\left[\tilde{A}, \tilde{B}\right]$ is to evaluate $\tilde{A}\,(\tilde{B}\,\psi) - \tilde{B}\,(\tilde{A}\,\psi)$.

2. Consider the two operators $\tilde{a}$ and $\tilde{a}^\dagger$ defined such that

$$\begin{aligned} \tilde{a}\,\psi(x) &= \frac{1}{2}\left(x + i\frac{d}{dx}\right)\,\psi(x) \\ \tilde{a}^\dagger\,\psi(x) &= \frac{1}{2}\left(x - i\frac{d}{dx}\right)\,\psi(x) \ . \end{aligned}$$

[10] The difference between $\phi(p)$ and $\phi(k)$ is the normalization. Thus $\phi(k)$ has a normalization of $1/\sqrt{2\pi}$, while $\phi(p)$ has a normalization of $1/\sqrt{2\pi\hbar}$.

(a) Show that the operators $\tilde{a}$ and $\tilde{a}^\dagger$ are linear operators.

(b) Calculate the commutation relation $\left[\tilde{a}, \tilde{a}^\dagger\right]$.

(c) Find the eigenvalues and eigenstates of the operator $\tilde{a} - \tilde{a}^\dagger$. How are these eigenvalues and eigenstates related to the eigenvalues and eigenstates of the momentum operator?

3. Solve the Schrödinger equation for a particle in a box with sides at $x = 0$ and $x = a$ with the boundary condition that

$$\psi(0) = \psi(a) = 0 \ .$$

(a) What are the eigenvalues and the normalized eigenstates?

(b) Use *Mathematica* to plot the wave function for the ground state and first excited state.

(c) Use *Mathematica* to calculate the probability of finding the particle in the first excited state between $x = 0$ and $x = a/2$.

(d) Does the wave function have a definite symmetry under the transformation $x \leftrightarrow -x$? Why?

4. Given: The one dimensional potential that represents a particle in a box is

$$V(x) = \begin{cases} +\infty & x < -a \\ 0 & -a < x < 0 \\ +\infty & x > 0 \end{cases} \ .$$

(a) Write the time independent Schrödinger equation for this potential in the different regions.

(b) What are the boundary conditions on the wave function?

(c) Solve the Schrödinger equation in the different regions and normalize the solution.

(d) What are the energies (i.e., eigenvalues) the particle can have in this potential?

(e) Does the wave function for the particle have a definite symmetry under the transformation $x \to -x$? Why?

(f) Plot the normalized wave function for the lowest three eigenstates, given the mass of the particle is one, and $a = 2$.

5. A particle is in the ground state of a box with sides at $x = \pm a$. Very suddenly the sides of the box are moved to $x = \pm b$ $(b > a)$.

(a) Write the ground state normalized wave function for the particle in the initial box with sides at $x = \pm a$.

(b) Write the ground state normalized wave function for a particle in the final box with sides at $x = \pm b$.

(c) What is the probability that the particle which is initially in the ground state of the box with sides at $\pm a$, will be found in the ground state for the final potential with sides at $\pm b$?

(d) What is the probability that the particle in the ground state of the box with sides at $x = \pm a$, will be found in the first excited state of the final box with sides at $x = \pm b$?

In the last case, the simple answer has a simple explanation. What is it?

Quantum Mechanics with Applications, 2011, 70-87

CHAPTER 5

Simple One Dimensional Problems

Abstract: This chapter is devoted to the application of the Schrödinger equation to one dimensional models of interesting phenomena. In this way we can get an understanding of quantum effects, and the role of symmetry, without the complexity of the mathematics encountered in three dimensional problems.

In this chapter we will consider the solution of the Schrödinger equation for simple one dimensional problems. The main motivation for considering systems in one-dimension is that the complexity of the mathematical formulation is reduced, yet the main ideas of a quantum system are retained. In particular, we will find that for a one-dimensional problem, with a time independent potential, the Schrödinger equation reduces to an ordinary second order differential equation. The formulation of quantum mechanics in one-dimension will allow us to consider the reflection and transmission of a wave by a potential well as a simple example of a scattering problem. In particular, we will test the concept of current conservation within the framework of a scattering experiment, and the relation between the scattering matrix and bound states in the potential.

5.1 Free-Particle

All scattering experiments involve a beam of particles, e.g., electrons, protons, and photons, incident on a target with the scattered particles being collected and analyzed by a detector. In Figure 5.1 we illustrate a typical experimental set-up where the incident beam is generated by an accelerator. This incident beam is then scattered from a target T, and the scattered particles are detected by the detector D.

Before the incident beam reaches the target it is traveling in a potential free zone (i.e., no external force on the particle), and therefore is described by the solutions of the

Iraj R. Afnan

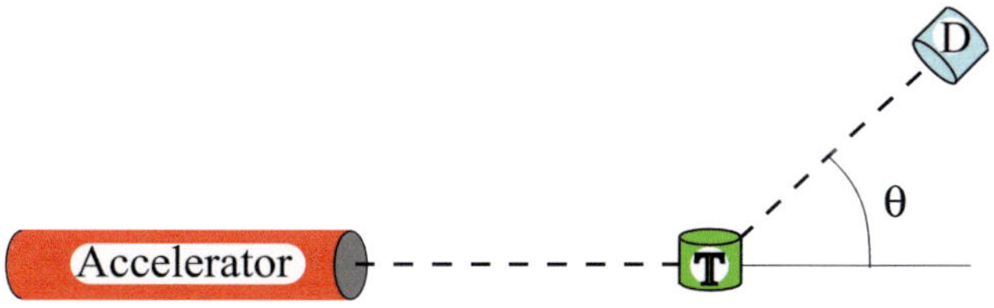

Figure 5.1: An illustration of a typical scattering experiment in which the incident beam is generated by an accelerator, is scattered by a target T, and is detected by the detector D. The scattering angle is θ.

Schrödinger equation for a free particle. The Hamiltonian for a free particle is given by [1]

$$H = -\frac{\hbar^2}{2m}\frac{d^2}{dx^2} , \tag{5.1}$$

and the Schrödinger equation takes the form

$$\frac{d^2\psi(x)}{dr^2} + k^2\psi(x) = 0 \qquad \text{with} \qquad k^2 = \frac{2mE}{\hbar^2} , \tag{5.2}$$

where E is the energy of the particles in the incident beam. The solution of Eq. (5.2) is given by

$$\psi_k^R(x) = e^{ikx} \qquad \psi_k^L(x) = e^{-ikx} . \tag{5.3}$$

Both of the above solutions satisfy the Schrödinger equation, Eq. (5.2). The corresponding time dependent solutions of the Schrödinger equation are

$$\Psi(x,t) - \psi(x)\ e^{-iEt/\hbar} .$$

If we now write the energy in terms of the the frequency, i.e., $E = \hbar\omega$, we can write the two time dependent solutions as

$$\Psi_k^R(x,t) = e^{i(kx-\omega t)} \quad \text{and} \quad \Psi_k^L(x,t) = e^{-i(kx+\omega t)} . \tag{5.4}$$

The most general solution could be a linear combination of the above two solutions. However, before we can consider such general solutions, let us examine the properties of the individual solutions. These solutions represent waves, and we would like to determine the direction of propagation of one of these waves, e.g., Ψ_k^R.

To determine the direction of propagation of the wave, we can think of a champion surfer that can stay at a specific point in front of the crest of a perfect wave for an

[1]From this point on we will drop the $\tilde{}$ from all operators and leave it to the reader to differentiate between operators and functions.

indefinite period of time. If at time t_0 he is at a position x_0, then his height above the surface of the sea is given by [2]

$$\Psi_k^R(x_0, t_0) = e^{i(kx_0 - \omega t_0)} .$$

At a time $t > t_0$, the surfer still maintains the same height above the surface of the sea, and his position x, will be determined by the condition:

$$\Psi(x, t) = \Psi(x_0, t_0) .$$

This is equivalent to the condition

$$kx_0 - \omega t_0 = kx - \omega t ,$$

or

$$x - x_0 = \frac{\omega}{k}(t - t_0) .$$

Thus for $t > t_0$ we have that $x > x_0$, and we can say the champion surfer, and therefore the wave, is traveling along the positive x-axis, i.e., to the right. This establishes the fact that the function $\Psi_k^R(x, t)$ represents a wave that is propagating to the right. In a similar manner we can show that $\Psi_k^L(x, t)$ represents a wave moving to the left.

This means that the time independent wave function for a wave propagating to the right is given by $\psi_k^R(x)$ while a wave proceeding to the left is denoted by $\psi_k^L(x)$ with

$$\psi_k^R(x) = e^{ikx} \quad \text{and} \quad \psi_k^L(x) = e^{-ikx} . \tag{5.5}$$

The general solution to the Schrödinger equation, Eq. (5.2), is now taken as a linear combination of a wave traveling to the right and one traveling to the left, i.e.,

$$\psi_k(x) = Ae^{ikx} + Be^{-ikx} . \tag{5.6}$$

The constants A and B are to be determined by the boundary conditions imposed by the experimental setup. These will be specified as we examine each individual problem in subsequent sections.

5.2 Potential Step

We would like now to make use of the general solution to the time independent Schrödinger equation as given in Eq. (5.6) to illustrate the difference between the classical problem and the corresponding quantum mechanical problem for the simplest system of a potential barrier and potential well. In this way we hope to also illustrate the problem of scattering

[2] The surface of the sea is taken to be the level of water when there are no waves.

from a barrier or a potential well, and how current conservation plays an important role in all scattering experiments.

As the simplest scattering problem, let us consider a plane wave incident from the left on a potential step, see Figure 5.2. The potential is taken to be of the form

$$V(x) = \begin{cases} 0 & x < 0 \\ V_0 & x > 0 \end{cases} . \tag{5.7}$$

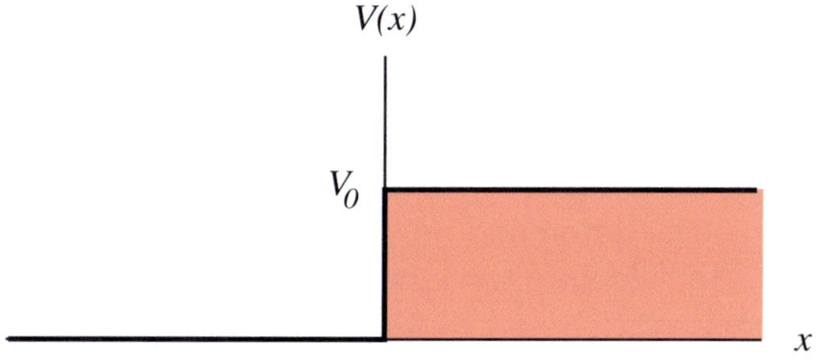

Figure 5.2: Plot for a potential barrier of height V_0.

Before we consider this problem within the framework of quantum mechanics, let us consider the classical problem. Here we recall that the force is given by

$$F = -\frac{dV}{dx} ,$$

which in this case is an impulse at $x = 0$ directed towards the negative x-axis. Let us first consider the case when $E < V_0$ and the particle is traveling to the right. Its momentum would be $p = \sqrt{2mE}$ for $x < 0$. When it gets to $x = 0$, it is reflected by the barrier, i.e., it bounces back with momentum $p' = -\sqrt{2mE}$. On the other hand, for $E > V_0$, the particle keeps going to the right but for $x > 0$, its momentum is reduced to $p' = \sqrt{2m(E - V_0)}$. The change in the momentum of the particle as it goes from $x < 0$ to $x > 0$ is a result of the impulse at $x = 0$.[3]

To get the corresponding quantum mechanical solution, we need to solve the time independent Schrödinger equation for the potential defined in Eq. (5.7). Here again we consider a beam of particles incident from the left, then, as in the classical case, we expect a reflected beam and a transmitted beam. Classically, the magnitude of the reflected and transmitter beam was determined by Newton's second law. In this case we assume a reflected and transmitted wave and let the Schrödinger equation determine the amplitude

[3]Recall that Newton's second law is given by $F = dp/dt$.

of the reflected and transmitted wave. This suggests that the most general solution to the Schrödonger equation for the potential in Eq. (5.7) be of the form

$$\psi(x) = \begin{cases} 1\, e^{ikx} + R\, e^{-ikx} & \text{for} \quad x < 0 \\ \\ T\; e^{ik'x} & \text{for} \quad x > 0 \end{cases} , \tag{5.8}$$

where the wave number for $x < 0$ is k, while the wave number for $x > 0$ is k'. These wave numbers are given in terms of the incident particle's energy E, and barrier height V_0 as

$$k = \sqrt{\frac{2mE}{\hbar^2}} \quad \text{and} \quad k' = \sqrt{\frac{2m}{\hbar^2}(E - V_0)} \ . \tag{5.9}$$

In writing the solution of the Schrödinger equation in Eq. (5.8), we have taken the incident wave to have unit amplitude while the reflected wave has an amplitude R. For the transmitted wave, the amplitude is taken to be T.

We now turn to current conservation and demonstrate that it is conserved at both the classical and quantum level. To calculate the incident, reflected and transmitted currents, and thus examine current conservation at $x = 0$, we will make use of the definition of the current as given in Eq. (3.50), i.e.,

$$j(x) = \frac{\hbar}{2im}\left\{\psi^* \frac{d\psi}{dx} - \frac{d\psi^*}{dx}\psi\right\} .$$

In this case the current for $x < 0$ is given, using the wave function in Eq. (5.8), by

$$\begin{aligned} j(x) &= \frac{\hbar}{2mi}[(e^{-ikx} + R^* e^{ikx})(ike^{ikx} - ikR\, e^{-ikx}) - \text{ complex conjugate}] \\ &= \frac{\hbar}{2mi}[2ik(1 - |R|^2)] \\ &= \frac{\hbar k}{m}(1 - |R|^2) \ . \end{aligned} \tag{5.10}$$

This current consists of two parts, one corresponds to the incident beam and is given by

$$\frac{\hbar k}{m} = \frac{p}{m} = v \ ,$$

which is the velocity of the incident particle. The second component of the current for $x < 0$ is the reflected current which is proportional to $|R|^2$, where R is the amplitude of the reflected wave.

On the other hand the current in the region $x > 0$ is given in terms of the amplitude for the transmitted wave, i.e., T, and can be written using the definition of the current as

$$j(x) = \frac{\hbar k'}{m}|T|^2 \ . \tag{5.11}$$

Now the conservation of current at $x = 0$ requires that the current for $x < 0$ be equal to the current for $x > 0$, and since these currents, as given in Eqs. (5.10) and (5.11) do not depend on x, we have

$$\frac{\hbar k}{m}(1 - |R|^2) = \frac{\hbar k'}{m}|T|^2 \quad \text{or} \quad \frac{\hbar k}{m} = \frac{\hbar k}{m}|R|^2 + \frac{\hbar k'}{m}|T|^2 \ . \tag{5.12}$$

In other words, the sum of the reflected and transmitted current is equal to the incident current. At this stage we note that current conservation does not determine the amplitude of the reflected and transmitted wave, it only puts a constraint on R and T in the form of Eq. (5.12), which holds for the classical case as well.

We therefore turn to the boundary condition and the properties of the Schrödinger equation to determine the constants R and T. Since the Schrödinger equation is a second order linear differential equation, we require that the wave function and its derivative be continuous at all points, and in particular at $x = 0$. Thus from the continuity of the wave function we have that

$$1 + R = T \ . \tag{5.13}$$

On the other hand the continuity of the derivative of the wave function requires that

$$k(1 - R) = k'T \ . \tag{5.14}$$

We now can solve these two linear algebraic equations (i.e., Eqs. (5.13) and (5.14)) resulting from the requirement of continuity of the wave function and its derivative, to determine the amplitude of the reflected and transmitted waves R and T, to be

$$R = \frac{k - k'}{k + k'} \ , \tag{5.15}$$

and

$$T = \frac{2k}{k + k'} \ . \tag{5.16}$$

If we now compare these results with the classical result we find that:

1. For $E > V_0$ we have a reflected as well as a transmitted wave, while classically we had no reflected particle.

2. The transmitted current $j_T = \frac{\hbar k'}{m}|T|^2$ and the reflected current $j_R = \frac{\hbar k}{m}|R|^2$ add up to give the incident current $j_I = \frac{\hbar k}{m}$, i.e.,

$$\frac{\hbar k}{m} = \frac{\hbar k}{m}|R|^2 + \frac{\hbar k'}{m}|T|^2 \ .$$

   This result, which is a statement of current conservation, is consistent with the amplitudes for the reflected and transmitted current as given in Eqs. (5.15) and (5.16).

3. In the limit, when the energy of the incident particles is much greater than the height of the barrier, i.e., $E \gg V_0$, the change in the momentum of the particle at $x = 0$ is expected to be small and $k' \approx k$. In this case the amplitude of the reflected wave is much smaller than the amplitude of the transmitted wave, $R \ll T$, and the reflected wave is negligible, which is close to the classical limit.

4. For the case when the energy of the incident beam is less than the barrier height, i.e., $E < V_0$, the wave number for $x > 0$ is given by

$$\begin{aligned} k' &= \sqrt{\frac{2m}{\hbar^2}(E - V_0)} \\ &= i\kappa \ , \end{aligned} \tag{5.17}$$

where κ is real and given by the expression

$$\kappa = \sqrt{\frac{2m}{\hbar^2}\,(V_0 - E)}$$

i.e., the momentum $p' = \hbar k' = i\hbar\kappa$ becomes imaginary. This means that for $x > 0$, the wave function for the transmitted wave is given by,

$$\psi(x) = T\ e^{-\kappa x} \quad \text{for} \quad x > 0\ . \tag{5.18}$$

This wave function decays exponentially, and the corresponding current is zero at a detector place at a point $x \gg 0$. In other words we have a decaying solution of the Schrödinger equation similar to the one we encountered when we had a particle bound in a well. The amplitude of the reflected wave in this case is given by,

$$R = \frac{k - i\kappa}{k + i\kappa}\ , \tag{5.19}$$

and the corresponding reflected current is given by

$$j_R = \frac{\hbar k}{m}|R|^2 = \frac{\hbar k}{m} = j_I\ . \tag{5.20}$$

In other words all of the incident current is reflected by the barrier despite the fact that the wave function is not zero for $x > 0$. The fact that the wave function is not zero for $x > 0$ is an indication that the particle penetrates this region. In fact, the amplitude of the transmitted wave T is not zero and is given by

$$T = \frac{2k}{k + i\kappa}\ . \tag{5.21}$$

To understand this apparent lack of current conservation, we should first recall that the wave in the region $x > 0$ is decaying exponentially, and the wave number k' in this region is given by

$$k' = i\kappa \quad \text{where} \quad \kappa = \sqrt{\frac{2m}{\hbar^2}(V_0 - E)}\ .$$

However, despite the fact that the wave function is not zero for $x > 0$, a calculation of the current in this region using Eq. (3.50) gives a zero current, proving the conservation of current. This is a result of the fact that we have a decaying wave function which goes to zero. The rate at which it goes to zero is governed by the value of κ. The larger the height of the barrier, the faster the decay of the wave function, and therefore the smaller the penetration depth. This non-zero penetration of the particle is a purely quantum mechanical effect. We will come back to this problem of barrier penetration and current conservation when the barrier is of finite length.

5.3 Potential Well

In the last section we considered scattering from a potential step, and when we compared the classical and quantum mechanical results we found two new features:

1. For $E > V_0$ there was a reflected wave.

2. For $E < V_0$ the particle penetrated the classically forbidden region $x > 0$. In particular, we found that in this region, the wave function decays exponentially, and in fact does not look like a wave any more.

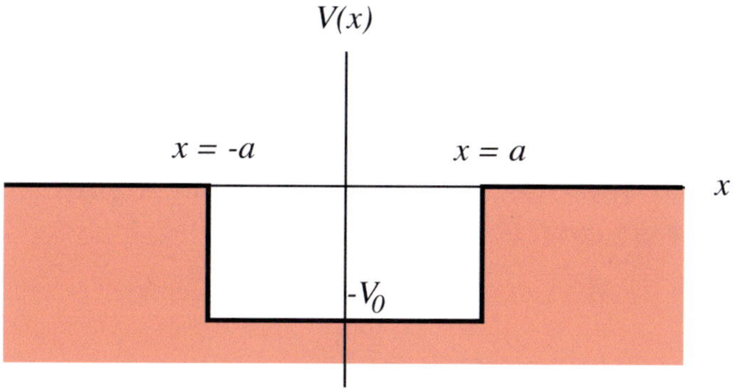

Figure 5.3: Plot for an attractive potential well of depth V_0 and width $2a$.

To see if these effects are present in other systems, and in particular if a particle when placed in a potential well does in fact penetrate the classically forbidden region, let us consider a potential well of the form illustrated in Figure 5.3, i.e.,

$$V(x) = \begin{cases} 0 & \text{for} \quad |x| > a \\ \\ -V_0 & \text{for} \quad |x| < a \end{cases} , \tag{5.22}$$

where $V_0 > 0$ is the depth of the potential well.

5.3.1 Bound state problem

Let us first consider the case when $E < 0$, in which case we expect the particle to be bound inside the well. The Schrödinger equation now takes the form

$$-\frac{\hbar^2}{2m}\frac{d^2\psi(x)}{dx^2} = E\,\psi(x) \qquad \text{for} \qquad |x| > a\ , \tag{ 5.23a}$$

while

$$-\frac{\hbar^2}{2m}\frac{d^2\psi(x)}{dx^2} - V_0\,\psi(x) = E\,\psi(x) \qquad \text{for} \qquad |x| < a\ . \tag{ 5.23b}$$

These equations can be written, after multiplication by $2m/\hbar^2$, as

$$\begin{aligned} \frac{d^2\psi(x)}{dx^2} - \alpha^2\psi(x) &= 0 \quad \text{where} \quad \alpha^2 = \frac{2m|E|}{\hbar^2} \quad \text{for} \quad |x| > a \\ &. \\ \frac{d^2\psi(x)}{dx^2} + k^2\psi(x) &= 0 \quad \text{where} \quad k^2 = \frac{2m}{\hbar^2}(V_0 - |E|) \quad \text{for} \quad |x| < a \end{aligned} \tag{5.24}$$

The general solution to these two equations is of the form

$$\psi(x) = \begin{cases} A\,e^{-\alpha x} + B\,e^{\alpha x} & \text{for} \quad |x| > a \\ \\ C\sin kx + D\cos kx & \text{for} \quad |x| < a \end{cases} . \tag{5.25}$$

To determine the constants A, B, C and D we need to make use of the boundary conditions for this problem. These boundary condition are: (i) The wave function and its derivative should be continuous for all values of x, and in particular at $x = \pm a$. (ii) For the wave function of a bound state to be normalizable, we require that the wave function go to zero at $x \to \pm\infty$. Let us first consider the region outside the well, i.e., $|x| > a$. The boundary condition that the wave function be normalizable requires that for $x < -a$ we have $A = 0$, while for $x > a$ we have $B = 0$, i.e.,

$$\psi(x) = \begin{cases} B\,e^{\alpha x} & \text{for} \quad x < -a \\ \\ C\sin kx + D\cos kx & \text{for} \quad -a < x < a \\ \\ A\,e^{-\alpha x} & \text{for} \quad x > a \end{cases} . \tag{5.26}$$

To determine the other constants, we make use of the continuity of the wave function and its first derivative at $x = \pm a$. Thus at $x = -a$ we have

$$B\ e^{-\alpha a} = -C \sin ka + D \cos ka$$

and

$$\alpha B\ e^{-\alpha a} = kC \cos ka + kD \sin ka\ ,$$

while at $x = +a$ we have

$$A\ e^{-\alpha a} = C \sin ka + D \cos ka$$

and

$$-\alpha A\ e^{-\alpha a} = kC \cos ka - kD \sin ka\ .$$

From these four equations we get

$$\begin{aligned} \alpha &= k\ \frac{C \cos ka + D \sin ka}{-C \sin ka + D \cos ka} \\ &= k\ \frac{D \sin ka - C \cos ka}{D \cos ka + C \sin ka} \end{aligned}$$

which, after cross multiplication, can be written as

$$(D \cos ka - C \sin ka)(D \sin ka - C \cos ka) = (D \cos ka + C \sin ka)(D \sin ka + C \cos ka)\ .$$

This can be simplified to give us the condition that

$$CD = -CD \quad \text{or} \quad CD = 0\ . \tag{5.27}$$

This means that either $D = 0$ or $C = 0$, i.e., the solution for $|x| < a$ is either $\sin kx$ or $\cos kx$. These two solutions differ by the fact that $\sin(-kx) = -\sin(kx)$, while $\cos(-kx) = \cos(kx)$, i.e., one solution is an odd solution while the second is an even solution.

5.3.2 Symmetry of the Hamiltonian

At this stage we should realize that the Hamiltonian for this potential is invariant under the transformation $x \to -x$. This means that the parity operator P commutes with the Hamiltonian, i.e.,

$$[H, P] = 0\ , \tag{5.28}$$

and the eigenstates of H, i.e., the wave function $\psi(x)$, should also be an eigenstate of the parity operator P. This implies that

$$P\,\psi(x) = \pi\,\psi(x)\ , \tag{5.29}$$

where $\pi = \pm 1$. For the case when $\pi = +1$, the wave function does not change under the transformation $x \to -x$. This solution, known as the even solution, is given by,

$$\psi_{\pi=+1}(x) = D\cos kx \qquad \text{for} \qquad -a < x < a\ , \tag{5.30}$$

where we have labeled the wave function $\psi(x)$ by its parity of $\pi = +1$. For $\pi = -1$, the wave function $\psi(x)$ should change sign under the transformation $x \to -x$. This solution, known as the odd solution, is given by

$$\psi_{\pi=-1}(x) = C\sin kx \qquad \text{for} \qquad -a < x < a\ . \tag{5.31}$$

The even solution corresponding to $\pi = +1$ must now satisfy the boundary conditions at $x = \pm a$, i.e., the wave function and its derivative must be continuous, so that at $x = -a$,

$$\begin{aligned} B\ e^{-\alpha a} &= D\cos ka \\ \alpha B\ e^{-\alpha a} &= kD\sin ka\ , \end{aligned}$$

while for $x = +a$ the boundary condition gives us

$$\begin{aligned} A\ e^{-\alpha a} &= D\cos ka \\ -\alpha A\ e^{-\alpha a} &= -kD\sin ka\ . \end{aligned}$$

In both cases, if we divide the second equation by the first equation, we get

$$\alpha = k\tan ka\ . \tag{5.32}$$

On the other hand, for the odd solution, i.e., $\pi = -1$, we have for the boundary condition at $x = -a$

$$\begin{aligned} -B\ e^{-\alpha a} &= C\sin ka \\ \alpha B\ e^{-\alpha a} &= kC\cos ka\ . \end{aligned}$$

In a similar way we get a set of boundary conditions for $x = a$. Both of the boundary conditions give the same expression, as was the case for the even solution, to be

$$-\ \alpha = k\cot ka\ . \tag{5.33}$$

Both of these transcendental equations, Eqs. (5.32) and (5.33), when solved will give us values for the energy E for which we can have a solution to the Schrödinger equation. In the case of bound state, i.e., $E < 0$, only certain energies will be allowed and these will be the eigenstates of the Hamiltonian.

5.3.3 Eigenstates of the Hamiltonian

To see how we can solve these transcendental equations, let us consider the case of the even parity solution i.e., $\pi = +1$. In this case the wave number inside the well is given by

$$k^2 = \frac{2m}{\hbar^2}(V_0 - |E|) \ , \tag{5.34}$$

while the corresponding quantity outside the well is

$$\alpha^2 = \frac{2m}{\hbar^2}|E| \ . \tag{5.35}$$

This allows us to write a relation between k and α of the form

$$k^2a^2 = \eta^2 - \alpha^2a^2 \ , \tag{5.36}$$

where η is given in terms of the potential depth V_0, and the potential width $2a$, by

$$\eta^2 = \frac{2mV_0a^2}{\hbar^2} \ . \tag{5.37}$$

With this relation between the wave number k and α, we can rewrite the transcendental Eq. (5.32) in a form that will allow us to get a graphical solution, i.e.,

$$\sqrt{\frac{\eta^2 - y^2}{y^2}} = \tan y \ , \tag{5.38}$$

where $y = ka$. To solve this equation graphically we need to plot the right and left hand side of Eq. (5.38), i.e.,

$$F(y) = \tan y \qquad \text{and} \qquad F(y) = \sqrt{\frac{\eta^2 - y^2}{y^2}} \ . \tag{5.39}$$

The points at which the two curves intersect give those values of y for which we have a solution that gives the allowed energies for the positive parity solutions. (See left plot in Figure 5.4). In a similar way we can write Eq. (5.33) as

$$F(y) = -\cot y \qquad \text{and} \qquad F(y) = \sqrt{\frac{\eta^2 - y^2}{y^2}} \ . \tag{5.40}$$

These are plotted on the right hand side in Figure 5.4, and the intercept gives the eigenvalues for the negative parity states.

Since $|E| < V_0$, we have that

$$F(y) = \sqrt{\frac{\eta^2 - y^2}{y^2}} = \begin{cases} 0 & \text{for} \quad y = \eta \\ \to \infty & \text{for} \quad y \to 0 \end{cases} \ , \tag{5.41}$$

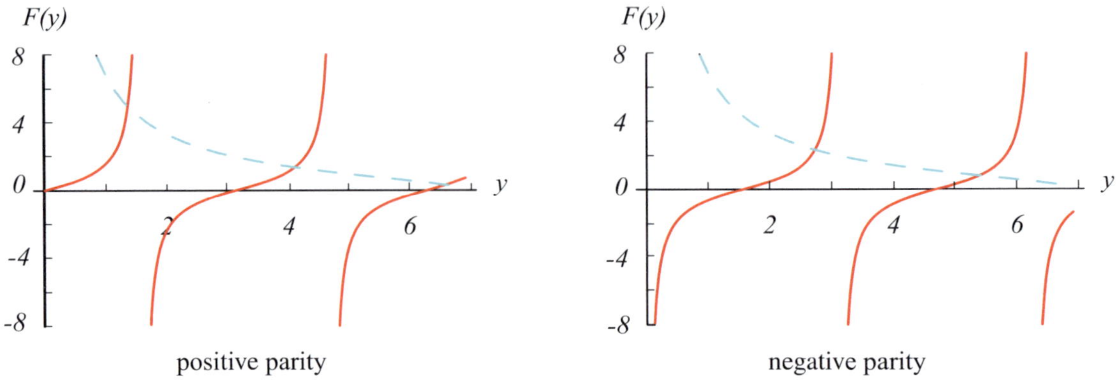

Figure 5.4: A plot of $F(y)$ for the positive parity (on the left), and for the negative parity solutions (on the right).

and thus an increase in η will lead to intercepts at greater values of y. We may then conclude that as η increases, the number of solutions increase and therefore the number of bound states increases. From Eq. (5.41) and Figure 5.4, we can draw the following general conclusions:

1. For the even solutions we have that
$$F(y) = \sqrt{\frac{\eta^2 - y^2}{y^2}} = \tan y$$
and from the graph on the left in Figure 5.4 we see that we always have at least one solution, and therefore one bound state.

2. For $\eta < \pi/2$, we have
$$V_0 a^2 < \frac{\hbar^2}{2m}\frac{\pi^2}{4} \ .$$
In this case Eq. (5.38) has a solution, see graph on the left-hand side of Figure 5.4. This bound state corresponds to the even solution, $\pi = +1$. For the odd solution, $\pi = -1$, the equation
$$F(y) = \sqrt{\frac{\eta^2 - y^2}{y^2}} = -\cot y \ ,$$
has no solution, see graph on the right-hand side of Figure 5.4. As a result, for $\eta < \pi/2$ there is only one solution which is even under the parity transformation.

3. The ground state of the system corresponds to an even solution, while the first excited state corresponds to the odd parity solution.

If we compare the bound states problem for this potential with that of a particle in a box as discussed in the last section, we find that:

1. The wave function for a potential well extends into the region $|x| > a$ which was not the case for a particle in a box. On the other hand the shape of the wave functions are similar. This is mainly due to the fact that the potential in both examples have the same symmetry.

2. For a potential well, we have only a finite number of bound states in contrast to the particle in a box where we had an infinite number of bound states. However as we will see, for a potential well we have scattering states which were not present in the case of a particle in a box.

5.3.4 The scattering problem

We now turn to the case when the energy $E > 0$, and consider a beam of particles of unit amplitude incident from the left. Making use of the general results of the potential step discussed at the beginning of this chapter, we observe that for $x < -a$ we have an incident wave from the left and a possible reflected wave. On the other hand, for $x > a$ we have only a transmitted wave. In the region in between, i.e $-a < x < a$, we expect waves to travel both to the right and to the left. We therefore can write our wave function as:

$$\psi(x) = \begin{cases} e^{ikx} + R\, e^{-ikx} & \text{for} \quad x < -a \\ A\, e^{ik'x} + B\, e^{-ik'x} & \text{for} \quad -a < x < a \\ T\, e^{ikx} & \text{for} \quad x > a \end{cases} , \tag{5.42}$$

where the wave number outside and inside the well are given by $k^2 = 2mE/\hbar^2$ and $k'^2 = 2m(E + V_0)/\hbar^2$, respectively. Here again we have taken the amplitude of the incident wave to be one, while that of the reflected and transmitted wave are taken to be R and T, respectively. We now can calculate the current in the three regions to be

$$j(x) = \begin{cases} \frac{\hbar k}{m}(1 - |R|^2) & \text{for} \quad x < -a \\ \frac{\hbar k'}{m}(|A|^2 - |B|^2) & \text{for} \quad -a < x < a \\ \frac{\hbar k}{m}|T|^2 & \text{for} \quad x > a \end{cases} , \tag{5.43}$$

and since the current in the three regions do not depend on the position x, the conservation of current will now require that

$$\frac{\hbar k}{m}(1 - |R|^2) = \frac{\hbar k'}{m}(|A|^2 - |B|^2) = \frac{\hbar k}{m}|T|^2 . \tag{5.44}$$

Here we note that $|R|^2+|T|^2 = 1$, i.e., the transmitted current plus the reflected current is equal to the incident current, i.e., no particles were created or destroyed in the scattering off the attractive square well. This is a statement of conservation of particle number or *unitarity*. We will come back to this conservation of particle number in a more general discussion of scattering in Chapter 14.

We note here that although the potential has reflection symmetry, the boundary condition that the incident beam is from the left has broken this symmetry and as a result the solution does not have the reflection symmetry we had in the bound state problem. Therefore, the constants R, A, B and T have to be determined by the boundary conditions that the wave function, and its first derivative be continuous at $x = \pm a$. In fact these boundary conditions give

$$e^{-ika} + R\ e^{ika} = A\ e^{-ik'a} + B\ e^{ik'a}\ ,$$

and

$$ik(e^{-ika} - R\ e^{ika}) = ik'(A\ e^{-ik'a} - B\ e^{ik'a})$$

for $x = -a$, while for $x = +a$, we have that

$$T\ e^{ika} = A\ e^{ik'a} + B\ e^{-ik'a}\ ,$$

and

$$ikT\ e^{ika} = ik'(A\ e^{ik'a} - B\ e^{-ik'a})\ .$$

The boundary conditions on the wave function have given us exactly four linear equations that can be solved for the four unknown constants R, A, B and T. In particular, we have for R and T

$$R = i\ e^{-2ika}\ \frac{(k'^2 - k^2)\sin 2k'a}{2kk'\cos 2k'a - i(k'^2 + k^2)\sin 2k'a}\ , \tag{5.45}$$

and

$$T = e^{-2ika}\ \frac{2kk'}{2kk'\cos 2k'a - i(k'^2 + k^2)\sin 2k'a}\ . \tag{5.46}$$

It is now a simple exercise to show that the above expression for the reflected and transmitted amplitudes satisfy the current conservation condition

$$|R|^2 + |T|^2 = 1\ .$$

If we now define the cross section for reflection and transmission as

$$\sigma_R = \frac{j_R}{j_I} = |R|^2 \quad \text{and} \quad \sigma_T = \frac{j_T}{j_I} = |T|^2 \tag{5.47}$$

then the sum of the two cross section is one, i.e.,

$$\sigma_R + \sigma_T = 1\ . \tag{5.48}$$

Note that if $E \gg V_0$, then $k'^2 \approx k^2$ and the amplitude for the reflected wave is considerably reduced. On the other hand, as the energy $E \rightarrow 0$, the wave number $k \rightarrow 0$, and the amplitude of the transmitted wave $T \rightarrow 0$. However if $\sin 2k'a = 0$, then $\cos 2k'a = 1$, and T does not go to zero as $k \rightarrow 0$. In this case

$$2k'a = n\pi \quad \text{and} \quad \lambda = \frac{2\pi}{k'} = \frac{4a}{n} \; . \tag{5.49}$$

Or, if the width of the well

$$2a = \frac{\lambda}{2} n \; , \tag{5.50}$$

then we get 100% transmission when the width of the well is an integer multiple of half a wave length. In this limit, the reflected wave has an amplitude of zero, i.e., the wave goes through the potential well as if it is not there. In other words, the system represented by the potential is transparent.

This effect, which corresponds to an increase in the transmission cross section with a corresponding reduction in the reflection cross section, is observed in noble gases and is known as the the Ramsauer-Townsend effect. This is illustrated in Figure 5.5

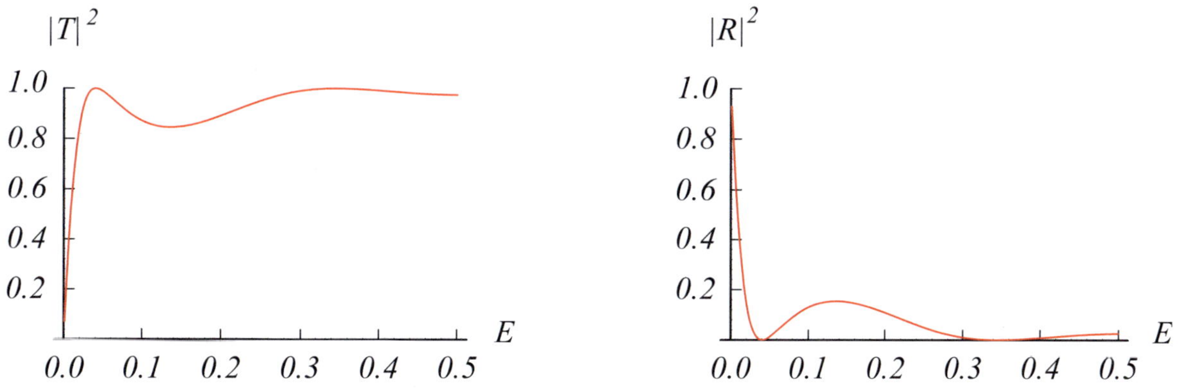

Figure 5.5: The transmission and reflection probability for an attractive square well. Note that at a specific energy the reflection probability goes to zero indicating zero scattering.

This effect of being able to change the transmission of a current by changing the width can have practical applications in switching devices at a microscopic level. Although in the above discussion we concentrated on changing the width of the potential well to change the transmission, we could have instead changed the potential depth and as a result the wave length of the particle in the well, to change the transmission amplitude. Note that the fact that we are considering a one dimensional system is not an approximation for a switching device since most macroscopic switches are wires which are one dimensional. In fact we expect the next generation of computer chips to rely on solid state devices that are based on this principle. The problem is how do we change the width or depth of the well in a controlled manner.

5.4 Problems

1. The wave function for a particle is given by

$$\psi(x) = A\, e^{ikx} + B\, e^{-ikx} \ .$$

   What is the corresponding current?

2. Show that Eq. (5.12) is valid if we take the amplitude for the reflected and transmitted wave, R and T, from Eqs. (5.15) and (5.16).

3. Consider the one dimensional potential

$$V(x) = \begin{cases} +V_0 & \text{for} \quad x < 0 \\ 0 & \text{for} \quad x > 0 \end{cases} \ .$$

   (a) Write the most general solution of the Schrödinger equation for $x > 0$ and for $x < 0$, assuming $E > V_0$.

   (b) Does this solution have reflection symmetry?, i.e., symmetry under $x \to -x$. Why?

   (c) Consider the experimental situation when there is an incident beam of particles from the left (i.e., from $x = -\infty$). Calculate the amplitude for the reflected and transmitted waves, given the amplitude for the incident wave is equal to one.

   (d) Show that the incident current is equal to the sum of the reflected current and the transmitted current.

4. Consider the square well potential

$$V(x) = \begin{cases} 0 & \text{for} \quad x < -a \\ -V_0 & \text{for} \quad -a < x < +a \\ 0 & \text{for} \quad x > a \end{cases}$$

   (a) Given the well size $a = 5 \times 10^{-15}$ m $\equiv$ 5 fm, find the well depth V_0, in MeV, that will support five bound states.

   (b) Use the value of V_0, determined in (a), to determine the energy and parity of each of the bound states.

   (c) Plot using *Mathematica* the normalized wave function for the ground state (i.e., the lowest energy state), and the first excited state.

We are given that $\hbar c = 197.3$ MeV-fm, and $mc^2 = 940$ MeV. Here, c is the velocity of light.

Hint: First use the graphical solution of the transcendental equation (5.32) and (5.33) to determine the well depth. Then use *Mathematica* to solve the transcendental equation for the energy of each of the bound states.

5. For a potential localized over a finite part of the x-axis, the solution of the Schrödinger equation to the left and to the right of the potential are given in the figure below.

The S-matrix relates the amplitude of the out going wave to the amplitude of the incoming waves and is defined in this case to be

$$\begin{pmatrix} C \\ B \end{pmatrix} = \begin{pmatrix} S_{11} & S_{21} \\ S_{12} & S_{22} \end{pmatrix} \begin{pmatrix} A \\ D \end{pmatrix} .$$

Show that current conservation requires that:

$$\begin{aligned} |S_{11}|^2 + |S_{21}|^2 &= 1 \\ |S_{12}|^2 + |S_{22}|^2 &= 1 \\ S_{11}^* S_{12} + S_{21}^* S_{22} &= 0 . \end{aligned}$$

This result is equivalent to the statement that the scattering matrix

$$S = \begin{pmatrix} S_{11} & S_{12} \\ S_{21} & S_{22} \end{pmatrix}$$

is unitary.

6. Calculate the elements of the scattering matrix, S_{11}, S_{12}, S_{21}, and S_{22} defined in the previous problem for the potential

$$V(x) = \begin{cases} 0 & \text{for} \quad x < -a \\ V_0 & \text{for} \quad -a < x < a \\ 0 & \text{for} \quad x > a \end{cases}$$

and show that the scattering matrix is unitary.

CHAPTER 6

Application of Quantum Mechanics

Abstract: Here we turn to some applications of quantum mechanics to systems in atoms, nuclei and solids that can be modeled in terms of one dimensional quantum systems.

In the present chapter we will develop simple one dimensional models of physical systems in atomic, nuclear and solid state physics. In all cases we try to simplify the physical problem by constructing a one dimensional quantum mechanical model. In each case, the solution of the Schrödinger equation in one dimension will illustrate the major features of the system under consideration. Although the results of such models may not be quantitative to the extent that they reproduce detailed experimental results, the main features are exhibited to get a qualitative understanding of the behavior of the system. In most cases further improvement will require an order of magnitude increase in computational power, which often renders such calculations beyond the scope of a first course in quantum mechanics at the undergraduate level.

6.1 Barrier Penetration and α-decay

In 1896, Becquerel [25] accidentally discovered that uranium, when placed near unexposed photographic plates, invariably results in the fogging of the plates when developed. Further examination of the phenomena led to the idea that some naturally occurring elements are radioactive, i.e., they emit radiation. This radiation, which we now know emanates from the nucleus of an atom, was found to be of three kinds:

1. α-**rays** – These are nuclei with two protons and two neutrons, i.e., they are the nuclei of one of the isotopes of He. The nucleus that emits α particles loses two protons and two neutrons and as a result its atomic number is reduced by four, while its charge decreases by two units of positive charge.

Iraj R. Afnan

2. β**-rays** – These are electrons or positrons that are emitted by the nucleus. These electrons or positrons are not present in the nucleus, but are created at the instant the nucleus decays. In this case the charge on the nucleus changes by one unit depending on the charge of the the β-ray. If the nucleus emits an electron, the charge on the nucleus increases by one unit. However, if the β-ray is a positron, then the charge on the nucleus decreases by one unit.[1] In the case of β decay, the number of neutrons or protons changes by one, while the atomic number of the nucleus does not change.

3. γ**-rays** – These are electromagnetic radiation, or photons, that are emitted by the nucleus. In this case the number of neutrons and protons in the nucleus does not change.

In all these decay modes, a nucleus in an excited state emits radiation and as a result ends in a state of lower energy. This process continues until the system, in this case a nucleus, is in its ground state, or lowest energy state and cannot lose any further energy. We then say that the nucleus is stable. In Figure 6.1 we illustrate the decay mode of ${}^{211}_{83}$Bi which has an atomic number of 211 with 83 protons, and therefore 128 neutrons. This nucleus has two possible decay modes (i.e., ways of losing energy). The first is by α decay to ${}^{207}_{81}$Tl, the second by β^--decay to ${}^{211}_{84}$Po. In turn, both ${}^{207}_{81}$Tl and ${}^{211}_{84}$Po decay by β^-- or α-decay to ${}^{207}_{82}$Pb. This is a simple example of how a system in an excited energy state loses its energy by any means it can to attain the lowest energy state, the ground state.

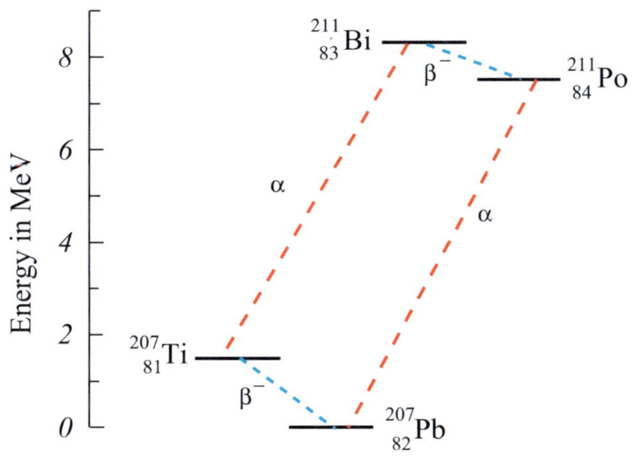

Figure 6.1: A bimodal decay of ${}^{211}_{83}$Bi in which the daughter nuclei decay to the same final nucleus ${}^{207}_{82}$Pb.

In the present section we would like to examine how a nucleus loses energy by α-decay. It was another 30 years after the discovery of radioactivity before Gamow [26], Condon

[1] The magnitude of the charge on the proton is identical to the magnitude of the charge on the electron.

and Gurney [27] in 1928 gave a quantitative explanation of α-decay in terms of a quantum mechanical barrier penetration. This model for α-decay has since been used to study the decay of nuclei, and in particular, the large variation in the lifetime (τ), or the probability for decay, of some of these nuclei, e.g.,

$$\begin{array}{llll} {}^{232}_{90}\mathrm{Th} & \longrightarrow & {}^{228}_{88}\mathrm{Ra} + \alpha & \tau = 2.03 \times 10^{10} \\ {}^{212}_{84}\mathrm{Po} & \longrightarrow & {}^{208}_{82}\mathrm{Pb} + \alpha & \tau = 4.3 \times 10^{-7} \end{array}$$

To understand this large variation in lifetime we will first consider the problem of barrier penetration in one dimension. We will then modify this simple model to give a representation of the decay of a nucleus by α-emission.

6.1.1 Potential Barrier

Consider the barrier of height V_0 and width $2a$ centered about the origin, see Figure 6.2. The corresponding potential is given by

$$V(x) = \begin{cases} 0 & \text{for} \quad |x| > a \\ +V_0 & \text{for} \quad |x| < a \end{cases} \tag{6.1}$$

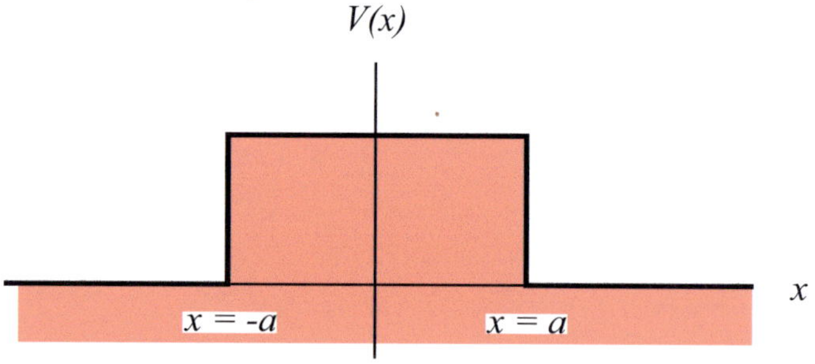

Figure 6.2: A plot of a potential barrier of height V_0 and width $2a$.

Since this potential is repulsive, we have no bound state, but as in the case of a well considered in the last chapter, we still have two possible domains for the energy E. These two energy domains correspond to: (i) Energies greater than the barrier height. (ii) Energies less than the barrier height. Classically, we expect the particle to be reflected off the barrier when the energy is less than the barrier height, and to go through the barrier when the energy E is greater than the barrier height. In the quantum case we will find that there is a transmitted and reflected wave in both cases. To consider these two possibilities, we first consider the Schrödinger equation for $|x| < a$, i.e.,

$$\frac{d^2\psi(x)}{dx^2} + \frac{2m}{\hbar^2}(E - V_0)\psi(x) = 0 \ , \tag{6.2}$$

The solution in this region is given by

$$\psi(x) = \begin{cases} A\ e^{ik'x} + B\ e^{-ik'x} & \text{for} \qquad E > V_0 \\ \\ A\ e^{-\kappa x} + B\ e^{\kappa x} & \text{for} \qquad E < V_0 \end{cases} , \tag{6.3}$$

where

$$\kappa^2 = \frac{2m}{\hbar^2}(V_0 - E) \quad \text{while} \quad k'^2 = \frac{2m}{\hbar^2}(E - V_0)\ . \tag{6.4}$$

The solution of the Schrödinger equation for $|x| > a$ is identical to that of the potential well considered in the last chapter (see e.g., Eq. (5.42)). Thus for $E > V_0$, the wave function for the potential barrier is the same as that of the potential well, the only difference being the value of k'. For the potential barrier, $k' < k$, while for the potential well, $k' > k$.

The interesting aspect of this potential occurs in the case when $0 < E < V_0$. The classical solution for this problem of a beam of particles incident from the left is that all the particles are reflected, because none of the particles have enough kinetic energy to overcome the potential energy barriers in the region $|x| < a$. However, the quantum mechanical problem leads to a solution of the Schrödinger equation of the form

$$\psi(x) = \begin{cases} e^{ikx} + R\ e^{-ikx} & \text{for} \quad x < -a \\ \\ A\ e^{-\kappa x} + B\ e^{\kappa x} & \text{for} \quad -a < x < a \\ \\ T\ e^{ikx} & \text{for} \quad x > a \end{cases} \tag{6.5}$$

To get the four constants (R, A, B and T) fixed, we need to make use of the facts that $\psi(x)$ and its derivative are continuous at $x = \pm a$. The result in this case is the same as the potential well if we make the substitution

$$k' \to i\kappa \tag{6.6}$$

in Eq. (5.46), i.e., the transmission amplitude T, in this case is given by

$$T = e^{-2ika} \frac{2ik\kappa}{2ik\kappa \cos 2i\kappa a - i(k^2 - \kappa^2)\sin 2i\kappa a}$$

but we have that

$$\cos 2i\kappa a = \frac{1}{2}(e^{-2\kappa a} + e^{2\kappa a}) = \cosh 2\kappa a$$

and

$$\sin 2i\kappa a = \frac{1}{2i}(e^{-2\kappa a} - e^{2\kappa a}) = i \sinh 2\kappa a\ .$$

This allows us to write the amplitude for the transmitted wave as

$$T(k, i\kappa) = e^{-2ika} \frac{2k\kappa}{2k\kappa \cosh 2\kappa a - i(k^2 - \kappa^2)\sinh 2\kappa a}\ . \tag{6.7}$$

This means the transmitted current, which is defined as $j_T = \frac{\hbar k}{m}|T|^2$, is given by

$$j_T = \left(\frac{\hbar k}{m}\right) \frac{(2k\kappa)^2}{(k^2+\kappa^2)^2 \sinh^2 2\kappa a + (2k\kappa)^2} \neq 0 \ , \tag{6.8}$$

i.e., the particles in the incident beam, or at least some of them, get through the barrier even though the energy of each particle is *less* than the barrier height, i.e., $E < V_0$. This effect is known as *tunneling*, and is purely a quantum effect and has no classical analog. It is this quantum effect that was required to understand α-decay as we will see in the next section.

Before we proceed with the application of this tunneling effect to α-decay, we should have a better understanding of how the tunneled or transmitted current depends on the barrier height. To illustrate the relation between the transmitted current and the barrier height, we consider the limit of $\kappa a \gg 1$, i.e., $a \gg 1$, or $V_0 \gg E$. In this case we have

$$\sinh 2\kappa a \longrightarrow \frac{1}{2}\, e^{2\kappa a} \ .$$

In this limit the transmitted current becomes

$$j_T = 4\left(\frac{\hbar k}{m}\right)\left(\frac{2k\kappa}{k^2+\kappa^2}\right)^2 e^{-4\kappa a} \ , \tag{6.9}$$

which is what we expected, i.e., that as the height (V_0) or thickness (a) of the barrier increases, the transmitted current drops to zero exponentially. This in turn means that the transmitted current is a very sensitive function of the a and V_0.

6.1.2 α-decay

There are several applications of tunneling in nature. The oldest and most famous is α-particle decay which was considered as the tunneling of α-particle [2] from the inside of the nucleus to the outside. From inside the nucleus, the α-particle sees a potential that can be approximated by a square well, while from the outside, the α-particle sees the Coulomb potential due to a positively charged nucleus and the fact that the charge on the α-particle is also positive. This allows us to write an approximation to the potential the α-particle sees that is of the form:

$$V(r) = \begin{cases} -V_0 & \text{for} \quad r < R \\ \\ +\dfrac{ZZ'e^2}{r} & \text{for} \quad r > R \end{cases} \ , \tag{6.10}$$

where R is approximately the radius of the nucleus, Z and Z' are the charge of the final nucleus and α-particle, and e is the charge on the proton.

[2] This was first considered by Gamow [26], Condon and Gurney[27] in 1928.

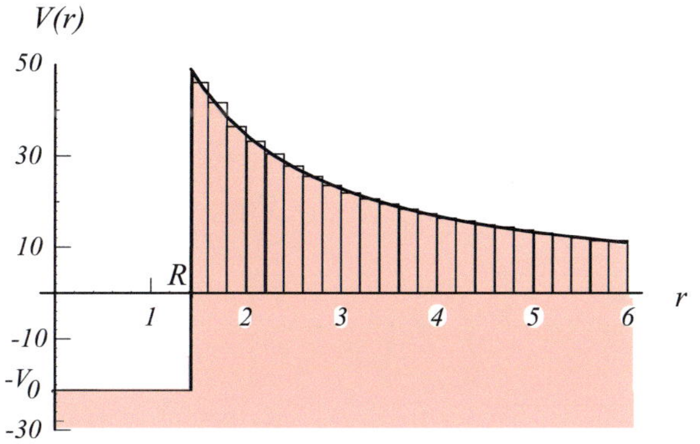

Figure 6.3: The nuclear potential as seen by an α-particle. In the figure we have illustrated how we can replace the Coulomb barrier by a set of rectangular barriers which would correspond to replacing the integral in Eq. (6.11) by a sum, and therefore the total transmission $|T|$ by a product of transmission for the different strips.

For the potential barrier we considered in Sec. 6.1.1, the κ in Eq. (6.9) is given by $\kappa = \sqrt{\frac{2m}{\hbar^2}(V_0 - E)}$, and is a constant since the potential was of constant height. However, for the more realistic nuclear situation in α-decay, the potential is not a constant but depends on the radial distance from the center of the nucleus r. As a result, the κ for the potential in Eq. (6.10) is now dependent on r. This means the factor of $2\kappa a$ in the amplitude for the transmitted wave through the barrier should be replaced by the integral of $\kappa(x)$ over the barrier, i.e., we can write the exponential in the amplitude of the transmitted wave as [3]

$$|T| \sim \exp\left\{-\int_{-a}^{+a} dx\, \kappa(x)\right\}$$

For α-decay the potential barrier is not constant in height, and the exponential factor is

$$|T| \sim \exp\left\{-\int_{R}^{b} dr \sqrt{\frac{2m}{\hbar^2}(V(r) - E)}\right\} . \tag{6.11}$$

[3] We can replace the barrier in α-decay as given in Figure 6.3 by a series of barriers strips of thickness Δx. In this case the total transmission amplitude is, i.e.,

$$|T| = |T_1|\,|T_2|\,|T_3|\,\ldots\,|T_N|^2$$

where $|T_i|$ is the transmission amplitude for the ith strip in Figure 6.3, i.e.,

$$|T_i| = e^{-\kappa_i \Delta x} .$$

Here, b is the point at which $V(b) = E$ outside the potential, i.e.,

$$E = \frac{ZZ'e^2}{b} \quad \text{or} \quad b = \frac{ZZ'e^2}{E} \; .$$

This allows us to write the transmitted amplitude T as

$$\begin{aligned} |T| &\sim \exp\left\{ -\sqrt{\frac{2m}{\hbar^2} ZZ'e^2} \int_R^b dr \sqrt{\frac{1}{r} - \frac{1}{b}} \right\} \\ &= \exp\left\{ -\sqrt{\frac{2m}{\hbar^2} ZZ'e^2}\, \sqrt{b} \left[\cos^{-1}\left(\frac{R}{b}\right)^{1/2} - \left(\frac{R}{b} - \frac{R^2}{b^2}\right)^{1/2} \right] \right\} \; . \end{aligned}$$

For $R \ll b$ this result simplifies to

$$\begin{aligned} |T| &\sim \exp\left\{ -\frac{\pi}{2} \sqrt{\frac{2m}{\hbar^2} ZZ'e^2 b} \right\} \\ &= \exp\left\{ -\frac{\pi ZZ'e^2}{2\hbar} \sqrt{\frac{2m}{E}} \right\} \; . \end{aligned} \tag{6.12}$$

The probability for transmission through the barrier is equal to the rate the α-particle presents itself at the barrier, $\frac{v}{2R}$, times the probability for transmission through the barrier $|T|^2$. This gives us, for the lifetime τ, the result [4]

$$\begin{aligned} \tau^{-1} &= \frac{v}{2R} |T|^2 \\ &\approx \frac{v}{2R} \exp\left\{ -\frac{4Z}{\sqrt{E(\text{MeV})}} \right\} \; , \end{aligned}$$

where we have taken the charge on the α-particle to be two, i.e., $Z' = 2$. This allows us to write the lifetime as

$$\log_{10} \tau^{-1} \approx C_1 - C_2 \frac{Z}{E^{1/2}(\text{MeV})} \; . \tag{6.13}$$

where C_1 and C_2 are constants. This result is in remarkable agreement with the experimental lifetime of α emitting nuclei.

[4] In writing this result we have made use of the fact that

$$\hbar c = 197.3\,\text{MeV fm} \qquad e^2 = 1.44\,\text{MeV fm}$$

and the mass of the α particle is approximately four times the mass of the proton, which is

$$m_p c^2 = 938\,\text{MeV} \; .$$

Here c is the velocity of light.

At this stage we observe that as the energy of the emitted α-particle increases, the probability for transmission through the barrier increases. This is due to the fact that the barrier width has decreased with increasing energy. Since the lifetime is inversely proportional to the probability for transmission $|T|^2$, the lifetime of the α-emitting nucleus decreases. This result based on this simple one-dimensional model is in good agreement with experiment.

6.2 The Deuteron

The deuteron is the simplest nuclear system we can consider. It consists of a bound state of a proton and a neutron with a binding energy of 2.2246 MeV. The properties of this system can be used to examine the properties of the force between the neutron and proton. In the present section, we would like to determine how the binding energy of the deuteron can be used to determine the depth of the potential that binds the proton-neutron system.

Before we examine the solution of the Schrödinger equation for the deuteron, let us assume that the force between the proton and the neutron is due to the exchange of a π meson (*pion*) of mass $m_\pi \approx 140$ MeV. From the energy-time uncertainty relation we have

$$\Delta E \Delta t \simeq \hbar$$

and therefore

$$\Delta t \simeq \frac{\hbar}{\Delta E} .$$

If we take the velocity of the pion to be approximately the velocity of light c, then the distance the pion can travel in the time Δt will be approximately the range of the potential, i.e.,

$$a \simeq c\,\Delta t = \frac{\hbar c}{\Delta E} \sim \frac{\hbar c}{m_\pi c^2} \simeq \frac{200}{140} \sim 1.5 \text{ fm} ,$$

where we have taken $\hbar c \approx 200$ MeV fm. Given the range of the force and the binding energy of the deuteron we need to estimate the depth of the potential. To determine the kinetic energy of the proton-neutron system we make use of the position momentum uncertainty, i.e.,

$$\Delta p \Delta x = \hbar .$$

Then taking Δx to be the range of the potential, we can determine Δp as

$$\Delta p \approx \frac{\hbar}{a} .$$

Assuming the relative momentum of the proton-neutron system is Δp, the kinetic energy can now be written in terms of the uncertainty in the momentum as

$$\frac{(\Delta p)^2}{2m} = \frac{\hbar^2}{2ma^2} = \frac{\hbar^2 c^2}{2(mc^2)\,a^2} = \frac{(200)^2}{10^3 \times (1.5)^2} = \frac{4 \times 10^4}{2 \times 10^3} \simeq 20 \text{ MeV} .$$

Here we have taken mc^2 to be the reduced mass of the neutron-proton system. Since the mass of the neutron is approximately equal to the mass of the proton, this reduced mass is half the the mass of the proton which is approximately 1000 MeV, i.e., $mc^2 \approx 500$ MeV. The total energy of a system is the sum of the kinetic energy (a positive number) and the potential energy (usually a negative number). Given the energy of the system to be ≈ -2.2 MeV,[5] and the kinetic energy is approximately 20 MeV, we expect the potential energy to be about -22.2 MeV.

6.2.1 The binding energy of the deuteron

We now have to construct a one dimensional model for the neutron-proton system. The first approximation is to take the potential between a neutron and a proton to be a square well of depth V_0 and width a. This is illustrated in Figure 6.4. Since the distance between the neutron and proton is $r > 0$, we need only consider the values of $x > 0$ when we convert the problem from three- to one-dimension. As a result our model for the neutron proton system is the one-dimensional potential given by

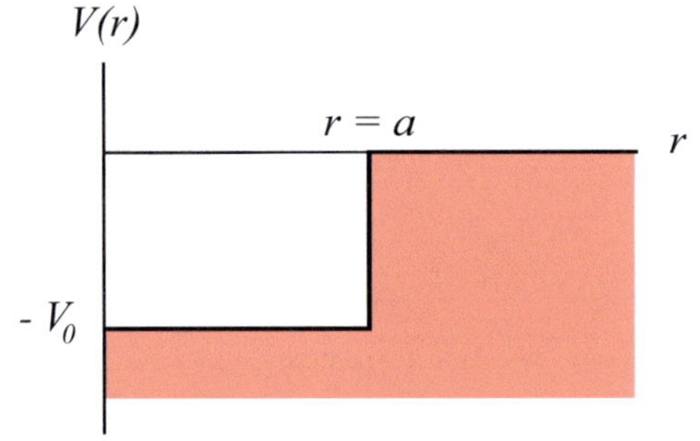

Figure 6.4: The radial potential between the neutron and proton in the deuteron.

$$V(x) = \begin{cases} +\infty & \text{for} \quad x < 0 \\ -V_0 & \text{for} \quad 0 < x < a \\ 0 & \text{for} \quad x > a \end{cases} , \tag{6.14}$$

For this potential, the solution of the Schrödinger equation for $x \leq 0$ is zero i.e., $\psi(x) = 0$ for $x \leq 0$. This leaves the domain $x > 0$ to play the role of the variable r which is the

[5]The binding energy of a system is a positive number, while the energy of a bound state is taken to be a negative number. Therefore, the binding energy of the system is the magnitude of the energy of that system, i.e., for a bound state we have:

$$\text{Binding energy of a system} = -\text{Energy of the system}$$

radial distance from the proton to the neutron. Thus for $x > 0$, the Schrödinger equation takes the form

$$\frac{d^2\psi(x)}{dx^2} = -\begin{cases} \frac{2m}{\hbar^2}(E + V_0)\ \psi(x) & \text{for} \quad 0 < x < a \\ \\ \frac{2m}{\hbar^2} E\ \psi(x) & \text{for} \quad x > a \end{cases}, \tag{6.15}$$

where m is the reduced mass of the proton-neutron system, i.e., $m = m_p m_n/(m_p + m_n)$ where m_p is the mass of the proton and m_n is the mass of the neutron. Since the deuteron is a bound state of a proton and a neutron, we will consider the solution of the above differential equation for $E < 0$. As the original problem is a three dimensional problem in which we have replaced the radial variable in spherical polar coordinates by the one dimension we are considering, we will replace x by r from this point on. The solution of our Schrödinger equation is now of the form

$$\psi(r) = \begin{cases} A \sin kr + B \cos kr & \text{for} \quad 0 < r < a \\ \\ C\ e^{-\alpha r} + D\ e^{\alpha r} & \text{for} \quad r > a \end{cases}. \tag{6.16}$$

Here the parameters α and k are given in terms of the energy E and the well depth V_0 by the relations

$$\alpha = \sqrt{\frac{2m}{\hbar^2}|E|} \qquad \text{and} \qquad k = \sqrt{\frac{2m}{\hbar^2}(V_0 - |E|)}\ . \tag{6.17}$$

The constants A, B, C and D are to be determined by the boundary conditions on the wave function and its derivative, and the normalization of this wave function. The boundary condition at the origin requires that the wave be zero, i.e., $\psi(0) = 0$, This condition is satisfied by requiring that $B = 0$. For a bound state, the wave function $\psi(r)$ should go to zero as $r \to \infty$. This second boundary condition is satisfied if we take $D = 0$. The solution of the Schrödinger equation with these two boundary conditions satisfied can now be written as

$$\psi(r) = \begin{cases} A \sin kr & \text{for} \quad 0 < r < a \\ \\ C\ e^{-\alpha r} & \text{for} \quad r > a \end{cases}. \tag{6.18}$$

To determine the remaining constants A, C and the allowed energies or eigenvalues E, we make use of the requirement that the wave function and its derivative to be continuous at $r = a$, i.e.,

$$\begin{aligned} A \sin ka &= C\ e^{-\alpha a} \\ kA \cos ka &= -\alpha C\ e^{-\alpha a}\ . \end{aligned} \tag{6.19}$$

Dividing the second equation by the first equation we eliminate the constants A and C. The resulting transcendental equation is similar to that encountered in the last chapter for an attractive potential well (see Eq. (5.33)), i.e.,[6]

$$k \cot ka = -\alpha \ . \tag{6.20}$$

We can solve this equation graphically by plotting the functions

$$F(y) = \cot \sqrt{y^2 - \alpha^2 a^2} \quad \text{and} \quad F(y) = -\sqrt{\frac{\alpha^2 a^2}{y^2 - \alpha^2 a^2}} \ , \tag{6.21}$$

where

$$y^2 = \frac{2ma^2 V_0}{\hbar^2} \ , \tag{6.22}$$

with y, α and k related by the equation

$$k^2 a^2 = y^2 - \alpha^2 a^2 \ . \tag{6.23}$$

The graphical solution of Eq. (6.21) is given in Figure 6.5.

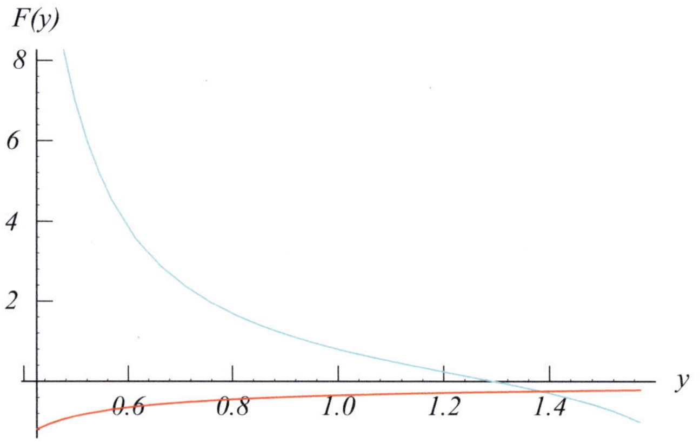

Figure 6.5: The graphic solution of the transcendental equation. The intercept of the two curves gives the solution.

The deuteron is a neutron-proton system with a binding energy of 2.2246 MeV. The question is: can we find a well depth V_0 that will give us a bound state with such a binding energy? Since there is only one such bound state, there is only one solution to the above transcendental equation. Let us call this solution y_0. This value of y_0, when used in Eq. (6.22), gives the potential depth V_0 provided we know the range of the potential a. We have estimated above, that for a range $a \approx 1.5$ fm, the well depth $V_0 \approx 22.2$ MeV. Since the binding energy is 2.2246 MeV, we have that $V_0 \gg |E|$. From the definition of y and α, we can establish that

$$y \gg \alpha^2 a^2$$

[6]Note this equation is identical to the case of the odd solution encountered in the last chapter. This is a result of the fact that we imposed the boundary condition that the wave function be zero at the origin.

so that

$$\cot\sqrt{y^2-\alpha^2a^2}\approx\cot y \qquad \text{and} \qquad \sqrt{\frac{\alpha^2a^2}{y^2-\alpha^2a^2}}\ll 1\ .$$

In this case $F(y)$, as defined in Eq. (6.21) is approximately zero, i.e.,

$$F(y)=\cot y\approx 0 \qquad \text{and} \qquad y\approx\frac{\pi}{2}\ .$$

In Figure 6.5 we have a plot of the Eq. (6.21) for a deuteron binding energy of 2.224 MeV, and find the graphical solution is at $y = 1.38$ radians. From this we can extract a value of the well depth given a value for the range of the potential a. For example, we observe here that the binding energy of the deuteron does not determine both the depth and range of our square well potential. It only determines the product V_0a^2. To determine the range and strength or depth of the potential we will need further information about the scattering of neutrons from the proton.

6.2.2 The deuteron wave function

We are now in a position to determine the relative wave function of the neutron and proton in the deuteron. Having determined the binding energy $|E|$, we now know the wave number k, and from the continuity of the wave function at $r = a$, we have that

$$A\sin ka = C\,e^{-\alpha a}\ .$$

This allows us to determine the constant C in terms of A, i.e.,

$$C = A\ e^{\alpha a}\sin ka\ . \tag{6.24}$$

We now have the wave function up to an overall constant A. This wave function is given by

$$\psi(r)\ = A\begin{cases}\sin kr & \text{for}\quad 0<r<a\\ \sin ka\ e^{-\alpha(r-a)} & \text{for}\quad r>a\end{cases}\ . \tag{6.25}$$

Finally, to determine the overall amplitude of our wave function A, we have to make use of the fact that the wave function for a bound state has to be normalized if its square is to be interpreted as a probability, i.e.,[7]

$$\int_0^{+\infty} dr\,\psi^*(r)\,\psi(r) = 1\ . \tag{6.26}$$

Making use of the wave function given in Eq. (6.25), we get

$$A^2\left\{\int_0^a dr\sin^2 kr + \sin^2 ka\,e^{2\alpha a}\int_a^\infty dr\ e^{-2\alpha r}\right\} = 1\ .$$

[7]Note in three dimensions the radial integral in Eq. (6.26) should have an angle integration which would give a factor of 4π. We have chosen to ignore this for simplicity.

Using the tabulated values of the integrals [8] we find that A is given by

$$A^2 \left\{ \frac{a}{2} - \frac{1}{4k} \sin 2ka + \frac{\sin^2 ka}{2\alpha} \right\} = 1$$

or

$$A = \left\{ \frac{a}{2} - \frac{1}{4k} \sin 2ka + \frac{\sin^2 ka}{2\alpha} \right\}^{-1/2} . \tag{6.27}$$

Given the parameters of the potential, we can determine the binding energy E, and the normalized wave function for the deuteron.

6.3 The δ-Function Potential

Having considered two examples of the application of the one-dimensional Schrödinger equation to nuclear physics problems, we turn our attention to atoms and solids. Here again we find cases where some of the basic properties of molecules and solids can be modeled in terms of the one dimensional solution of the Schrödinger equation. Since both solids and molecules are made up of atoms, we need to first consider a model for the atom that is simple enough to lend itself to analytic solutions. This model for an atom is then used to construct first a model of a diatomic molecule, and then a one dimensional solid. In both cases we find that our model has some of the basic symmetries of the more complicated system. As a result, those features of the physical system that are due to symmetry, are preserved in the simple one dimensional models we consider.

In a first step towards developing a one dimensional model of a molecule or a solid, we turn our attention to the case of a potential of the form

$$\frac{2m}{\hbar^2} V(x) = -\frac{\lambda}{a} \delta(x) . \tag{6.28}$$

This may be considered as the potential that determines the behavior of the valence electron in an atom. The Schrödinger equation for the bound state ($E < 0$) in this potential takes the form

$$\frac{d^2\psi(x)}{dx^2} - \kappa^2 \psi(x) = -\frac{\lambda}{a} \delta(x) \psi(x) , \tag{6.29}$$

[8] We have made use of the integrals

$$\int_0^a dx\ \sin^2 kx = \frac{1}{k} \left\{ \frac{ka}{2} - \frac{1}{4} \sin 2ka \right\}$$

and

$$\int_a^\infty dx\ e^{-2\alpha x} = +\frac{e^{-2\alpha a}}{2\alpha} .$$

where $\kappa^2 = \frac{2m}{\hbar^2}|E|$. The interest in this potential stems from the fact that one can combine two such potentials to get a simple model of a molecule with one electron being shared by the two nuclei in the molecule. Also, we can consider an infinite number of such potentials that are equally spaced along the x-axis. This is a simple model of a one dimensional lattice, and has the feature of being the potential the electron sees in a one dimensional solid or a one dimensional polymer.

The new feature of the Schrödinger equation for this potential is that it is singular at $x = 0$, and as a result the first derivative of the wave function is not continuous at $x = 0$. In fact if we integrate the above equation between $-\epsilon$ and $+\epsilon$ and take the limit as $\epsilon \to 0$, we get

$$\left.\frac{d\psi}{dx}\right|_{-\epsilon}^{+\epsilon} = \left.\frac{d\psi}{dx}\right|_{+\epsilon} - \left.\frac{d\psi}{dx}\right|_{-\epsilon} = -\frac{\lambda}{a}\psi(0) \ . \tag{6.30}$$

We will need to use this boundary condition to determine the constants in the solution of the Schrödinger equation for this potential.

The solution of the Schrödinger equation for $x \neq 0$ is that of an equation with no potential, i.e., for $E < 0$ it is given by

$$\psi(x) = \begin{cases} A\ e^{\kappa x} & \text{for} \quad x < 0 \\ B\ e^{-\kappa x} & \text{for} \quad x > 0 \end{cases} \ . \tag{6.31}$$

The continuity of the wave function at $x = 0$ requires that $A = B$, while the discontinuity in the first derivative of the wave function should satisfy Eq. (6.30). This gives us

$$\lim_{\epsilon \to 0} \left[-A\kappa\ e^{-\kappa\epsilon} - A\kappa\ e^{\kappa\epsilon}\right] = -\frac{\lambda}{a}A \ . \tag{6.32}$$

Taking into consideration the fact that the exponentials inside the brackets are one in the limit as $\epsilon \to 0$, we get

$$2\kappa = \frac{\lambda}{a} \ . \tag{6.33}$$

From this result we see that this δ-function potential supports one, and only one bound state with the binding energy given by

$$|E| = \frac{\hbar^2}{2m}\kappa^2 = \frac{\hbar^2\lambda^2}{8ma^2} \ . \tag{6.34}$$

The wave function for this bound state is given in Eq. (6.31) with $A = B$ and κ given by Eq. (6.33). The overall normalization constant A is now determined by the fact that the wave function has to be normalized, i.e.,

$$\int_{-\infty}^{+\infty} dx\ |\psi(x)|^2 = 2A^2 \int_0^{\infty} dx e^{-2\kappa x} = \frac{A^2}{\kappa} = 1 \ .$$

This gives, for the normalization constant, $A = \sqrt{\kappa}$, and the normalized wave function as

$$\psi(x) = \sqrt{\kappa} \begin{cases} e^{\kappa x} & \text{for} \quad x < 0 \\ \\ e^{-\kappa x} & \text{for} \quad x > 0 \end{cases} . \tag{6.35}$$

This wave function is illustrated in Figure 6.6, where the wave function is scaled so that it has a unit amplitude at $x = 0$, which would correspond to $\kappa = 1$. This wave function is similar to the wave function of a valence electron for large separation between the electron and the nucleus of the atom. This is the most important feature of the wave function of an electron in an atom when it comes to constructing either molecules or solids. In this way we have preserved that feature of the atomic wave function needed for molecular and solid state physics. In the atomic case, the binding energy for this potential will be the ionization energy of the valence electron.

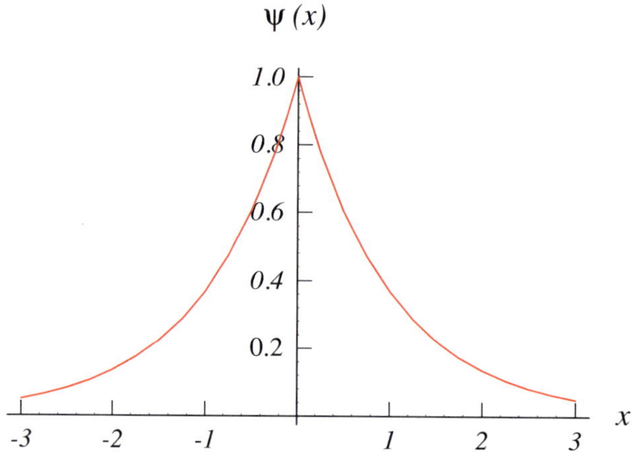

Figure 6.6: The wave function for a δ-function potential.

6.4 The Diatomic Molecule

In a diatomic molecule we have two nuclei a distance R apart with one or more electrons being shared by the two atoms in the molecule. These electrons are in a potential with two centers. The simplest such molecule is the hydrogen molecular ion. In this case we have one electron and two protons. Since the protons are much heavier than the electron, to first approximation we can assume the two protons are stationary a distance R apart, while the electron is shared by the two centers. In this way we can replace the three-body problem (the electron and two protons) by an electron moving in an average field of the two protons. This problem can be approximated by a one dimensional problem in which we have one proton stationary at $x = a$ while the second proton is at $x = -a$. This means that R, the distance between the two protons, is $2a$. The potential energy the electron sees is the Coulomb potential due to the two protons, which is attractive. This we can approximate by the potentials used in the last section for the valence electron in an atom, i.e., the potential is the sum of two δ-functions.

Let us now consider the problem of two δ-function potentials placed symmetrically with respect to the origin, e.g., at $x = \pm a$. This potential can be written as

$$\frac{2m}{\hbar^2} V(x) = -\frac{\lambda}{a}\{\delta(x+a) + \delta(x-a)\} \ , \tag{6.36}$$

and is invariant under inversion i.e., $x \to -x$. As a result of this symmetry, the solutions are eigenstates of the parity operator which means they are either symmetric (even) or antisymmetric (odd). We first consider the symmetric solution which is given by

$$\psi_e(x) = \begin{cases} C\, e^{\kappa x} & \text{for} \quad x < -a \\ A \cosh \kappa x & \text{for} \quad -a < x < a \\ C\, e^{-\kappa x} & \text{for} \quad x > a \end{cases} \ . \tag{6.37}$$

Note, $\cosh \kappa x = \frac{1}{2}(e^{\kappa x} + e^{-\kappa x})$ and is a symmetric function, i.e., it is an even function of x. The continuity of the wave function at $x = a$ gives

$$A \cosh \kappa a = C\, e^{-\kappa a} \ . \tag{6.38}$$

On the other hand, the discontinuity in the derivative of the wave function at $x = a$ gives (see Eq. (6.30))

$$-\kappa\, C\, e^{-\kappa a} - \kappa\, A \sinh \kappa a = -C\, \frac{\lambda}{a} e^{-\kappa a}$$

or

$$A \sinh \kappa a = C \left(\frac{\lambda}{\kappa a} - 1\right) e^{-\kappa a} \ . \tag{6.39}$$

Combining the results of Eqs. (6.38) and (6.39), we get

$$\tanh \kappa a = \left(\frac{\lambda}{\kappa a} - 1\right) \ . \tag{6.40}$$

Since $\tanh \kappa a$ is positive for κa positive, we expect the solution of this transcendental equation to be restricted to those values of κ that satisfy the relation $(\frac{\lambda}{\kappa a} - 1) > 0$ or $\frac{\lambda}{\kappa a} > 1$, and therefore $\kappa < \frac{\lambda}{a}$. On the other hand since $\tanh \kappa a < 1$, the solution of the transcendental equation is in the domain where $(\frac{\lambda}{\kappa a} - 1) < 1$ or $\frac{\lambda}{\kappa a} < 2$ and therefore $\kappa > \frac{\lambda}{2a}$. In other words, the solution of Eq. (6.40) is restricted to values of κ in the domain

$$\frac{\lambda}{2a} < \kappa < \frac{\lambda}{a} \ .$$

In Figure 6.7 we have a graphical solution of Eq. (6.40). Here again we have only one bound state with positive parity. At this point we note that $\kappa > \frac{\lambda}{2a}$, i.e., the even solution has a binding energy greater than that of one δ-function potential as given in Eq. (6.33).

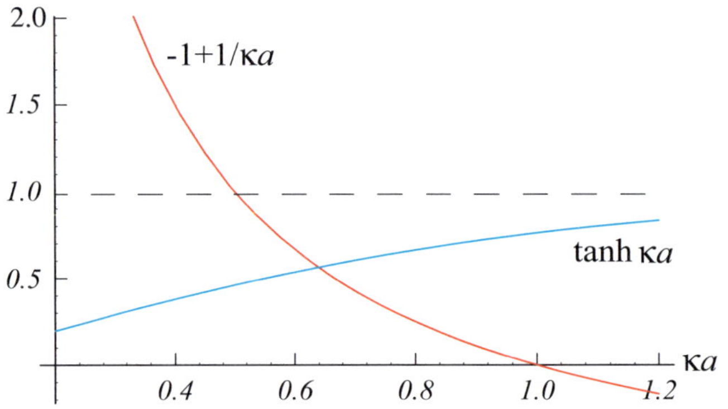

Figure 6.7: This is a graphical determination of the solution to the equation $\tanh \kappa a = \frac{1}{\kappa a} - 1$. The numerical solution is $\kappa = 0.639$.

We now turn to the antisymmetric (odd) solution. This solution is given by

$$\psi_o(x) = \begin{cases} -D\, e^{\kappa x} & \text{for} \quad x < -a \\ B \sinh \kappa x & \text{for} \quad -a < x < +a \\ D\, e^{-\kappa x} & \text{for} \quad x > a \end{cases} . \tag{6.41}$$

In this case, the boundary condition that the wave function is continuous at $x = a$ gives us

$$B \sinh \kappa a = D\, e^{-\kappa a} , \tag{6.42}$$

while the boundary condition on the derivative of the wave function at $x = a$ gives us

$$-\kappa\, D\, e^{-\kappa a} - \kappa B \cosh \kappa a = -D\, \frac{\lambda}{a}\, e^{-\kappa a} ,$$

which can be written as

$$B \cosh \kappa a = D \left(\frac{\lambda}{\kappa a} - 1 \right) e^{-\kappa a} . \tag{6.43}$$

Combining the results of Eqs. (6.42) and (6.43), we get

$$\coth \kappa a = \left(\frac{\lambda}{\kappa a} - 1 \right)$$

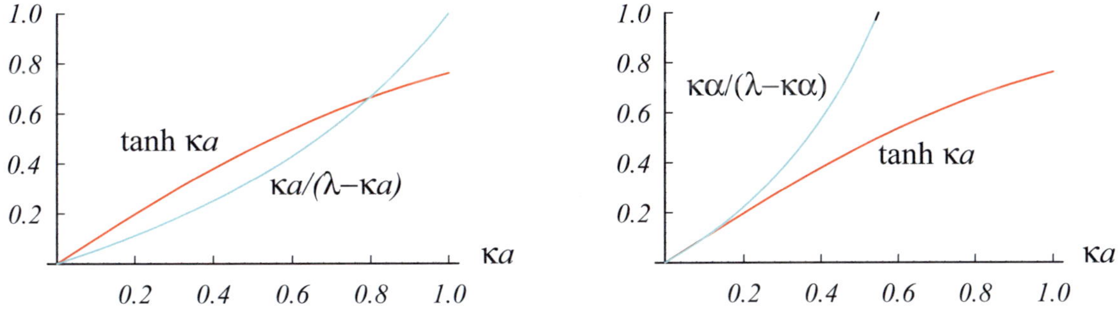

Figure 6.8: The graphical determination of the solution to the equation $\tanh \kappa a = \frac{\kappa a}{\lambda - \kappa a}$ for $\lambda = 2$ (left graph), and $\lambda = 1.1$ (right graph).

or

$$\tanh \kappa a = \left(\frac{\lambda}{\kappa a} - 1\right)^{-1} = \frac{\kappa a}{\lambda - \kappa a} \ . \tag{6.44}$$

In this case we are not always guaranteed a solution. If the slope of $\kappa a/(\lambda - \kappa a)$ is greater than the slope of $\tanh \kappa a$ at $\kappa a = 0$ there is no solution (see Figure 6.8 on the right), and therefore there is no bound state for the antisymmetric case. If there is a solution to the transcendental equation it should be for values of κ that satisfy the relation

$$\frac{\kappa a}{\lambda - \kappa a} < 1 \ ,$$

or

$$2\kappa a < \lambda \quad \text{and therefore} \quad \kappa < \frac{\lambda}{2a} \ . \tag{6.45}$$

This means the binding energy of this state is less than that of the symmetric solution. The wave function for the two solutions is given below in Figure 6.9. Here we note that the behavior of the wave function for $|x| > a$ is completely determined by the binding energy. This behavior is preserved if we go to a more realistic model for the molecule. On the other hand, the behavior of the wave function for $|x| < a$ is determined for the two solutions by the symmetry of the system under reflection, i.e., $x \to -x$. In the case of the real molecule this symmetry corresponds to exchanging the two protons in the molecule. Since the two protons are identical, the exchange of the two protons is a good symmetry of the system and therefore the Hamiltonian. As a result, the qualitative behavior of the two solutions for $|x| < a$, which is governed by the symmetry of the system, will hold true for the real molecule as well. In particular, we expect the even solution (blue) to correspond to a state which is more bound than that of the odd solution (red). We note here that we applied our boundary conditions at $x = +a$ and symmetry of the system took care of the boundary condition at $x = -a$.

To get a better understanding as to why the symmetric solution is more bound than the antisymmetric solution, we write the total energy of the molecule as the sum of the

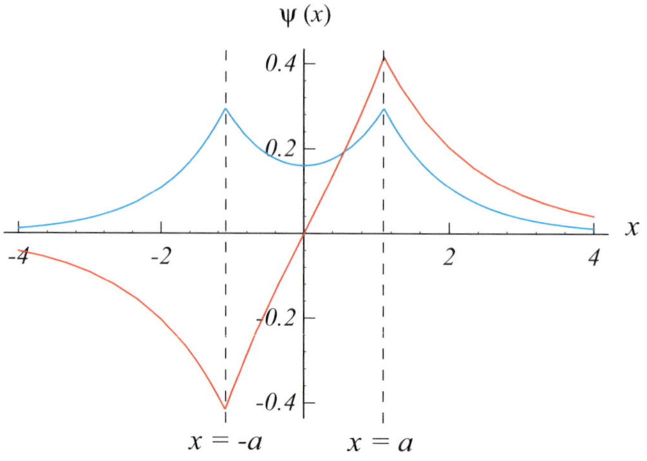

Figure 6.9: This is a graph of the wave functions of the ground state (blue) and first excited state (red) for the potential with two attractive δ-functions at $x = \pm a$.

kinetic energy and the potential energy i.e.,

$$E = K.E. + P.E. \; . \tag{6.46}$$

We know that the potential energy is negative for attractive potentials while the kinetic energy is always positive. To get a bound state we need to have the energy $E < 0$. The potential energy is given by the matrix element of the potential, i.e.,

$$P.E = \int dx\, \psi^*(x)\, V(x)\, \psi(x) = \int dx\, |\psi(x)|^2\, V(x) \; . \tag{6.47}$$

Since $V(x) < 0$ for attractive potentials, and $|\psi_S(x)|^2 > |\psi_A(x)|^2$ because $\psi_A(x)$ goes through zero, the $P.E$ for the symmetric wave function is larger in magnitude than the $P.E.$ for the antisymmetric wave function, i.e.,

$$(P.E)_S = \int dx\, |\psi_S(x)|^2\, V(x) < \int dx\, |\psi_A(x)|^2\, V(x) = (P.E)_A \; . \tag{6.48}$$

On the other hand, the $K.E$ is given by

$$\begin{aligned} K.E &= \int dx\, \psi^*(x) \left(-\frac{\hbar^2}{2m} \frac{d^2}{dx^2} \right) \psi(x) \\ &= \frac{\hbar^2}{2m} \int dx \left| \frac{d\psi}{dx} \right|^2 > 0 \; . \end{aligned} \tag{6.49}$$

But from Figure 6.9 it is clear that

$$\left| \frac{d\psi_S}{dx} \right| < \left| \frac{d\psi_A}{dx} \right| \; ,$$

for $-a < x < +a$, and therefore

$$(K.E)_S < (K.E)_A \ . \tag{6.50}$$

This allows us to compare the energy of the symmetric and the antisymmetric states, with the result that

$$\begin{aligned} E_S &= (K.E)_S + (P.E)_S < (K.E)_A + (P.E)_S \\ &< (K.E)_A + (P.E)_A = E_A \ . \end{aligned} \tag{6.51}$$

Thus our simple model provides us with a simple explanation for why the H_2^+ molecule has a symmetric wave function for the ground state. We observe that this simple model, that has the basic symmetry of the real H_2^+, has given us a good qualitative understanding of this diatomic molecule. This is just an example of the powerful role symmetry plays in quantum systems.

6.5 The Atomic Clock

We now consider the above diatomic molecule in which we have two states with different energies, one corresponding to an even solution the other due to an odd solution. We now demonstrate how this system can be used to construct a frictionless oscillator that can be used as a clock.

Consider the above two-center potential with an initial condition in which the particle is localized to one of the two centers. Clearly this state is going to be a linear combination of the symmetric (even) and antisymmetric (odd) wave functions, i.e.,

$$\psi_i(x) = \psi_S(x) + \alpha\, \psi_A(x) \ . \tag{6.52}$$

The constant α is determined by the requirement that the probability of finding the particle along the negative x-axes (i.e., the nucleus on the left) should be zero, i.e.,

$$\int_{-\infty}^{0} dx |\psi_i(x)|^2 = 0 \ . \tag{6.53}$$

The time dependence of this initial wave function is given by

$$\begin{aligned} \Psi_i(x,t) &= \psi_S(x)\; e^{-iE_S t/\hbar} + \alpha\; e^{-iE_A t/\hbar}\psi_A(x) \\ &= e^{-E_S t/\hbar}\left\{\psi_S(x) + \alpha\; e^{i(E_S - E_A)t/\hbar}\psi_A(x)\right\} \ . \end{aligned} \tag{6.54}$$

Since the symmetric and antisymmetric states have different energies, the time evolution of the two eigenstates will be different. As a result, the relative phase of $\psi_S(x)$ and $\psi_A(x)$ changes with time. Thus if we take α at $t = 0$ so the wave function is that of an electron localized to the right hand nucleus ($|\Psi(x,0)|^2 = 0$ for $x < 0$), then at a later time $t_1 > 0$, we have

$$e^{i(E_S - E_A)t_1/\hbar} = -1 \ ,$$

and this corresponds to the electron being localized to the left hand side nucleus ($|\Psi(x,t_1)|^2 = 0$ for $x > 0$). This oscillation of the electron between the left and right nucleus keeps going on with time, i.e., the electron hops from one nucleus to the other and back with a frequency

$$\omega = 2(E_S - E_A)/\hbar \ . \tag{6.55}$$

This frequency depends on the energy difference between the symmetric state and the antisymmetric state. The period of this oscillation is directly related to the time it takes for the state bound on one nucleus in the molecule to penetrate the region between the two nuclei. Such a system can be used as a very accurate clock and since there is no friction in the system, the frequency does not change with time and the clock is stable.

6.6 One Dimensional Solids

Having demonstrated the success of predicting some of the features of a diatomic molecule in terms of a potential with two δ-function in one dimension, we turn our attention to the possibility of determining some of the features of a solid by extending the above model to solids. In a solid, the atoms are placed at regular sites with fixed separation between the atoms. In a diatomic molecule the valence electrons are shared by the two atoms, while in a solid the valence electrons are shared by more than one atom. In fact, we will find that the electrons move through the solid almost as if they were a gas of free electrons. The atoms, or ions,[9] in the solid are fixed in space. The most the ions can do is vibrate about their equilibrium position. For the present analysis we will assume that the atoms or ions are fixed, and concentrate on the motion of the electron. Since the ions are fixed in space we can model the solid in terms of potential wells placed at regular intervals. In one dimension this would correspond to atoms placed at a regular spacing of length a. Thus if we neglect the ends of our one dimensional solid, the solid has the symmetry that a translation by an amount a along the x-axis leaves the solid unchanged. This symmetry under translation will be present in the potential the electron feels, and should be included in the analysis of the problem of the motion of the electron in the solid. This translational symmetry is also present in two and three dimensional crystals. Thus in a three dimensional crystal, we have translational symmetry along three axes. We will first consider the constraint this translational symmetry places on the wave function. We will see that with the help of this symmetry, a problem that looks very difficult becomes almost as simple as that of a diatomic molecule.

[9]Since each atom has lost its valence electron, what is left behind is a positively charged ion.

6.6.1 Translational symmetry – Bloch's Theorem

In solids, and metals in particular, the atoms are fixed to certain positions in space, while the electrons are free to move. In fact, the valence electrons in metals behave like a free gas of electrons with a background potential that is periodic. This periodicity or symmetry can be used to derive the general form of the wave function that insures the symmetry of the system is satisfied. This was first established by Felix Bloch and is known as the Bloch's Theorem [28].

In one dimension, this periodicity can be stated as a symmetry in the potential of the form

$$V(x) = V(x+a) \ , \tag{6.56}$$

and the Hamiltonian has the property that

$$H(x) = H(x+a) \ . \tag{6.57}$$

This means the wave function should also have this symmetry.[10]

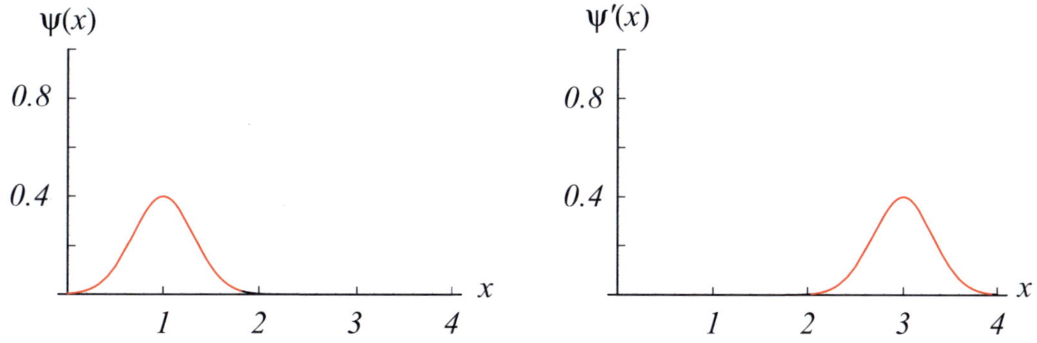

Figure 6.10: The wave function $\psi(x)$, peaked at $x = x_0 = 1$, and the translated wave function $\psi'(x)$, peaked $x = x_0 + \lambda = 3$.

Let us examine this symmetry in more detail to determine the operator for this symmetry. Consider a system described by the wave function $\psi(x)$. If we displace this system by the transformation

$$x \to x + \lambda \ , \tag{6.58}$$

then the new system will be described by a new wave function $\psi'(x)$. This new wave function is related to the old wave function $\psi(x)$ by the action of a displacement operator, i.e.,

$$\psi'(x) = D(\lambda)\ \psi(x) \ , \tag{6.59}$$

[10]Although solids are three dimensional systems, it is possible to form polymers that are chains of molecules that form one dimensional periodic systems.

where $D(\lambda)$ is the operator that moves the system by the distance λ. From Figure 6.10 it is clear that

$$\begin{aligned} \psi'(x) &= \psi(x-\lambda) \\ &= \psi(x) - \lambda\frac{d}{dx}\psi(x) + \frac{\lambda^2}{2!}\frac{d^2}{dx^2}\psi(x) - \cdots \\ &= \left\{1 - \lambda\frac{d}{dx} + \frac{1}{2!}\lambda^2\frac{d^2}{dx^2} - \cdots\right\}\psi(x)\ . \end{aligned} \tag{6.60}$$

In writing the second line of Eq. (6.60), we have made use of the Taylor series expansion for $\psi(x-\lambda)$ about $\psi(x)$. Since the momentum operator in coordinate space is given by $p = -i\hbar\frac{d}{dx}$, we can write the last line of Eq. (6.60) as

$$\begin{aligned} \psi'(x) &= \left\{1 - \frac{i\lambda p}{\hbar} - \frac{1}{2!}\left(\frac{i\lambda p}{\hbar}\right)^2 - \cdots\right\}\psi(x) \\ &= e^{-i\lambda p/\hbar}\ \psi(x)\ . \end{aligned} \tag{6.61}$$

From this result we can read off the operator $D(\lambda)$ to be [11]

$$D(\lambda) = e^{-i\lambda p/\hbar}\ . \tag{6.62}$$

If the Hamiltonian does not change under the transformation $x \longrightarrow x + \lambda$, then the Hamiltonian should commute with the operator $D(\lambda)$,[12] i.e.,

$$[H, D(\lambda)] = 0\ . \tag{6.63}$$

[11]Note the operator $D(\lambda)$ moves the system by a distance λ to the right. This is equivalent to moving the coordinates system a distance λ to the left, see Figure 6.10.

[12]Given the time dependent Schrödinger equation

$$i\hbar\frac{\partial\Psi}{\partial t} = H\,\Psi\ ,$$

we can operate on this equation with the translation operator $D(\lambda)$ to get

$$\begin{aligned} i\hbar\frac{\partial}{\partial t}D(\lambda)\Psi \equiv i\hbar\frac{\partial\Psi'}{\partial t} &= D(\lambda)\,H\,\Psi \\ &= D(\lambda)\,HD^{-1}(\lambda)\,D(\lambda)\,\Psi \\ &= D(\lambda)\,H\,D^{-1}(\lambda)\,\Psi'\ . \end{aligned}$$

Here, we have defined $\Psi' \equiv D(\lambda)\,\psi$. Invariance of the equation under the transformation $D(\lambda)$ (i.e., the shape of the wave function does not change) requires that both Ψ and Ψ' satisfy the same equation. This can be achieved only if

$$D(\lambda)\,H\,D^{-1}(\lambda) = H \quad \text{or} \quad D(\lambda)\,H = H\,D(\lambda)\ .$$

Because $D(\lambda)$ is a function to the momentum operator p, $D(\lambda)$ is defined in terms of a power series in the momentum p (see Eq. (3.60)). Therefore for Eq. (6.63) to be valid, the momentum operator has to commute with the Hamiltonian, i.e.,

$$[H, p] = 0 \; , \tag{6.64}$$

This implies that the momentum is a constant of the motion, and that we can construct simultaneous eigenstates of H and p, i.e.,

$$H\,\psi = E\,\psi \tag{6.65}$$

$$p\,\psi = \hbar k\,\psi \; . \tag{6.66}$$

For solids, $\lambda = a$ is the spacing between the atoms, while in free space, λ can take on any value. We now assume that the wave function of a one dimensional periodic system is of the form

$$\psi_k(x) = e^{ikx} U_k(x) \; , \tag{6.67}$$

where $U_k(x)$ is known as the *Bloch function*. Then the wave function at $x + a$ is given as

$$\begin{aligned} D^{-1}(a)\psi_k(x) &= \psi_k(x+a) \\ &= e^{ik(x+a)} U_k(x+a) \\ &= e^{ika}\left\{ e^{ikx} U_k(x+a) \right\} \; . \end{aligned}$$

But since $p\psi_k = \hbar k \psi_k$ and $D^{-1}(a) = e^{iap/\hbar}$, we have that

$$D^{-1}(a)\,\psi_k(x) = e^{ika}\,\psi_k(x) = e^{ika}\,e^{ikx}\,U_k(x) \; .$$

We therefore have that

$$U_k(x) = U_k(x+a) \; , \tag{6.68}$$

and

$$D(a)\,\psi_k(x) = e^{-ika}\,\psi_k(x) \; , \tag{6.69}$$

i.e., $\psi_k(x)$ is also an eigenstate of the displacement operator $D(a)$. At the same time we can calculate the wave function at $x + a$, given the wave function at x, by the application of $D^{-1}(a)$, i.e.,

$$\psi_k(x+a) = D^{-1}(a)\,\psi_k(x) \; .$$

We emphasize here that the operator $D(\lambda)$ was constructed to move the physical system represented by a wave function and not the coordinate system defined by the observer.

6.6.2 The Kronig-Penney Model

Let us now turn to a specific example of a one dimensional solid first suggested by Kronig and Penney and referred to in the literature as the Kronig-Penney Model [29]. In this model we can explicitly determine the Bloch wave function $U_k(x)$ and the corresponding eigenstates or spectrum. Consider the potential that is the sum of δ-functions potentials with the same strength and at a regular spacing of a. Each of these δ-functions would correspond to the potential an electron feels due to one of the ions in our one dimensional crystal. This potential is of the form

$$V(x) = -\frac{\hbar^2}{2m}\frac{\lambda}{a}\sum_{n=-\infty}^{+\infty}\delta(x-na)\ . \tag{6.70}$$

This potential consists of a series of attractive δ-functions (spikes) with spacing a between the spikes. The Schrödinger equation for this system is

$$-\frac{\hbar^2}{2m}\frac{d^2\psi(x)}{dx^2} + V(x)\psi(x) = E\psi(x)\ . \tag{6.71}$$

For $E > 0$, this equation can be written as

$$\frac{d^2\psi(x)}{dx^2} + k_0^2\psi(x) = -\frac{\lambda}{a}\sum_{n=-\infty}^{+\infty}\delta(x-na)\psi(x)\ , \tag{6.72}$$

where $k_0^2 = \frac{2m}{\hbar^2}E$.

From the symmetry of the problem, we know that the wave function can be written as

$$\psi(x) = e^{ikx}U_k(x)\ , \tag{6.73}$$

with $U_k(x+a) = U_k(x)$. This implies that

$$\psi(x+a) = e^{ika}\psi(x)\ . \tag{6.74}$$

For $-a < x < 0$, the potential is zero and we can write the solution of the Schrödinger equation as

$$\psi_1(x) = A\ e^{ik_0x} + B\ e^{-ik_0x} \quad \text{for} \quad -a < x < 0\ , \tag{6.75}$$

while for $0 < x < a$, we can write the wave function, making use of Eq. (6.74), as

$$\begin{aligned}\psi_2(x) &= e^{ika}\psi_1(x-a)\\ &= e^{ika}\left\{A\ e^{ik_0(x-a)} + B\ e^{-ik_0(x-a)}\right\} \quad \text{for} \quad 0 < x < a\ .\end{aligned} \tag{6.76}$$

As illustrated in Figure 6.11, the wave function ψ_1 is defined over the interval $-a < x < 0$, while the wave function ψ_2 is defined over the interval $0 < x < +a$.

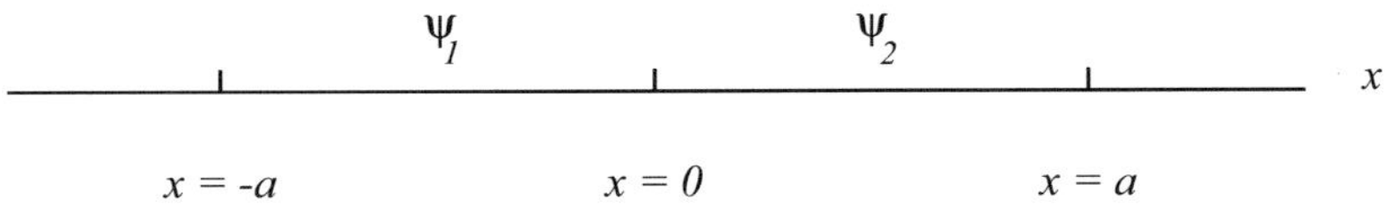

Figure 6.11: The domain for the wave functions $\psi_1(x)$ and $\psi_2(x)$.

The boundary condition at $x = 0$ for the wave function is now given as

$$\psi_1(-\epsilon) = \psi_2(\epsilon) \quad \text{as} \quad \epsilon \longrightarrow 0 \,, \tag{6.77}$$

while that on the derivative of the wave function is

$$\psi_2'(\epsilon) - \psi_1'(-\epsilon) = -\frac{\lambda}{a}\,\psi_1(-\epsilon) \quad \text{as} \quad \epsilon \longrightarrow 0 \,. \tag{6.78}$$

Using our wave functions, as given in Eqs. (6.75) and (6.76), in these boundary conditions, we get

$$A + B = e^{ika}\,\left\{A\,e^{-ik_0a} + B\,e^{ik_0a}\right\} \,, \tag{6.79}$$

and

$$ik_0\,e^{ika}\left\{A\,e^{-ik_0a} - B\,e^{ik_0a}\right\} - ik_0(A - B) = -\frac{\lambda}{a}(A + B) \,. \tag{6.80}$$

These two equations resulting from the application of the boundary conditions at $x = 0$, can be rewritten as

$$A\left\{1 - e^{i(k-k_0)a}\right\} + B\left\{1 - e^{i(k+k_0)a}\right\} = 0 \,, \tag{6.81}$$

and

$$A\left\{\frac{\lambda}{ik_0a} - 1 + e^{i(k-k_0)a}\right\} + B\left\{\frac{\lambda}{ik_0a} + 1 - e^{i(k+k_0)a}\right\} = 0 \,. \tag{6.82}$$

For these two homogeneous equations to have a solution, the determinant of coefficients must be zero, i.e.,

$$\begin{aligned}\left\{\frac{\lambda}{ik_0a} + 1\right. & - \left. e^{i(k+k_0)a} - \frac{\lambda}{ik_0a}e^{i(k-k_0)a} - e^{i(k-k_0)a} + e^{2ika}\right\} \\ & - \left\{\frac{\lambda}{ik_0a} - 1 + e^{i(k-k_0)a} - \frac{\lambda}{ik_0a}e^{i(k+k_0)a} + e^{i(k+k_0)a} - e^{2ika}\right\} = 0\end{aligned}$$

or

$$2e^{ika}\left(e^{-ika} + e^{ika}\right) - 2e^{ika}\left(e^{ik_oa} + e^{-ik_0a}\right) + \frac{\lambda e^{ika}}{ik_0a}\left(e^{ik_0a} - e^{-ik_0a}\right) = 0 \,.$$

This result can be written as a transcendental equation of the form

$$\cos ka = \cos k_0a - \frac{\lambda}{2k_0a}\sin k_0a \,. \tag{6.83}$$

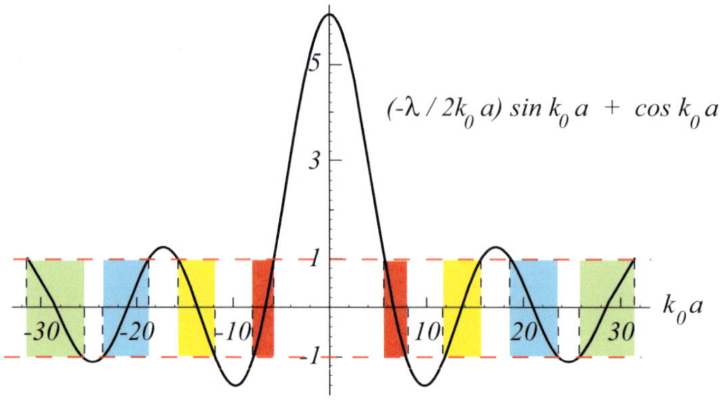

Figure 6.12: This is a graphical solution of the equation $\cos ka = \frac{\lambda}{2k_0 a}\sin k_0 a + \cos k_0 a$. Since $|\cos ka| \leq 1$, only certain ranges of values of $k_0 a$ are allowed. These correspond to the energy bands in our one dimensional solid.

This is similar in form to Eq. (6.20) for the bound state of a particle in a square well (e.g., the deuteron), and can be solved graphically as illustrated in Figure 6.12.

The left hand side of Eq. (6.83) can take values between -1 and $+1$ for all values of the momentum k. On the other hand, the right hand side of this equation can have values greater than $+1$ and less than -1 as demonstrated in Figure 6.12. Thus the energy $E = \frac{\hbar^2}{2m}k_0^2$ can take only those values for which the right hand side of Eq. (6.83) is between -1 and $+1$. This suggests that there are bands of energies allowed, and within each band we have continuous values for the energy E.

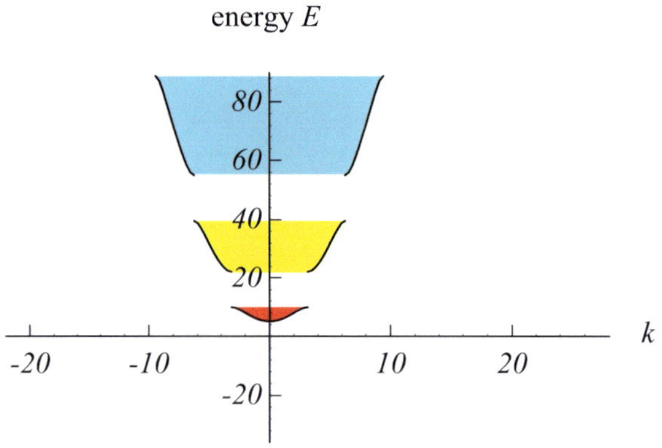

Figure 6.13: The energy-wave vector plot for a one dimensional Kronig-Penney model.

From the wave function of the electron as given in Eqs. (6.75) and (6.76) for a one dimensional solid, we expect the electron to behave as a free particle. This can be confirmed by the relation between the energy E and the momentum k. In Figure 6.13 we present a plot of the energy E verses the momentum k. It is clear from this figure that for those energies allowed the relation between the energy and momentum is almost that of a free particle. This suggests that the electrons within a given band behave like a gas of free non-interacting electrons, and the mobility of these electrons determines the conductivity of the solid. The detailed properties of the band, i.e., the relation between the energy and momentum, are determined by the crystal structure. Thus a central problem in solid state physics is to find crystal structures for which the band, and therefore the properties of the solid, are those useful for certain devices. Here again we find that a simple model that has the correct symmetry gives the general properties of the solid in the form of energy bands and the fact that the electrons within a band behave like a free gas. A more detailed and therefore more realistic model for a solid will give the Bloch wave function $U_k(x)$, and the relation between the energy E and momentum k. To achieve this will require more complex computational methods than those considered here.

6.7 Problems

1. Consider the potential

$$V(r) = \begin{cases} -V_0 & \text{for} \quad r < R_0 \\ \frac{V_1}{R_0}(R_1 - r) & \text{for} \quad R_0 < r < R_1 \\ 0 & \text{for} \quad r > R_1 \end{cases}$$

as a model for α-decay.

(a) Calculate the lifetime when $R_0 = 5$ fm., $R_1 = 20$ fm., $V_0 = 50$ MeV, and $V_1 = 10$ MeV. The energy of the final α particle is 1 MeV.

(b) Use *Mathematica* to calculate the lifetime of the nucleus as a function of the energy of the α particle.

2. Given: the binding energy of the deuteron is 2.2246 MeV and the range of the neutron proton potential is 1.5 fm:

(a) Assuming the potential is a square well, find the well depth V_0.

(b) Use *Mathematica* to plot the wave function of the deuteron.

3. Consider the δ-function potential

$$V(x) = \frac{\hbar^2}{2m}\frac{\lambda}{a}\delta(x)$$

(a) Show that the scattering matrix (S-matrix as defined in problem 5 of Chapter 5) is given in this case by

$$S(E) = \frac{1}{c - \lambda}\begin{pmatrix} c & \lambda \\ \lambda & c \end{pmatrix},$$

where

$$c = 2ika \qquad \text{and} \qquad k^2 = \frac{2mE}{\hbar^2} \; .$$

(b) Prove that this matrix is unitary.

(c) Show that for $\lambda < 0$ the S-matrix becomes infinite when E is the bound state energy for the potential.

4. Consider the Kronig-Penney potential

$$\frac{2m}{\hbar^2} V(x) = \frac{\lambda}{a} \sum_{n=-\infty}^{n=+\infty} \delta(x - na)$$

with $\lambda = 3\pi$.

(a) Make a detailed plot of

$$\cos x + \frac{\lambda}{2} \frac{\sin x}{x}$$

as a function of $x = k_0 a$.

(b) Show that forbidden energy bands start just above $k_0 a = n\pi$.

(c) Show that the allowed energy bands get narrower as λ increases.

(d) Plot the energy $\hbar^2 k_0^2 / 2m$ as a function of k.

CHAPTER 7

Molecular Vibration – The Harmonic Oscillator

Abstract: In the present chapter we turn to the one-dimensional harmonic oscillator potentials for which the Schrödinger equation admits analytic solutions. This will allow us to examine the vibrational spectrum of quantum systems, and in particular, diatomic molecules.

Vibrational modes of excitation are a common feature in many quantum, as well as classical systems. Invariably these vibrational modes are the result of the displacement of the system from equilibrium, with these deviations being small in magnitude. In quantum systems, these vibrational excitations are present in molecules, nuclei, solids and molecular clusters. Usually, such vibrational excitations have the common feature that the energy spectrum has the unique signature whereby the spacing between the levels is constant and can be represented by $\hbar\omega$. In the present chapter, we will consider vibration in diatomic molecules, but the formulation can be extended to more complicated systems such as the vibrational spectrum of nuclei or the vibration of an atom in a solid about its equilibrium position. The choice of diatomic molecules is dictated by the fact that we can model this vibration in terms of a one dimensional harmonic oscillator.

In the last two chapters we restricted our analysis to systems for which the corresponding Schrödinger equation has the simple form

$$\left(\frac{d^2}{dx^2} \pm k^2\right) \psi(x) = 0 \ .$$

This equation had simple solutions, and to that extent, we avoided the question of how to solve the Schrödinger equation as a second order linear differential equation. We are now at a stage to examine more complicated potentials and as a result need to consider the problem of solving the Schrödinger equation for these more complicated potentials. In

Iraj R. Afnan

practical terms there are only a handful of potentials for which we have analytic solutions. The more common practice is either to solve the differential equation numerically on a computer, or to use one of the standard packages such as *Mathematica* [39] to extract the numerical or analytic solutions. One of the motivations for considering the solution of the Schrödinger equation for the harmonic oscillator is to determine the mechanism by which we end up with a spectrum that is quantized.

7.1 The One Dimensional Harmonic Oscillator

The harmonic oscillator problem in Physics is a very good approximation to all phenomena which result from the system deviating by a small amount from equilibrium. A simple example of such a situation is the vibrational excitation of a diatomic molecule. Here the potential between the two atoms in a diatomic molecule is attractive at large distances and is the result of each atom polarizing the other atom, i.e., the interaction is a dipole-dipole interaction (see Chapter 13 Sec. 13.4). On the other hand, as the atoms get closer and their electronic clouds overlap, this interaction is highly repulsive. In other words the potential is of the form of the Lennard Jones potential [30] (see Figure 7.1)

$$V(r) = 4\varepsilon\left[\left(\frac{\sigma}{r}\right)^{12} - \left(\frac{\sigma}{r}\right)^{6}\right] \tag{7.1}$$

where ε and σ are parameters of the potential that are adjusted to fit either scattering data, or the thermodynamic properties of the system. The lowest energy state of the molecule corresponds to an average separation of r_0, where r_0 is the distance at which the potential is a minimum. If we now consider the lowest excitation of the molecule, then they are of two kinds: (i) the vibration of the two atoms about the equilibrium distance r_0. (ii) the rotation of the molecule as a rigid object. For the vibrational motion, provided the amplitude of vibration is small, the potential energy $V(r)$ can be expanded about the equilibrium separation of the two atoms, i.e.,

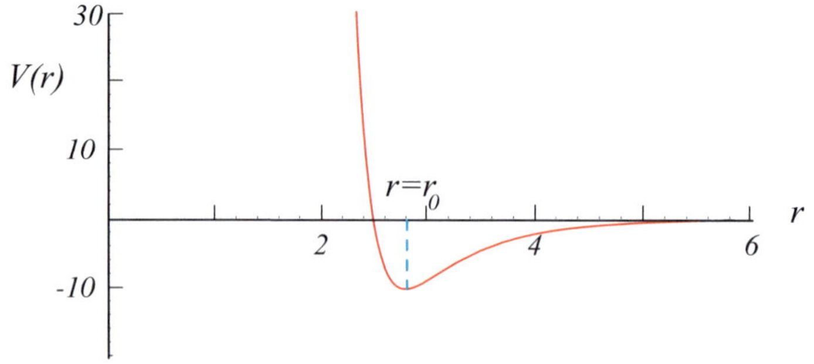

Figure 7.1: The Lennard-Jones potential for the *He*-*He* system.

$$V(r) = V(r_0) + (r - r_0)\left(\frac{dV}{dr}\right)_{r_0} + \frac{1}{2}(r - r_0)^2\left(\frac{d^2V}{dr^2}\right)_{r_0} + \dots \,. \tag{7.2}$$

Since $V(r)$ is minimum at $r = r_0$, then $\left(\frac{dV}{dr}\right)_{r_0} = 0$, and if we take our zero of energy to be such that $V(r_0) = 0$ then the potential that describes the oscillation about equilibrium is given to lowest order by

$$V(x) = \frac{1}{2}m\omega^2 x^2 \,, \tag{7.3}$$

where $x = (r - r_0)$ and $\left(\frac{d^2V}{dr^2}\right)_{r_0} = m\omega^2$. Thus to examine the low energy vibrational spectrum of molecules we need to examine the potential $V(x)$ given above. The other motivation for considering this potential is that the Schrödinger equation for this potential is slightly more complicated than the examples we have considered so far. In fact the Schrödinger equation takes the form

$$-\frac{\hbar^2}{2m}\frac{d^2\psi(x)}{dx^2} + \frac{1}{2}m\omega^2 x^2\psi = E\psi(x) \,, \tag{7.4}$$

or after multiplication by $\frac{2}{\hbar\omega}$ we have

$$-\frac{\hbar}{m\omega}\frac{d^2\psi(x)}{dx^2} + \frac{m\omega}{\hbar}x^2\psi = \frac{2E}{\hbar\omega}\psi(x) \,. \tag{7.5}$$

To simplify the differential equation, we define new variables η and a rescale energy ϵ as

$$\eta = \sqrt{\frac{m\omega}{\hbar}}x \quad \text{and} \quad \epsilon = \frac{2E}{\hbar\omega} \,. \tag{7.6}$$

Our Schrödinger equation now takes the simple form

$$\frac{d^2\psi(\eta)}{d\eta^2} + (\epsilon - \eta^2)\psi(\eta) = 0 \,. \tag{7.7}$$

To solve this second order differential equation, we first consider the asymptotic solutions, i.e., in the limit as $|x| \to \infty$. Considering the fact that $V(x) \to \infty$ as $|x| \to \infty$ we may conclude that the wave function should go to zero as $|x| \to \infty$. To get the form of the wave function in this region, we will first consider the differential equation for $\eta \gg \epsilon$. In this limit, the Schrödinger equation reduces to

$$\frac{d^2\psi(\eta)}{d\eta^2} - \eta^2\,\psi(\eta) = 0 \,. \tag{7.8}$$

We now change the variable in this differential equation to $z = \eta^2$ so that

$$\frac{d\psi}{d\eta} = \frac{dz}{d\eta}\frac{d\psi}{dz} = 2\eta\,\frac{d\psi}{dz}$$

and

$$\begin{aligned}\frac{d^2\psi}{d\eta^2} &= 2\frac{d\psi}{dz} + 2\eta\frac{d}{d\eta}\left(\frac{d\psi}{dz}\right) \\ &= 2\frac{d\psi}{dz} + 4\eta^2\frac{d^2\psi}{dz^2} \\ &= 2\frac{d\psi}{dz} + 4z\frac{d^2\psi}{dz^2} \ .\end{aligned}$$

Our differential equation is now given by

$$\frac{d^2\psi(z)}{dz^2} + \frac{1}{2z}\frac{d\psi(z)}{dz} - \frac{1}{4}\psi(z) = 0 \tag{7.9}$$

or for $z \to \infty$ we have

$$\frac{d^2\psi(z)}{dz^2} - \frac{1}{4}\psi(z) = 0 \ . \tag{7.10}$$

In this way we have reduced the Schrödinger equation to the form we are familiar with, and the solution for large z is given by

$$\psi = e^{-z/2} = e^{-\eta^2/2} \ . \tag{7.11}$$

We now make use of our knowledge of the asymptotic solution for large η to write the general solution to the Schrödinger equation for the harmonic oscillator potential as

$$\psi(\eta) = u(\eta)e^{-\eta^2/2} \ , \tag{7.12}$$

with the hope that $u(\eta)$ is some simple polynomial in η. To determine the equation $u(\eta)$ satisfies, we substitute the above expression for $\psi(\eta)$ in terms of $u(\eta)$ in the Schrödinger equation for this potential, i.e., Eq. (7.7). Making use of the fact that

$$\frac{d^2\psi(\eta)}{d\eta^2} = e^{-\eta^2/2}\left\{\frac{d^2u(\eta)}{d\eta^2} - 2\eta\,\frac{du(\eta)}{d\eta^2} + (\eta^2 - 1)u(\eta)\right\}$$

we can write the equation for $u(\eta)$ as

$$\frac{d^2u(\eta)}{d\eta^2} - 2\eta\,\frac{du(\eta)}{d\eta} + (\epsilon - 1)u(\eta) = 0 \ . \tag{7.13}$$

At this stage it appears that we have replaced the Schrödinger equation by a more complicated differential equation. However, we will find that the above equation will give polynomial solutions. In fact we will find that the requirement that the wave function goes to zero as $|x| \to \infty$ will admit us not only a polynomial solution, but also that only certain energies are allowed, i.e., we will get a discreet spectrum of energy levels. To solve the second order differential equation in Eq. (7.13), we use the *method of series* to write a series solution for the unknown function $u(\eta)$ of the form

$$u(\eta) = \sum_{n=0}^{\infty} a_n \eta^n \, , \tag{7.14}$$

with the constants a_n to be determined from the differential equation (7.13). To achieve this we need to determine the series for the first and second derivative of the function $u(\eta)$. For the first derivative we have

$$\begin{aligned} \frac{du}{d\eta} &= \sum_{n=0}^{\infty} n \, a_n \, \eta^{n-1} \\ &= \sum_{n=0}^{\infty} (n+1) \, a_{n+1} \, \eta^n \, . \end{aligned}$$

In writing the second line above, we have changed the sum over n to run from $1 \to \infty$ since the first term in the series in the first line is zero. We then replace n by $(n+1)$ and rewrite the sum to run over $0 \to \infty$. We now can write the second term in our differential equation as

$$\begin{aligned} 2\eta \frac{du}{d\eta} &= \sum_{n=0}^{\infty} 2(n+1) \, a_{n+1} \, \eta^{n+1} \\ &= \sum_{n=1}^{\infty} 2n \, a_n \, \eta^n \, . \end{aligned}$$

In a similar manner we can write the series for the second derivative of $u(\eta)$ as

$$\begin{aligned} \frac{d^2u}{d\eta^2} &= \sum_{n=0}^{\infty} (n+1) \, n \, a_{n+1} \, \eta^{n-1} \\ &= \sum_{n=0}^{\infty} (n+2) \, (n+1) \, a_{n+2} \, \eta^n \, . \end{aligned}$$

We now can substitute the series for the function $u(\eta)$ it's first and second derivative in the differential equation, Eq. (7.13) to get

$$\sum_{n=0}^{\infty} (n+2) \, (n+1) \, a_{n+2} \, \eta^n - \sum_{n=0}^{\infty} 2n \, a_n \, \eta^n + (\epsilon - 1) \sum_{n=0}^{\infty} a_n \, \eta^n = 0 \, . \tag{7.15}$$

Since the right hand side of this equation is zero for any η, the coefficients of η^ℓ, for any ℓ, must also be zero i.e., for

$$\eta^0 \quad : \qquad 2a_2 + (\epsilon - 1)a_0 = 0 \quad \Longrightarrow \quad a_2 = \frac{1}{2}(1 - \epsilon)a_0$$

$$\eta^1 \quad : \qquad 6a_3 - 2a_1 + (\epsilon - 1)a_1 = 0 \quad \Longrightarrow \quad a_3 = \frac{1}{6}(3-\epsilon)a_1$$

$$\eta^2 \quad : \qquad 12a_4 - 4a_2 + (\epsilon - 1)a_2 = 0 \quad \Longrightarrow \quad a_4 = \frac{1}{12}(5-\epsilon)a_2$$

$$\vdots$$

$$\eta^\ell \quad : \quad (\ell+2)(\ell+1)a_{\ell+2} - 2\ell a_\ell + (\epsilon - 1)a_\ell = 0 \ .$$

In this way we have made use of the differential equation to set up relations between the coefficients a_n. These relations are unique to this differential equation and will allow us to write the solution with two undetermined constants, a_0 and a_1. This is expected considering the fact that the solution of a second order linear differential equation has two arbitrary constants. These constants are to be determined by the boundary condition of the physical system under consideration.

From the above relation between the constants a_n, we can deduce a general relation between a_ℓ and $a_{\ell+2}$ that is of the form

$$a_{\ell+2} = \frac{1}{(\ell+2)(\ell+1)}\left\{(2\ell+1) - \epsilon\right\} a_\ell \ . \tag{7.16}$$

Thus given a_0 and a_1, we can generate all the other coefficients of the power series. The fact that we have two series, one for $n =$ odd, the other for $n =$ even, is the result of the fact that the odd n series corresponds to a final solution that is odd under parity, while the even n solution gives a wave function that is even under parity. Furthermore these two solutions do not mix. This is the result of the fact that $V(x)$ does not change under parity, i.e., $x \to -x$.

For arbitrary values of ϵ, the series is not finite, and for large N we have that

$$a_{N+2} = \frac{2}{N} a_N \quad \text{for} \quad N \gg 1$$

For the even solution, we can write the series as

$$\begin{aligned} u(\eta) &= \{\text{polynomial in } \eta \text{ of order } (N-2)\} \\ &\qquad + a_N\left\{\eta^N + \frac{2}{N}\eta^{N+2} + \frac{2}{N}\frac{2}{N+2}\eta^{N+4} + \cdots\right\} \\ &= \{\text{polynomial in } \eta \text{ of order } (N-2)\} \\ &\qquad + a_N\eta^2\left\{(\eta^2)^{\frac{N}{2}-1} + \frac{(\eta^2)^{\frac{N}{2}}}{\frac{N}{2}} + \frac{(\eta^2)^{\frac{N}{2}+1}}{\frac{N}{2}(\frac{N}{2}+1)} + \cdots\right\} \end{aligned}$$

$$
\begin{aligned}
= & \ \{\text{polynomial in } \eta \text{ of order } (N-2)\} \\
& + a_N \eta^2 \left(\frac{N}{2}-1\right)! \left\{ \frac{(\eta^2)^{\frac{N}{2}-1}}{(\frac{N}{2}-1)!} + \frac{(\eta^2)^{\frac{N}{2}}}{(\frac{N}{2})!} + \frac{(\eta^2)^{\frac{N}{2}+1}}{(\frac{N}{2}+1)!} + \cdots \right\} .
\end{aligned}
$$

But we have that [1]

$$
\left\{ \frac{(\eta^2)^{\frac{N}{2}-1}}{(\frac{N}{2}-1)!} + \frac{(\eta^2)^{\frac{N}{2}}}{(\frac{N}{2})!} + \frac{(\eta^2)^{\frac{N}{2}+1}}{(\frac{N}{2}+1)!} + \cdots \right\} = e^{\eta^2} - \{ \text{ polynomial in } \eta \} .
$$

If we now make use of this solution in Eq. (7.12), we find that our solution for $\psi(\eta)$ is of the form

$$
\psi(\eta) = \{ \text{ polynomial in } \eta \} \ e^{-\eta^2/2} + a_N \, \eta^2 \left(\frac{N}{2}-1\right)! \ e^{\eta^2/2} . \tag{7.17}
$$

This solution goes to ∞ as $\eta \to \infty$ and does not satisfy the boundary condition that the wave function should go to zero as $\eta \to \infty$. Thus to satisfy the boundary condition and have a wave function that goes to zero as $\eta \to \infty$, we need to guarantee that $u(\eta)$ is a finite order polynomial in η. Since the coefficients in the power series are related by the recursion relation

$$
a_{\ell+2} = \frac{(2\ell+1-\epsilon)}{(\ell+2)(\ell+1)} a_\ell ,
$$

to turn this infinite series to a polynomial in η, we need to have $a_{\ell+2} = 0$ when $a_\ell \neq 0$. This condition can be satisfied only if

$$
\epsilon = 2\ell + 1 , \tag{7.18}
$$

where ℓ is an integer. This allows us to write the energy as

$$
\begin{aligned}
E_\ell &= \frac{\hbar\omega}{2} \epsilon \\
&= \hbar\omega(\ell + \frac{1}{2}) \quad \text{where} \quad \ell = 0, 1, 2, \cdots .
\end{aligned} \tag{7.19}
$$

Here we observe that the requirement that the wave function goes to zero as $\eta \to \infty$ results in a constraint on the energy of the system. This condition essentially states that only certain energies are allowed, i.e., we have generated a quantized spectrum.

---

[1] We have made use of the fact that

$$
e^x = 1 + x + \frac{x^2}{2!} + \frac{x^3}{3!} + \cdots
$$

where

$$
n! = n(n-1)(n-2)\cdots 1 .
$$

We are now in a position to write the allowed energies and corresponding wave function for our harmonic oscillator potential. The lowest energy state corresponds to taking the coefficient $a_2 = 0$, and this corresponds to $\ell = 0$. In this case the energy, which is the ground state energy, is given by

$$E_0 = \frac{1}{2}\hbar\omega \tag{7.20}$$

with the corresponding ground state wave function being

$$\psi_0 = a_0 e^{-\eta^2/2} \ . \tag{7.21}$$

The next eigenstate corresponds to $a_1 \neq 0$ while $a_3 = 0$. Here $\ell = 1$ and the energy is given by

$$E_1 = \hbar\omega(1 + \frac{1}{2}) = \frac{3}{2}\hbar\omega \ , \tag{7.22}$$

while the wave function for this state is given by

$$\psi_1 = a_1 \eta e^{-\eta^2/2} \ . \tag{7.23}$$

We now observe that the ground state wave function with energy $E_0 = \frac{1}{2}\hbar\omega$ is even under parity, while the wave function of the first excited state with energy $E_1 = \frac{3}{2}\hbar\omega$ is odd under parity. For the second excited state we have that $a_2 \neq 0$ while $a_4 = 0$. Here, a_2 is given in terms of a_0 by the relation

$$a_2 = \frac{1-\epsilon}{2} a_0 = -\ell a_0 \ .$$

In this case, $\ell = 2$ and the energy is given by

$$E_2 = \frac{5}{2}\hbar\omega \ , \tag{7.24}$$

while the wave function is given by

$$\psi_2 = a_0(1 - 2\eta^2)\ e^{-\eta^2/2} \ . \tag{7.25}$$

In this way we can continue to generate all the eigenvalues and eigenstates of the harmonic oscillator Hamiltonian.

An alternative way of generating the eigenstates is to examine the form of the differential equation for $u(\eta)$ when we take $\epsilon = 2\ell + 1$. This gives us a differential equation of the form

$$\frac{d^2 H_\ell(\eta)}{d\eta^2} - 2\eta \frac{dH_\ell(\eta)}{d\eta} + 2\ell H_\ell(\eta) = 0 \ . \tag{7.26}$$

This equation is known as the Hermite equation, and $H_\ell(\eta)$ is the Hermite polynomial. Special cases of the Hermite polynomial are

$$\begin{aligned} H_0(\eta) &= 1 \\ H_1(\eta) &= 2\eta \\ H_2(\eta) &= 4\eta^2 - 2 \\ H_3(\eta) &= 8\eta^3 - 12\eta \\ H_4(\eta) &= 16\eta^4 - 48\eta^2 + 12 \end{aligned}$$

These polynomials have the recursion relation

$$2\eta H_\ell = H_{\ell+1} + 2\ell H_{\ell-1} \ . \tag{7.27}$$

The general solution for the one dimensional harmonic oscillator Hamiltonian can now be written as

$$\psi_\ell(\eta) = (\sqrt{\pi}\, 2^\ell\, \ell!)^{-\frac{1}{2}}\, H_\ell(\eta)\, e^{-\eta^2/2} \ , \tag{7.28}$$

where $\eta = \sqrt{\frac{m\omega}{\hbar}}\, x$. The corresponding energies are

$$E_\ell = \hbar\omega(\ell + 1/2) \ . \tag{7.29}$$

These wave functions given in Eq. (7.28) are orthonormal, i.e.,

$$\int_{-\infty}^{+\infty} d\eta\, \psi_\ell(\eta)\, \psi_{\ell'}(\eta) = \delta_{\ell\ell'} \ , \tag{7.30}$$

and form a complete set of states.

The above solution of the Schrödinger equation for the harmonic oscillator potential can be used in many applications where the system is not far from equilibrium as is the case of the vibration of a diatomic molecule or the vibration of the lattice in a solid. The three dimensional harmonic oscillator has been used extensively in Nuclear Physics as a shell model of nuclei in analogy with the atomic shell model. We will find that the method we used to solve the Schrödinger equation in this case is very similar to the solution of the corresponding radial equation in three dimensional problems. This will be the subject of Chapter 9 in which we consider the three-dimensional square well as well as the problem of an electron in the Coulomb field of a nucleus.

7.2 Vibrational Spectrum of Diatomic Molecule

Having established the spectrum of the one dimensional harmonic oscillator, we turn our attention to the application of these results to the case of a diatomic molecule. In particular we will consider the spectrum of H_2, HD and DD, where D is the isotope of hydrogen in which the nucleus consists of a proton and a neutron i.e., the deuteron.

We showed in the last section that the energy levels of the harmonic oscillator are given by

$$E_\ell = \hbar\omega\,(\ell + 1/2)$$

where the frequency ω is given by

$$\omega = \frac{1}{\sqrt{\mu}}\sqrt{\left(\frac{d^2V}{dr^2}\right)_{r_0}} \quad \text{where} \quad \mu = \frac{m_1\,m_2}{m_1 + m_2} \ . \tag{7.31}$$

Here μ is the reduced mass of the molecule, with m_1 and m_2 being the masses of the two atoms in the molecule. Since the potential between the atoms in the molecule depends on the wave function of the electron, and that, to first approximation, does not depend on the mass of the nucleus, we can assume that

$$\left(\frac{d^2V}{dr^2}\right)_{r_0} = C$$

and is the same for all the three molecules under consideration. The reduce masses for the three diatomic molecules are given by,

$$\mu_{H_2} = \frac{m_p}{2}\ , \qquad \mu_{HD} = \frac{2m_p}{3} = \frac{4}{3}\,\mu_{H_2}\ , \quad \text{and} \quad \mu_{DD} = m_p = 2\,\mu_{H_2}\ . \tag{7.32}$$

where we have assumed that the mass of D is twice the mass of the proton (m_p). This allows us to calculate the ratio of the frequency of vibration for these three molecules. Making use of the relation between the frequency ω and the reduced mass μ, we can calculate in this model the ratio of the frequencies to be

$$\frac{\omega_{H_2}}{\omega_{HD}} = \sqrt{\frac{4}{3}} \approx 1.1547 \qquad \text{and} \qquad \frac{\omega_{H_2}}{\omega_{DD}} = \sqrt{2} \approx 1.4142\ . \tag{7.33}$$

These numbers can be compared with the corresponding experimental values of

$$\frac{\omega_{H_2}}{\omega_{HD}} = 1.1514 \qquad \text{and} \qquad \frac{\omega_{H_2}}{\omega_{DD}} = 1.4096\ . \tag{7.34}$$

This agreement is impressive considering the fact that we have a very simple model for a diatomic molecule. These results clearly illustrate that this simple approximation works very well for simple diatomic molecules. The model also suggests a limitation on the application of the simple one dimensional harmonic oscillator. From Figure 7.1, we expect that for higher excitation, the difference between the harmonic oscillator and the real spectrum should deviate. Since the potential in Figure 7.1 is not the same as a harmonic oscillator potential, we expect a deviation in the spectrum from that of a simple harmonic excitation as verified by experiment. This problem will be examined in Chapter 13 when we study perturbation theory.

Simple one dimensional models for a vibrational spectrum can also be used to understand the specific heat of solids due to the vibration of atoms in the solid about their equilibrium position. This model of the solid is commonly referred to as the Einstein solid, and predicts the fact that the specific heat of a solid tends to zero as the temperature approaches zero. For nuclei, the vibrational motion is a more complex, and we need to extend this one dimensional model to five dimensions to describe the vibration on the surface of a liquid drop as a model of the nucleus. Here again, the agreement between the model and experiment is very good.

7.3 Problems

1. Use *Mathematica* to solve Eq. (7.13) for $\epsilon = 3, 5$ and compare the results with the analytic solution.

CHAPTER 8

Central Force Problem I

Abstract: In this chapter we commence with the solution of the Schrödinger equation in three dimensions and consider the case when the potential has a spherical symmetry. This requires the introduction of spherical polar coordinates and the reduction of the three dimensional partial differential equation to a set of ordinary differential equations. The solution of these equations for the angular dependence will allow us to examine the rotational spectrum of diatomic molecules.

So far we have concentrated our effort on one dimensional systems with the hope of gaining some understanding of the fundamental ideas of quantum mechanics while discussing application to nuclei, atoms, molecules and solids. Before we proceed to a formal development of quantum theory it may be useful to examine some simple systems in three dimensions. The hydrogen atom is in a sense the simplest example of a real atom, and it was the hydrogen atom that was first considered by the early developers of quantum theory. However, we will find through further studies, that a full understanding of the hydrogen atom will strain our theoretical understanding of physics to the limit.

The simplest model of the hydrogen atom involves the system consisting of a proton and an electron with the Coulomb potential being the force of attraction that binds the two particles together to form the hydrogen atom. The classical Hamiltonian for this two particle system consists of the kinetic energy of the electron and proton and the potential energy, the Coulomb attraction, i.e.,

$$H = \frac{p_e^2}{2m_e} + \frac{p_p^2}{2m_p} + V(\vec{r}_e - \vec{r}_p) \ , \tag{8.1}$$

where $\vec{p}_e$ ($\vec{r}_e$) and $\vec{p}_p$ ($\vec{r}_p$) are the momenta (position) of the electron and proton respectively, while m_e and m_p are the corresponding masses. The Coulomb potential is given by [1]

$$V(\vec{r}_e - \vec{r}_p) = -\frac{e^2}{|\ \vec{r}_e - \vec{r}_p\ |}\ , \tag{8.2}$$

with e being the magnitude of the charge of the proton or electron. The two are identical in magnitude.

At this stage we could quantize our system by the substitution

$$\vec{p} \rightarrow -i\hbar\vec{\nabla}\ , \tag{8.3}$$

where the gradient, ∇, in rectangular coordinates is given by

$$\vec{\nabla} = \hat{\imath}\frac{\partial}{\partial x} + \hat{\jmath}\frac{\partial}{\partial y} + \hat{k}\frac{\partial}{\partial z}\ . \tag{8.4}$$

Here $\hat{\imath}$, $\hat{\jmath}$ and $\hat{k}$ are unit vectors in the x-, y- and z-directions. The Schrödinger equation for this system now depends on the coordinates of both the electron and proton, i.e.,

$$H\Psi(\vec{r}_1, \vec{r}_2) = E\Psi(\vec{r}_1, \vec{r}_2)\ , \tag{8.5}$$

and is a partial differential equation in six variables. This problem can be considerably simplified if we introduce relative and center of mass coordinates. Since we are interested in the relative motion of the electron and proton, we hope, and rightly so, to reduce the problem to three dimensions, or variables. The relative and center of mass momenta $\vec{p}$ and $\vec{P}$ are defined as

$$\vec{p} = \frac{m_p\vec{p}_e - m_e\vec{p}_p}{m_p + m_e} \qquad \vec{P} = \vec{p}_e + \vec{p}_p\ , \tag{8.6}$$

while the relative and center of mass coordinates $\vec{r}$ and $\vec{R}$ are defined as

$$\vec{r} = \vec{r}_e - \vec{r}_p \qquad \vec{R} = \frac{m_e\vec{r}_e + m_p\vec{r}_p}{m_e + m_p}\ . \tag{8.7}$$

From the above definition of relative and center of mass variables we have

$$(m_p + m_e)\vec{p} + m_e\vec{P} = (m_e + m_p)\vec{p}_e$$

and

$$-(m_p + m_e)\vec{p} + m_p\vec{P} = (m_e + m_p)\vec{p}_p\ .$$

Therefore, we can write the momentum of the proton and electron in terms of the relative and center of mass momenta as

$$\vec{p}_e = \vec{p} + \frac{m_e}{m_e + m_p}\vec{P} \quad \text{and} \quad \vec{p}_p = -\vec{p} + \frac{m_p}{m_e + m_p}\vec{P}\ . \tag{8.8}$$

---

[1]We will use units in which $\frac{1}{4\pi\epsilon_0} = 1$. This corresponds to using Gaussian units which are more convenient for microscopic systems.

We now can write the kinetic energy of the electron and proton in terms of the relative and center of mass momenta as

$$\begin{aligned}
\frac{p_e^2}{2m_e} &= \frac{1}{2m_e}\left(p^2 + \frac{2m_e}{m_e+m_p}\vec{p}\cdot\vec{P} + \frac{m_e^2}{(m_e+m_p)^2}P^2\right) , \\
\frac{p_p^2}{2m_p} &= \frac{1}{2m_p}\left(p^2 - \frac{2m_p}{m_e+m_p}\vec{p}\cdot\vec{P} + \frac{m_p^2}{(m_e+m_p)^2}P^2\right) .
\end{aligned}$$

With these results in hand we can write the kinetic energy of the system as

$$\frac{p_e^2}{2m_e} + \frac{p_p^2}{2m_p} = \frac{p^2}{2\mu} + \frac{P^2}{2M} , \tag{8.9}$$

where the reduced mass μ and total mass M are given by

$$\mu = \frac{m_e m_p}{m_e + m_p} \qquad M = m_e + m_p . \tag{8.10}$$

Here we observe that the kinetic energy of the proton and electron can be written as the sum of the kinetic energy of the center of mass plus the relative kinetic energy. Since the potential energy depends on the relative position of the electron and proton in the atom, we can write the Hamiltonian in terms of the new coordinates as

$$H = \frac{P^2}{2M} + \left(\frac{p^2}{2\mu} + V(r)\right) , \tag{8.11}$$

where the first term describes the motion of the center of mass of the atom as a whole, while the second term describes the relative motion of the electron and proton which gives us the properties of the hydrogen atom. Since we are only interested in the internal structure of the atom, we can take the center of mass momentum $\vec{P}$ to be zero. The resultant coordinate system is known as the center of mass system.

For the hydrogen atom, $\frac{m_p}{m_e} \approx 2000$ and as a result

$$\mu = \frac{m_e}{1 + \frac{m_e}{m_p}} \approx m_e , \tag{8.12}$$

i.e., the reduced mass is basically the mass of the electron. This is equivalent to assuming that the inertial frame in which the proton is stationary is also the center of mass coordinate system.

From this point on, we will discuss the properties of the hydrogen atom and other two-body systems in the center of mass coordinate system where $\vec{P} = 0$. These properties are determined by the Schrödinger equation resulting from the quantization of the Hamiltonian in Eq. (8.11) when $\vec{P} = 0$, i.e.,

$$\left\{\frac{p^2}{2\mu} + V(r)\right\}\psi(\vec{r}) = E\,\psi(\vec{r}) , \tag{8.13}$$

where the momentum operator is given by

$$\vec{p} = -i\hbar\vec{\nabla} \ . \tag{8.14}$$

Although the above derivation of the Schrödinger equation was for the Coulomb potential, the derivation is valid for any potential between two particles, provided the potential is a function of the magnitude of the relative distance between the two particles that form the system. Such potentials are known as *central potentials*. The Schrödinger equation (8.13) for the potential $V(r)$ is now a partial differential equation in three variables.

At this stage we observe that in rectangular coordinates the kinetic energy operator in coordinate space is given as

$$\frac{p^2}{2\mu} = -\frac{\hbar^2}{2\mu}\nabla^2 = -\frac{\hbar^2}{2\mu}\left\{\frac{\partial^2}{\partial x^2} + \frac{\partial^2}{\partial y^2} + \frac{\partial^2}{\partial z^2}\right\} , \tag{8.15}$$

where ∇^2 is the Laplacian, while the potential energy operator for the case of the Coulomb potential takes the form

$$V(r) = -\frac{e^2}{[x^2 + y^2 + z^2]^{\frac{1}{2}}} \ . \tag{8.16}$$

The form of this potential suggests that the Schrödinger equation, Eq. (8.13), in rectangular coordinates is very difficult if not impossible to solve because of the square root in the potential energy. However, we note that the potential between the two particles depends on the magnitude of the relative distance between the two particles, and rotating our coordinate system does not change this potential. At the same time this rotation of the coordinate system does not change the relative kinetic energy. In other words, the Hamiltonian has a symmetry under rotation in three dimensional space. This suggests that the spherical polar coordinate system is the best choice for solving the Schrödinger equation. Before we write our differential equation in spherical polar coordinates, let us write p^2 in a form that will allow us to separate the radial from the angular dependence.

8.1 Cartesian Tensors

In this section we will introduce a tool for simplifying identities in vector calculus that is very powerful. Although all the results reported in this chapter can be derived using standard methods developed in several variable calculus, the power of the approach developed in this section can be applied in diverse fields from electromagnetic theory to general relativity to hydrodynamics, and to that extent is a very useful tool. For the present chapter, the Cartesian tensor approach is no more than a shorthand notation for products of a combination of vectors and gradients.

The angular momentum in classical mechanics is given by

$$\vec{L} = \vec{r} \times \vec{p} \ . \tag{8.17}$$

To define the angular momentum in quantum mechanics, we make use of the quantization rules in Table 3.1 to write the angular momentum in coordinate space as

$$\vec{L}_{op} = \vec{r} \times \vec{p}_{op} = -i\hbar\, \vec{r} \times \vec{\nabla} \ . \tag{8.18}$$

From this point on we will drop the '*op*' subscript on $\vec{L}$ and $\vec{p}$ with the result

$$L^2 = \vec{L} \cdot \vec{L} = (\vec{r} \times \vec{p}) \cdot (\vec{r} \times \vec{p}) \ . \tag{8.19}$$

Note the momentum operator $\vec{p}$ does not commute with the coordinate operator $\vec{r}$. In fact we have a generalization of Eq. (3.69) in the form

$$[r_i, p_j] = i\hbar\delta_{ij} \ , \tag{8.20}$$

with $i,\ j = 1, 2, 3$ labeling the components of the position and momentum vectors in rectangular or Cartesian coordinates, e.g. $\vec{r} = (r_1, r_2, r_3)$, with r_1 being the component of the position vector along the x-axis.

To calculate L^2, we introduce *tensor notation* that is very useful in calculating vector identities involving $\vec{\nabla}$ in vector products.[2] In this tensor notation, the $i^{\rm th}$ component of the angular momentum

$$\vec{L} = \vec{r} \times \vec{p} \ , \tag{8.21}$$

is written as

$$L_i = \epsilon_{ijk} r_j p_k \ , \tag{8.22}$$

where the indices $i, j, k = 1, 2, 3$ label the Cartesian components of a vector and ϵ_{ijk} is the *totally antisymmetric tensor* that has the property

$$\epsilon_{ijk} = \begin{cases} 0 & \text{if any two indices are equal} \\ +1 & \text{if } (ijk) \text{ is an even permutation of } (123) \\ -1 & \text{if } (ijk) \text{ is an odd permutation of } (123) \end{cases} \ . \tag{8.23}$$

We have also made use of the notation whereby repeated indices in a product are summed over, e.g., the product of two vectors

$$\vec{A} = A_1\hat{\imath} + A_2\hat{\jmath} + A_3\hat{k} \quad \text{and} \quad \vec{B} = B_1\hat{\imath} + B_2\hat{\jmath} + B_3\hat{k}$$

can be written as

$$\vec{A} \cdot \vec{B} = \sum_{i=1}^{3} A_i B_i \equiv A_i B_i \ .$$

[2]This notation is very useful in continuum mechanics and relativity. It is also used in advanced electromagnetic theory.

In a similar manner, we can write the component of the angular momentum in Eq. (8.22) as,[3]

$$L_i = \sum_{jk} \epsilon_{ijk}\, r_j p_k \equiv \epsilon_{ijk}\, r_j p_k \ . \tag{8.24}$$

Finally, to simplify complicated expressions we need to know that the totally antisymmetric tensor ϵ_{ijk} has the property that

$$\sum_i \epsilon_{ijk}\, \epsilon_{i\ell m} \equiv \epsilon_{ijk}\, \epsilon_{i\ell m} = \delta_{j\ell}\delta_{km} - \delta_{jm}\delta_{k\ell} \ . \tag{8.25}$$

With these results, we can turn to the calculation of $\vec{L} \cdot \vec{L}$. In our new notation we have

$$\begin{aligned} L^2 = L_i\, L_i &= \epsilon_{ijk}\, r_j p_k\, \epsilon_{i\ell m} r_\ell p_m \\ &= \epsilon_{ijk}\, \epsilon_{i\ell m}\, r_j\, p_k\, r_\ell\, p_m \ . \end{aligned} \tag{8.26}$$

Since the position r_ℓ and momentum p_k operators do not commute, we will keep the order of these operators as they are in the original scalar product. We then have, making use of Eq. (8.25), that

$$\begin{aligned} L^2 &= (\delta_{j\ell}\delta_{km} - \delta_{jm}\delta_{k\ell})\, r_j\, p_k\, r_\ell\, p_m \\ &= r_j\, p_k\, r_j\, p_k - r_j\, p_k\, r_k\, p_j \ . \end{aligned} \tag{8.27}$$

Note, in writing the right hand side of this equation, we need to remember the the order in which we write the components of the position and momentum vectors is important as they do not commute. We also need to make sure that we use different indices for writing the two components of the angular momentum as there is an implicit sum assumed. Using the commutation relation of r_i and p_j as given in Eq. (8.20), we get

$$\begin{aligned} L^2 &= r_j\, (-i\hbar\delta_{jk} + r_j p_k)\, p_k - r_j\, (-i\hbar\delta_{kk} + r_k p_k)\, p_j \\ &= -i\hbar(\vec{r}\cdot\vec{p}) + r^2 p^2 + 3i\hbar(\vec{r}\cdot\vec{p}) - r_j r_k p_k p_j \\ &= 2i\hbar(\vec{r}\cdot\vec{p}) + r^2p^2 - r_k(i\hbar\delta_{jk} + p_k r_j)p_j \\ &= r^2p^2 + i\hbar(\vec{r}\cdot\vec{p}) - (\vec{r}\cdot\vec{p})\,(\vec{r}\cdot\vec{p}) \ . \end{aligned} \tag{8.28}$$

This result allows us to write the kinetic energy for relative motion as

$$\frac{p^2}{2\mu} = \frac{1}{2\mu r^2}\left\{ (\vec{r}\cdot\vec{p})\,(\vec{r}\cdot\vec{p}) - i\hbar(\vec{r}\cdot\vec{p}) + L^2 \right\} \ . \tag{8.29}$$

Furthermore, since $\vec{p} = -i\hbar\vec{\nabla}$, we have

$$\vec{r}\cdot\vec{p} = -i\hbar r\, \frac{\partial}{\partial r} \ , \tag{8.30}$$

[3] As an exercise write the components of the angular momentum L_1, L_2 and L_3. Compare your answer with you favorite method of calculating vector product. Do they agree?

and we can be write the kinetic energy as

$$\frac{p^2}{2\mu} = -\frac{\hbar^2}{2\mu}\nabla^2 = -\frac{\hbar^2}{2\mu r^2}\left\{r\frac{\partial}{\partial r}\left(r\frac{\partial}{\partial r}\right) + r\frac{\partial}{\partial r} - \frac{L^2}{\hbar^2}\right\}$$
$$= -\frac{\hbar^2}{2\mu}\left\{\frac{1}{r^2}\frac{\partial}{\partial r}\left(r^2\frac{\partial}{\partial r}\right) - \frac{L^2}{\hbar^2 r^2}\right\} . \tag{8.31}$$

We will show in the next section that L^2 depends on the angle variable (θ, ϕ) in spherical polar coordinates. As a result, the above expression for the kinetic energy has effectively separated the radial variable from the angle variable. This we will see will allow us to solve the Schrödinger equation using separation of variables. We could have derived the above result by writing ∇^2 in spherical polar coordinates. The Hamiltonian for a central potential can now be written in spherical polar coordinates as:

$$H = -\frac{\hbar^2}{2\mu}\left\{\frac{1}{r^2}\frac{\partial}{\partial r}\left(r^2\frac{\partial}{\partial r}\right) - \frac{L^2}{\hbar^2\, r^2}\right\} + V(r) . \tag{8.32}$$

The corresponding Schrödinger equation takes the form

$$\left[-\frac{\hbar^2}{2\mu}\left\{\frac{1}{r^2}\frac{\partial}{\partial r}\left(r^2\frac{\partial}{\partial r}\right) - \frac{L^2}{\hbar^2\, r^2}\right\} + V(r)\right]\psi(\vec{r}) = E\,\psi(\vec{r}) . \tag{8.33}$$

The Hamiltonian in Eq. (8.32) is the sum of two parts – one depends on the radial variable r, and the second which includes L^2, depends on the angles θ and ϕ. As a result, we can use the method of separation of variables to write

$$\psi(\vec{r}) = R(r)\; Y(\theta, \phi) . \tag{8.34}$$

Furthermore, since the angular momentum square, L^2, is a linear Hermitian operator, we can take $Y(\theta, \phi)$ to be a solution of the eigenvalue problem

$$L^2 Y(\theta, \phi) = \hbar^2\lambda\; Y(\theta, \phi) , \tag{8.35}$$

where λ is a parameter to be determined by the boundary condition on the Schrödinger equation. We now can write the *radial Schrödinger equation* as:

$$\left[-\frac{\hbar^2}{2\mu}\left\{\frac{1}{r^2}\frac{d}{dr}\left(r^2\frac{d}{dr}\right) - \frac{\lambda}{r^2}\right\} + V(r)\right] R(r) = E\, R(r) . \tag{8.36}$$

In this way we have reduced the Schrödinger equation in three dimensions, to a radial equation in one dimension for $R(r)$, and a partial differential equation in the two angle variables that involves the eigenstates of L^2. It is important to note at this stage that the angular part of the solution $Y(\theta, \phi)$ does not depend on the choice of potential $V(r)$, and will be a valid angular solution to the Schrödinger equation for any potential that depends on the radial variable r. In other words, the angular solution is valid for any potential with spherical symmetry.

8.2 Angular Momentum in Quantum Mechanics

Since space is isotropic, rotational symmetry is present in all problems where external forces do not break this symmetry of space. As a result, all atoms, molecules, and nuclei in free space have spherical symmetry. To that extent, angular momentum in quantum mechanics plays a central role in most of the above problems.

In classical mechanics angular momentum is defined as

$$\vec{L} = \vec{r} \times \vec{p}\ , \tag{8.37}$$

and the corresponding quantity in quantum mechanics results from the substitution $\vec{p} \to -i\hbar\vec{\nabla}$, i.e.,

$$\vec{L} = -i\hbar\vec{r} \times \vec{\nabla} = \vec{r} \times \tilde{\vec{p}}\ . \tag{8.38}$$

The components of this angular momentum operator are given in rectangular coordinates as

$$\begin{aligned} L_x &= yp_z - zp_y = -i\hbar\left(y\frac{\partial}{\partial z} - z\frac{\partial}{\partial y}\right) \\ L_y &= zp_x - xp_z = -i\hbar\left(z\frac{\partial}{\partial x} - x\frac{\partial}{\partial z}\right) \\ L_z &= xp_y - yp_x = -i\hbar\left(x\frac{\partial}{\partial y} - y\frac{\partial}{\partial x}\right)\ . \end{aligned} \tag{8.39}$$

We have found that in quantum mechanics we cannot measure the position and momentum of a particle to any desired accuracy. This results in the position-momentum uncertainty relation, which can be represented mathematically by the commutation relation

$$[r_i, p_j] = i\hbar\delta_{ij}\ . \tag{8.40}$$

Using these commutation relations we can calculate the commutation relation between the components of the angular momentum, $\vec{L}$, e.g.

$$\begin{aligned} [L_x, L_y] &= L_xL_y - L_yL_x \\ &= (yp_z - zp_y)(zp_x - xp_z) - (zp_x - xp_z)(yp_z - zp_y) \\ &= (yp_zzp_x - yp_zxp_z - zp_yzp_x + zp_yxp_z) \\ &\quad -(zp_xyp_z - zp_xzp_y - xp_zyp_z + xp_zzp_y) \\ &= yp_x(p_zz - zp_z) + xp_y(zp_z - p_zz) \\ &= i\hbar(xp_y - yp_x) = i\hbar L_z\ . \end{aligned} \tag{8.41}$$

From this result we may conclude that one cannot measure simultaneously the x- and y-components of the angular momentum vector. Alternatively we can say that we cannot construct a state that is an eigenstate of two different components of the angular momentum vector $\vec{L}$. The above result can be written, in general, as

$$[L_i, L_j] = i\hbar\epsilon_{ijk}\, L_k \ , \tag{8.42}$$

where ϵ_{ijk} is the totally antisymmetric tensor used in the previous section. We have used an implicit sum over repeated indices on the right hand side of Eq. (8.42). We now define the angular momentum square as

$$L^2 = \vec{L}\cdot\vec{L} = L_x^2 + L_y^2 + L_z^2 \equiv \sum_{i=1}^{3} L_i^2 \ . \tag{8.43}$$

This allows us to calculate the commutation relation between the angular momentum square, L^2, and the component of the angular momentum, L_j, i.e.,

$$\begin{aligned}
[L^2, L_j] &= \sum_i [L_i^2, L_j] = \sum_i \left(L_i^2 L_j - L_j L_i^2\right) \\
&= \sum_i \left(L_i^2 L_j - L_i L_j L_i + L_i L_j L_i - L_j L_i^2\right) \\
&= \sum_{ik} \left(L_i i\hbar\epsilon_{ijk} L_k + i\hbar\epsilon_{ijk} L_k L_i\right) \\
&= i\hbar \sum_{ik} \left(\epsilon_{ijk} L_i L_k + \epsilon_{ijk} L_k L_i\right) \\
&= i\hbar \sum_{ik} \left(\epsilon_{ijk} L_i L_k + \epsilon_{kji} L_i L_k\right) = 0 \ ,
\end{aligned}$$

where we have made use of the fact that $\epsilon_{kji} = -\epsilon_{ijk}$. Thus we have established that

$$[L^2, L_j] = 0 \ . \tag{8.44}$$

In this case we may conclude that we can construct states that are eigenstates of both L^2 and one of the components of the angular momentum L_k. It is common practice to take L_k to be the z-component of the angular momentum. In other words we can write

$$L^2\, Y_{\lambda m}(\theta,\phi) = \hbar^2\lambda\, Y_{\lambda m}(\theta,\phi) \tag{8.45a}$$
$$L_z\, Y_{\lambda m}(\theta,\phi) = \hbar m\, Y_{\lambda m}(\theta,\phi) \ . \tag{8.45b}$$

We now have to find what the eigenvalues λ and m are, and what the eigenstate $Y_{\lambda m}(\theta,\phi)$ that determines the angular part of the wave function is. In other words, we need to find the solution of the differential equations (8.45). This is best achieved by changing variables to spherical polar coordinates, (see Figure 8.1) i.e.,

$$\begin{aligned}
x &= r\sin\theta\cos\phi \\
y &= r\sin\theta\sin\phi \\
z &= r\cos\theta \ .
\end{aligned} \tag{8.46}$$

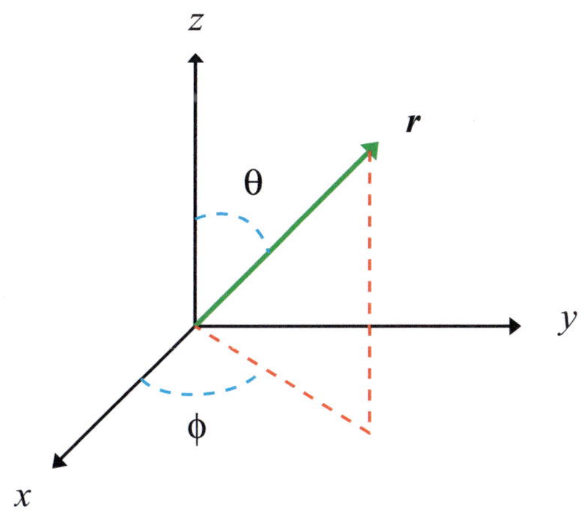

Figure 8.1: Spherical polar coordinates for central potentials.

Taking the differential of these equations, we get

$$\begin{aligned} dx &= \sin\theta\cos\phi\, dr + r\cos\theta\cos\phi\, d\theta - r\sin\theta\sin\phi\, d\phi \\ dy &= \sin\theta\sin\phi\, dr + r\cos\theta\sin\phi\, d\theta + r\sin\theta\cos\phi\, d\phi \\ dz &= \cos\theta\, dr - r\sin\theta\, d\theta\ . \end{aligned} \tag{8.47}$$

Solving for dr, $d\theta$, and $d\phi$, we get

$$\begin{aligned} dr &= \sin\theta\cos\phi\, dx + \sin\theta\sin\phi\, dy + \cos\theta\, dz \\ d\theta &= \frac{1}{r}(\cos\theta\cos\phi\, dx + \cos\theta\sin\phi\, dy - \sin\theta\, dz) \\ d\phi &= \frac{1}{r\sin\theta}(-\sin\phi\, dx + \cos\phi\, dy)\ . \end{aligned} \tag{8.48}$$

With these results we can write L_x, L_y and L_z in spherical polar coordinates making use of the fact that

$$\frac{\partial}{\partial x} = \frac{\partial r}{\partial x}\frac{\partial}{\partial r} + \frac{\partial \theta}{\partial x}\frac{\partial}{\partial \theta} + \frac{\partial \phi}{\partial x}\frac{\partial}{\partial \phi}\ ,$$

with similar expressions for the partial derivatives with respect to y and z. This gives, after some algebra,

$$\begin{aligned} L_x &= i\hbar\left(\sin\phi\frac{\partial}{\partial\theta} + \cot\theta\cos\phi\frac{\partial}{\partial\phi}\right) \\ L_y &= i\hbar\left(-\cos\phi\frac{\partial}{\partial\theta} + \cot\theta\sin\phi\frac{\partial}{\partial\phi}\right) \\ L_z &= -i\hbar\frac{\partial}{\partial\phi}\ , \end{aligned} \tag{8.49}$$

and

$$L^2 = L_x^2 + L_y^2 + L_z^2$$
$$= -\hbar^2 \left\{ \frac{1}{\sin\theta} \frac{\partial}{\partial\theta} \left(\sin\theta \frac{\partial}{\partial\theta} \right) + \frac{1}{\sin^2\theta} \frac{\partial^2}{\partial\phi^2} \right\} . \tag{8.50}$$

With L_z and L^2 written in terms of θ and ϕ, we can write Eq. (8.45) as partial differential equations. The second of these equations, Eq. (8.45b), is just a first order differential equation of the form

$$-i\hbar \frac{\partial Y_{\lambda m}}{\partial\phi} = \hbar m Y_{\lambda m} . \tag{8.51}$$

This equation can be integrated to give us [4]

$$Y_{\lambda m}(\theta, \phi) = e^{im\phi} P_{\lambda m}(\theta) . \tag{8.52}$$

In writing this solution we have taken the constant to be a function of the other angle in the problem, i.e., θ. The boundary condition that $Y_{\lambda m}(\theta, \phi + 2\pi) = Y_{\lambda m}(\theta, \phi)$ implies that m is an integer. We can now write the equation

$$L^2\, Y_{\lambda m}(\theta, \phi) = \hbar^2 \lambda\, Y_{\lambda m}(\theta, \phi) , \tag{8.53}$$

as

$$\left\{ \frac{1}{\sin\theta} \frac{\partial}{\partial\theta} \left(\sin\theta \frac{\partial}{\partial\theta} \right) + \frac{1}{\sin^2\theta} \frac{\partial^2}{\partial\phi^2} \right\} Y_{\lambda m}(\theta, \phi) = -\lambda\, Y_{\lambda m}(\theta, \phi) . \tag{8.54}$$

But we have that

$$\frac{\partial^2}{\partial\phi^2} Y_{\lambda m}(\theta, \phi) = -m^2\, Y_{\lambda m}(\theta, \phi) , \tag{8.55}$$

and therefore we have

$$\left\{ \frac{1}{\sin\theta} \frac{\partial}{\partial\theta} \left(\sin\theta \frac{\partial}{\partial\theta} \right) - \frac{m^2}{\sin^2\theta} \right\} P_{\lambda m}(\theta) = -\lambda\, P_{\lambda m}(\theta) . \tag{8.56}$$

We now change variables to

$$\omega = \cos\theta , \tag{8.57}$$

and write the derivative with respect to θ as

$$\frac{\partial}{\partial\theta} = \frac{\partial\omega}{\partial\theta} \frac{\partial}{\partial\omega} = -\sin\theta \frac{\partial}{\partial\omega}$$

---

[4] We can use the separation of variables, i.e., $Y_{\ell m}(\theta, \phi) = \Phi(\phi)\,\Theta(\theta)$ and Eq. (8.51) to write the ϕ dependence as

$$\Phi(\phi) = e^{im\phi} .$$

Here m has to be an integer for $\Phi(\phi) = \Phi(\phi + 2\pi)$, i.e the function is single valued.

or

$$\frac{\partial}{\partial\omega} = -\frac{1}{\sin\theta}\frac{\partial}{\partial\theta} \; .$$

The equation for $P_{\lambda m}$ can now be written as

$$\left\{ -\frac{\partial}{\partial\omega}\left(-\sin^2\theta\frac{\partial}{\partial\omega}\right) + \lambda - \frac{m^2}{\sin^2\theta} \right\} P_{\lambda m}(\theta) = 0$$

or

$$\left\{ \frac{d}{d\omega}\left((1-\omega^2)\frac{d}{d\omega}\right) + \left(\lambda - \frac{m^2}{1-\omega^2}\right) \right\} P_{\lambda m}(\omega) = 0 \; ,$$

since $\sin^2\theta = 1 - \cos^2\theta = 1 - \omega^2$. Therefore, we have

$$(1-\omega^2)\frac{d^2P(\omega)}{d\omega^2} - 2\omega\frac{dP(\omega)}{d\omega} + \left(\lambda - \frac{m^2}{1-\omega^2}\right) P(\omega) = 0 \; . \tag{8.58}$$

This is known as the *Legendre differential equation*, and can be solved by the methods of series. For $\lambda = \ell(\ell+1)$ with ℓ an integer and $|m| \leq \ell$, the solutions can be written in terms of a polynomial known as *Associate Legendre function.*[31] These functions are given by [5]

$$\begin{aligned} P_\ell^m(\omega) &= (1-\omega^2)^{m/2}\,\frac{d^m}{d\omega^m}P_\ell(\omega) \\ &= \frac{1}{2^\ell \ell!}\,(1-\omega^2)^{m/2}\,\frac{d^{\ell+m}}{d\omega^{\ell+m}}(\omega^2-1)^\ell \; . \end{aligned} \tag{8.59}$$

In the above expression we have introduced the functions $P_\ell(\omega) = P_\ell^0(\omega)$ which are known as the *Legendre polynomials.* These polynomials satisfy the orthogonality relation

$$\int_{-1}^{+1} d\omega\, P_\ell^m(\omega)P_{\ell'}^m(\omega) = \frac{2}{2\ell+1}\frac{(\ell+m)!}{(\ell-m)!}\,\delta_{\ell\ell'} \; . \tag{8.60}$$

They also satisfy the recursion relations

$$\begin{aligned} (2\ell+1)\omega P_\ell^m &= (\ell+1-m)P_{\ell+1}^m + (\ell+m)P_{\ell-1}^m \\ (1-\omega^2)\frac{d}{d\omega}P_\ell^m &= -\ell\omega P_\ell^m + (\ell+m)P_{\ell-1}^m \\ &= (\ell+1)\omega P_\ell^m - (\ell+1-m)P_{\ell+1}^m \end{aligned} \tag{8.61}$$

[5] We have introduced the notation that for $\lambda = \ell(\ell+1)$

$$P_\ell^m(\omega) \equiv P_{\lambda m}(\omega) \; .$$

Some special properties of these polynomials are

$$\begin{aligned}
P_\ell(1) &= 1\ ; \qquad P_\ell(-1) = (-)^\ell \\
P_\ell^m(1) &= P_\ell^m(-1) = 0 \quad \text{for} \quad m \neq 0 \\
P_0(\omega) &= 1\ ; \qquad P_1(\omega) = \omega \\
P_2(\omega) &= \frac{1}{2}(3\omega^2 - 1)\ ; \qquad P_1^1(\omega) = \sqrt{1-\omega^2} \\
P_2^1(\omega) &= 3\omega\sqrt{1-\omega^2}\ ; \qquad P_2^2(\omega) = 3(1-\omega^2)\ .
\end{aligned}$$

We now combine the solution for the ϕ and θ dependence to get the *spherical harmonics*, $Y_{\ell m}(\theta,\phi)$, given by

$$Y_{\ell m}(\theta,\phi) = (-)^m \left\{\frac{(2\ell+1)}{4\pi}\frac{(\ell-m)!}{(\ell+m)!}\right\}^{\frac{1}{2}} P_\ell^m(\cos\theta)\, e^{im\phi}\ . \tag{8.62}$$

This function, which is an eigenstate of L^2 and L_z, is normalized to the extent that

$$\int_{-1}^{+1} d(\cos\theta) \int_0^{2\pi} d\phi\, Y_{\ell m}^*(\theta,\phi)\, Y_{\ell' m'}(\theta,\phi) = \delta_{\ell\ell'}\,\delta_{mm'}\ . \tag{8.63}$$

The lowest order spherical harmonics are:

$$\begin{aligned}
Y_{00}(\theta,\phi) &= \frac{1}{\sqrt{4\pi}}\ ; \qquad Y_{10}(\theta,\phi) = \sqrt{\frac{3}{4\pi}}\cos\theta \\
Y_{1\pm1}(\theta,\phi) &= \mp\sqrt{\frac{3}{8\pi}}\sin\theta\, e^{\pm i\phi}\ ; \qquad Y_{20}(\theta,\phi) = \sqrt{\frac{5}{16\pi}}(3\cos^2\theta - 1) \\
Y_{2\pm1}(\theta,\phi) &= \mp\sqrt{\frac{15}{8\pi}}\sin\theta\cos\theta\, e^{\pm i\phi}\ ; \qquad Y_{2\pm2}(\theta,\phi) = \frac{1}{4}\sqrt{\frac{15}{2\pi}}\sin^2\theta\, e^{\pm 2i\phi}\ .
\end{aligned}$$

We thus have established that the components of the angular momentum $\vec{L}$ do not commute. In particular,

$$[L_i, L_j] = i\hbar\epsilon_{ijk}L_k\ . \tag{8.64}$$

However

$$[L^2, L_i] = 0\ , \tag{8.65}$$

which allows us to write eigenstates of L^2 and L_z as

$$\begin{aligned}
L^2\, Y_{\ell m}(\theta,\phi) &= \hbar^2\ell(\ell+1)\, Y_{\ell m}(\theta,\phi) \\
L_z\, Y_{\ell m}(\theta,\phi) &= \hbar m\, Y_{\ell m}(\theta,\phi)\ ,
\end{aligned} \tag{8.66}$$

with the eigenstates $Y_{\ell m}(\theta, \phi)$ being the spherical harmonics.

Here we observe that the magnitude of the angular momentum, $\sqrt{\langle L^2 \rangle}$ is given by

$$\left\{ \int d\Omega \, Y^*_{\ell m}(\theta, \phi) \, L^2 \, Y_{\ell m}(\theta, \phi) \right\}^{\frac{1}{2}} = \hbar \sqrt{\ell(\ell+1)}$$

where the integration is over all direction in space, i.e., $d\Omega = d\phi \, d\theta \, \sin\theta$. On the other hand, the maximum projection of $\vec{L}$ along the z-axis is given by

$$\int d\Omega \, Y^*_{\ell\ell}(\theta, \phi) \, L_z \, Y_{\ell\ell}(\theta, \phi) = \hbar\ell < \hbar\sqrt{\ell(\ell+1)} \; ,$$

i.e., the projection of the angular momentum along the z-axis is always less than the magnitude of the angular momentum. This suggests that the angular momentum vector cannot point along the z-axis. If we force $\vec{L}$ to be along the z-axis this would mean its components along the x and y axis are identically zero. This is not possible because the x- and y-components of the angular momentum do not commute, and from the uncertainty principle we know that we can not determine both the x- and y-components of the angular momentum to any desired accuracy, e.g. both being zero. This suggests that the angular momentum vector is precessing about the z-axis, and that this precession is purely a quantum mechanical effect resulting from the uncertainty principle.

8.3 Rotational Motion

In this section we will take advantage of the quantization of the angular momentum to examine the rotation of a diatomic molecule or a deformed nucleus. For excitations of a diatomic molecule that are small compared to the vibrational excitation $\hbar\omega$ (see Sec. 7.2), the distance between the atoms of a molecule are fixed, and the molecule can have rotational energy as if it is a rigid object. Classically, this rotational energy is the kinetic energy of rotation given by

$$\frac{1}{2}\mathcal{I}\omega^2 \; , \tag{8.67}$$

where $\mathcal{I}$ is the moment of inertia for the rigid body about the axis of rotation, and ω is the angular velocity of rotation. We can write this kinetic energy in terms of the classical angular momentum $\vec{L}$ as

$$H_{rot} = \frac{L^2}{2\mathcal{I}} \; . \tag{8.68}$$

Since the kinetic energy of rotation is also the total energy, we have defined this total rotational energy as H_{rot}.

We now have to quantize this classical Hamiltonian for rotational motion. From the results of the last section, we know that the angular momentum in quantum mechanics is given by Eq. (8.38), and in particular L^2 is the differential operator in Eq. (8.50). This means we can write the Hamiltonian for a quantized rigid body as

$$H_{rot} = -\frac{\hbar^2}{2\mathcal{I}}\left\{\frac{1}{\sin\theta}\frac{\partial}{\partial\theta}\left(\sin\theta\frac{\partial}{\partial\theta}\right) + \frac{1}{\sin^2\theta}\frac{\partial^2}{\partial\phi^2}\right\} . \tag{8.69}$$

and the corresponding Schrödinger equation for this quantized rigid body is

$$H_{rot}\,\psi_{rot} = E\,\psi_{rot} . \tag{8.70}$$

But we know that the spherical harmonics $Y_{\ell m}$ are the eigenstates of L^2. We therefore can establish that

$$\psi_{rot} = Y_{\ell m}(\theta, \phi) , \tag{8.71}$$

where the angles θ and ϕ give the orientation of the solid in space.[6] The corresponding eigenstates or possible energy of rotation is given by

$$E_\ell = \frac{\hbar^2}{2\mathcal{I}}\ell(\ell+1) . \tag{8.72}$$

Thus the energy spectrum for rotation is completely determined by the moment of inertia of the molecule.

To test the validity of this model for the rotational spectrum of a diatomic molecule, we consider the molecules H_2 and D_2. Since in both cases the electronic wave function is the same, we expect the distance between the two nuclei in the two molecules to be the same. Assuming the ground state energy of both molecules is zero, and the first excited state corresponds to $\ell = 2$, then the ratio of the energy of the first excited state for the two molecules is given by

$$\frac{E_2(H_2)}{E_2(D_2)} = \frac{\mathcal{I}(D_2)}{\mathcal{I}(H_2)} \tag{8.73}$$

Taking the moment of inertia of the molecule about an axis through the center of mass and perpendicular to the line joining the two nuclei in the molecule, we have

$$\mathcal{I}(H_2) = \frac{1}{2}m_p R_0^2 \qquad \text{and} \qquad \mathcal{I}(D_2) = \frac{1}{2}m_d R_0^2 = m_p R_0^2 . \tag{8.74}$$

where R_0 is the separation between the two atoms in the molecule, i.e., the bond length. Therefore, the ratio of the energy of the first excited state of the two molecules is

$$\frac{E_2(H_2)}{E_2(D_2)} = 2 . \tag{8.75}$$

[6]For the most general solid body, we need three angles to determine the orientation of the body in space. As a result, the motion of the most general solid object has a wave function that depends on three angles and not two. However, for diatomic molecules, where there is an axis of symmetry along a line joining the two nuclei, we only need two angles and therefore the wave function is a function of these two angles.

We can compare this result with the experimental ratio, which is 1.995. In a similar manner if we calculate the ratio of the moment of inertia of HD and H_2 molecules, then the model gives a value of 4/3, while experiment gives 1.3286. Here again we have good agreement, suggesting that our simple model for the rotational motion of a diatomic molecule is very good.

In the discussion above, we took the first excited state of H_2 to correspond to $\ell = 2$. The fact that there is no $\ell = 1$ or for that matter no odd ℓ state is a result of the fact that the H_2 molecule has a symmetry about a plane going through the center of mass of the molecule and perpendicular to the axis joining the two nuclei in the molecule. This same symmetry is present in O_2 and N_2, but is not satisfied for HD or $NaCl$. As a result, the latter molecules will have states with ℓ being both even and odd.

This same model has been extended to deformed nuclei which also exhibit a rotational spectrum similar to that observed in molecules.

8.4 The Radial Equation

Let us now consider potentials that are a function of the radial variables r, i.e., $V(\vec{r}) = V(r)$. These potentials are known as central potentials, and have spherical symmetry, i.e., they are invariant under rotation in three dimensions. The Hamiltonian for this class of potentials is of the form

$$H = T + V(r) \; , \tag{8.76}$$

where the kinetic energy, T, is given as

$$T = \frac{p^2}{2\mu} = T_r + \frac{L^2}{2\mu r^2} \; , \tag{8.77}$$

which can be written in terms of the radial variable r and the angular momentum operator square L^2 as given in Eq. (8.31). Since the angular momentum operator is a function of the angular variables θ and ϕ, then $[T_r, \vec{L}] = 0$. We also have shown that $[L^2, \vec{L}] = 0$, thus the kinetic energy T always commutes with the angular momentum $\vec{L}$. If in addition the potential is central, i.e.,

$$V(\vec{r}) = V(r) \; ,$$

then the Hamiltonian H commutes with the angular momentum $\vec{L}$, i.e.,

$$[H, \vec{L}] = 0 \; . \tag{8.78}$$

This result implies that the eigenstates of H can also be eigenstates of the angular momentum. But since the components of the angular momentum do not commute, i.e., $[L_x, L_y] \neq 0$, we cannot find an eigenstate of $\vec{L}$, but we can find eigenstates of L^2 and

L_z. Thus the solution of the Schrödinger equation for a central potential can also be an eigenstate of L^2 and L_z, i.e.,

$$H\,\psi_{n\ell m} = E_{n\ell m}\,\psi_{n\ell m} \tag{8.79a}$$

$$L^2\,\psi_{n\ell m} = \hbar^2\ell(\ell+1)\,\psi_{n\ell m} \tag{8.79b}$$

$$L_z\,\psi_{n\ell m} = \hbar m\,\psi_{n\ell m}\ , \tag{8.79c}$$

where n, ℓ and m are quantum numbers with ℓ being the angular momentum and m, the projection of the angular momentum along the z-axis. Here, n is an additional quantum number to be determined later. We will find that the definition of n in terms of the energy will depend on the choice of the central potential. Because the Hamiltonian can be divided into two parts, one depending on r, the other on θ and ϕ, the wave function can be written as a product, i.e.,

$$\psi_{n\ell m}(\vec{r}) = R_{n\ell}(r)\,Y_{\ell m}(\theta,\phi) \tag{8.80}$$

with the radial function $R_{n\ell}(r)$ satisfying the equation

$$-\frac{\hbar^2}{2\mu}\left\{\frac{1}{r^2}\frac{d}{dr}\left(r^2\frac{d}{dr}\right) - \frac{\ell(\ell+1)}{r^2}\right\} R_\ell(r) + V(r)R_\ell(r) = E\,R_\ell(r)\ . \tag{8.81}$$

We now write the radial wave function as

$$R_\ell(r) = \frac{u_\ell(r)}{r}\ , \tag{8.82}$$

and then we have that

$$\frac{dR_\ell(r)}{dr} = \frac{1}{r}\frac{du_\ell(r)}{dr} - \frac{u_\ell(r)}{r^2}$$

and

$$\frac{d}{dr}\left(r^2\frac{dR_\ell}{dr}\right) = \frac{d}{dr}\left(r\frac{du_\ell}{dr}\right) - \frac{du_\ell}{dr} = r\frac{d^2u_\ell}{dr^2}\ .$$

This allows us to write the radial Schrödinger equation for $u_\ell(r)$ as

$$-\frac{\hbar^2}{2\mu}\frac{d^2u_\ell(r)}{dr^2} + \left(V(r) + \frac{\hbar^2}{2\mu}\frac{\ell(\ell+1)}{r^2}\right)u_\ell(r) = Eu_\ell(r) \tag{8.83}$$

or, in terms of an 'effective potential', as

$$-\frac{\hbar^2}{2\mu}\frac{d^2u_\ell(r)}{dr^2} + V_{eff}(r)u_\ell(r) = Eu_\ell(r)\ , \tag{8.84}$$

where this angular momentum dependent effective potential is given in terms of the original potential $V(r)$ as:

$$V_{eff}(r) = V(r) + \frac{\hbar^2}{2\mu}\frac{\ell(\ell+1)}{r^2} . \qquad (8.85)$$

This is illustrated in Figure 8.2, where we have assumed $V(r)$ is a square well, and the resultant effective potential is represented by the dashed curve. We observe that Eq. (8.84) is identical in form to the one dimensional Schrödinger equation we considered in previous chapters, the only differences being:

1. We now have $r \geq 0$ only.

2. The potential $V(r)$ is replaced by $V_{eff}(r)$ which depends on ℓ, the angular momentum.

3. Near the origin of the wave function $u(r) \rightarrow 0$ as $r \rightarrow 0$, otherwise $R(r)$ will be infinite at the origin. In this way we have insured the radial probability density $|R(r)|^2$ is finite at $r = 0$. Thus at $r = 0$ we have a situation similar to the one dimensional case with $V(x) \rightarrow +\infty$ for $x \leq 0$.

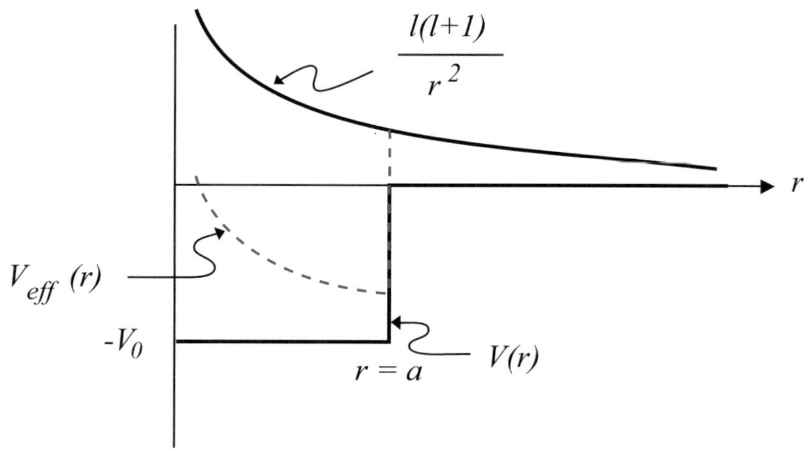

Figure 8.2: A comparison of the 'effective potential' V_{eff} and the square well $V(r)$.

We observe here that for $r > a$, the effective potential is repulsive for $\ell \neq 0$. This repulsion is due to an angular momentum barrier which keeps the particle away from the origin if the particle is outside the well, or possibly traps the particle if the particle is inside the well. This potential is illustrated in the Figure 8.2.

8.5 Problems

1. Use the tensor notation developed in this chapter to prove that

$$\nabla \times (\vec{A} \times \vec{B}) = \vec{A}(\nabla \cdot \vec{B}) - \vec{B}(\nabla \cdot \vec{A}) + (\vec{B} \cdot \nabla)\vec{A} - (\vec{A} \cdot \nabla)\vec{B} \ .$$

2. Given the change of variables from rectangular to spherical polar coordinates (see Eq. (8.46)), show that the Laplacian in spherical polar coordinates is given by

$$\nabla^2 = \frac{1}{r^2}\frac{\partial}{\partial r}\left(r^2\frac{\partial}{\partial r}\right) + \frac{1}{r^2 \sin\theta}\frac{\partial}{\partial\theta}\left(\sin\theta\frac{\partial}{\partial\theta}\right) + \frac{1}{r^2\sin^2\theta}\frac{\partial^2}{\partial\phi^2} \ .$$

3. The generating function for the Legendre polynomial is given by

$$\frac{1}{\sqrt{1-2ws+s^2}} = \sum_{\ell=0}^{\infty} P_\ell(w)\, s^\ell \ .$$

   Use this generating function to evaluate

$$\int_{-1}^{+1} dw\, P_\ell(w)\, P_{\ell'}(w) \ .$$

4. Using the definition of the components of the angular momentum L_x, L_y, L_z, and the commutation relation of the coordinate with the canonical momentum, i.e., $[r_i, p_j] = i\hbar\delta_{ij}$, show that:

   (a) The commutation relation of the component of the angular momentum operator with the component of the position and momentum operator are

$$[L_i, r_j] = i\hbar\varepsilon_{ijk}r_k \quad \text{and} \quad [L_i, p_j] = i\hbar\varepsilon_{ijk}p_k$$

   (b) The commutation relation of the component of the angular momentum are

$$[L_i,\, L_j] = L_iL_j - L_jL_i = i\hbar\varepsilon_{ijk}L_k$$

   and

$$\left[L_z,\, L^2\right] = 0 \ ,$$

   where

$$L^2 = L_x^2 + L_y^2 + L_z^2 \ .$$

5. A rigid system rotates freely about the z-axis with moment of inertia $\mathcal{I}$. By expressing the energy of the system in terms of the angular momentum, L_z, show that the possible energy levels of the system are

$$E_m = \frac{\hbar^2 m^2}{2\mathcal{I}}, \qquad m = 0, 1, 2, \ldots$$

with eigenfunctions

$$u_m(\phi) = e^{\pm im\phi} ,$$

where ϕ is the angle specifying the orientation of the system in the $x - y$ plane.

6. A diatomic molecule, e.g. H_2, rotates about its center of mass and in a plane perpendicular to the z-axis. The Hamiltonian for this system is given by

$$H = \frac{1}{2\mathcal{I}} L_z^2 ,$$

where $\mathcal{I}$ is the moment of inertia of the molecule.

(a) Assuming the mass of each of the atoms in the molecule is m, and the distance between the atoms is a, what is the moment of inertia of the molecule?

(b) Write the above Hamiltonian in spherical polar coordinates.

(c) What are the solutions of the Schrödinger equation for this Hamiltonian?

(d) What is the rotational energy spectrum for this diatomic molecule?

(e) What is the ratio of the energies of the first excited state of H_2 and D_2 molecules, given that the mass of D is twice the mass of H, and the separation between the nuclei in the two molecules is the same?

7. The rotational energy of a diatomic molecule may be obtained by regarding the molecule as a rigid system consisting of two point particles a fixed distance apart. This rigid system is free to rotate about its center of mass. By expressing the total energy in terms of the moment of inertia $\mathcal{I}$, and the total angular momentum, show that the rotational energy levels are

$$E_\ell = \frac{\hbar}{2\mathcal{I}} \ell(\ell + 1), \qquad \ell = 0, 1, 2, \ldots$$

and that the level specified by ℓ has $(2\ell + 1)$ different eigenstates, corresponding to the possible values of the z-component of the angular momentum

$$L_z = \hbar m \qquad m = 0, \pm 1, \pm 2 \ldots, \pm \ell .$$

8. Given the operator

$$L_- = L_x - iL_y \ ;$$

show that $L_-Y_{11}(\theta, \phi)$ is proportional to $Y_{10}(\theta, \phi)$, i.e.,

$$L_-Y_{11}(\theta, \phi) = aY_{10}(\theta, \phi) \ .$$

Determine the constant a.

9. Show that

$$|Y_{11}(\theta, \phi)|^2 + |Y_{10}(\theta, \phi)|^2 + |Y_{1-1}(\theta, \phi)|^2 = \frac{3}{4\pi} P_1(\cos\theta) \ .$$

This is a particular case of the theorem, which states that

$$\sum_m Y^*_{\ell m}(\theta, \phi)\, Y_{\ell m}(\theta, \phi) = \frac{2\ell + 1}{4\pi} P_\ell(\cos\theta) \ .$$

This result shows that when all m states corresponding to a given ℓ are occupied, the system has spherical symmetry.

10. The parity operator P is defined by the relation

$$P\,\psi(\vec{r}) = \psi(-\vec{r}) \ .$$

Show that

$$P\,Y_{\ell m}(\theta, \phi) = (-1)^\ell\, Y_{\ell m}(\theta, \phi) \ .$$

Show that the solutions of the Schrödinger equation for central potentials have definite parity.

CHAPTER 9

Central Force Problem II

Abstract: This chapter is devoted to the solution of the radial Schrödinger equation for three different spherically symmetric potentials whose solutions are used to discuss the approximation methods in atomic, molecular and nuclear problems in Chapter 13.

In the last Chapter we started with the simple model of the hydrogen atom and showed that in general the problem of two particles interacting via a potential that depends on the relative distance between them can be divided into two parts. The angular part of the wave function, which was the same for all central potentials, and a radial wave function that depended on the form of the potential. In the last chapter we considered in detail the angular part of the Schrödinger equation for the central potential. In this Chapter we turn to the radial part of the Schrödinger equation and start by considering the case of a square well in three-dimensions, then proceed to the harmonic oscillator, and finally go to the Coulomb problem which was the initial motivation for considering the Schrödinger equation for central potentials. In the process we introduce the special functions needed for scattering theory, and which form the basis for modeling more complex systems.

9.1 The Three-Dimensional Square Well

In the last Chapter we showed that the solution to the Schrödinger equation for a central potential can be reduced to solving the differential equation (8.81) for the radial wave function $R_\ell(r)$. To solve this radial equation we need to specify the potential $V(r)$. In the present section we will consider the case of a simple square well potential given by

$$V(r) = \begin{cases} -V_0 & \text{for} \quad r < a \\ \\ 0 & \text{for} \quad r > a \end{cases} , \tag{9.1}$$

Iraj R. Afnan

where a is the radius of the well. For this potential, the radial equation for $r < a$ takes the form

$$\frac{d^2R_\ell(r)}{dr^2} + \frac{2}{r}\frac{dR_\ell(r)}{dr} + \left(\alpha^2 - \frac{\ell(\ell+1)}{r^2}\right) R_\ell(r) = 0 \ , \tag{9.2}$$

where

$$\alpha^2 = \frac{2m}{\hbar^2}\left(E + V_0\right) \ . \tag{9.3}$$

If we now define $\rho = \alpha r$, the radial equation reduces to

$$\frac{d^2R_\ell(\rho)}{d\rho^2} + \frac{2}{\rho}\frac{dR_\ell(\rho)}{d\rho} + \left(1 - \frac{\ell(\ell+1)}{\rho^2}\right) R_\ell(\rho) = 0 \ . \tag{9.4}$$

This equation is known as the *spherical Bessel equation.* To relate it to the ordinary Bessel's differential equation. i.e.,

$$\frac{d^2J_p(\rho)}{d\rho^2} + \frac{1}{\rho}\frac{dJ_p(\rho)}{d\rho} + \left(1 - \frac{p^2}{\rho^2}\right) J_p(\rho) = 0 \ , \tag{9.5}$$

we have to write the radial wave function as

$$R_\ell(\rho) = \rho^{-1/2}\, J(\rho) \ .$$

Substituting this in Eq. (9.4), we find that $J(\rho)$ satisfies the equation

$$\frac{d^2J(\rho)}{d\rho^2} + \frac{1}{\rho}\frac{dJ(\rho)}{d\rho} + \left(1 - \frac{(\ell+\frac{1}{2})^2}{\rho^2}\right) J(\rho) = 0 \tag{9.6}$$

which is basically Eq. (9.5) for $p = \ell + 1/2$. We therefore can write the radial wave function in terms of the half-integer Bessel function as

$$R_\ell(\rho) = \sqrt{\frac{\pi}{2\rho}}\, J_p(\rho) \qquad p = \pm\left(\ell + \frac{1}{2}\right) \ . \tag{9.7}$$

We now introduce the *spherical Bessel function,* $j_\ell(\rho)$ as

$$j_\ell(\rho) = \sqrt{\frac{\pi}{2\rho}}\, J_{\ell+\frac{1}{2}}(\rho) \tag{9.8}$$

and the *spherical Neumann function,* $n_\ell(\rho)$ as

$$n_\ell(\rho) = (-1)^\ell \sqrt{\frac{\pi}{2\rho}}\, J_{-\ell-\frac{1}{2}}(\rho) \ , \tag{9.9}$$

Alternatively, we can define the *spherical Hankel function* of the first kind $h_\ell^{(+)}(\rho)$, and the second kind $h_\ell^{(-)}(\rho)$ as

$$h_\ell^{(\pm)}(\rho) = n_\ell(\rho) \pm i\, j_\ell(\rho) \ . \tag{9.10}$$

From the above results, we may deduce that there are two independent solutions to the radial equation. These are $j_\ell(\rho)$ and $n_\ell(\rho)$, or the Hankel functions $h_\ell^{(\pm)}$. The existence of two independent solutions is a consequence of the fact that we have a second order linear differential equation. Here, j_ℓ and n_ℓ are real solutions and are the convenient choice for the application of the boundary condition at $r = 0$. On the other hand, $h_\ell^{(\pm)}$ are complex solutions, which we will use to impose the boundary condition for $r \to \infty$. For the case of $\ell = 0$ and 1, these functions take the form of [1]

$$\begin{aligned}
j_0(\rho) &= \frac{\sin\rho}{\rho}\,, \qquad j_1(\rho) = \frac{\sin\rho}{\rho^2} - \frac{\cos\rho}{\rho}\,, \\
n_0(\rho) &= \frac{\cos\rho}{\rho}\,, \qquad n_1(\rho) = \frac{\cos\rho}{\rho^2} + \frac{\sin\rho}{\rho}\,, \\
h_0^{(\pm)}(\rho) &= \frac{e^{\pm i\rho}}{\rho}\,, \qquad h_1^{(\pm)}(\rho) = \left(\frac{1}{\rho^2} \mp \frac{i}{\rho}\right) e^{\pm i\rho}\,.
\end{aligned} \tag{9.11}$$

The asymptotic forms of these functions for $\rho \to \infty$ are

$$\begin{aligned}
j_\ell(\rho) &\to \frac{1}{\rho}\sin\left(\rho - \frac{\pi\ell}{2}\right) \\
n_\ell(\rho) &\to \frac{1}{\rho}\cos\left(\rho - \frac{\pi\ell}{2}\right) \\
h_\ell^{(\pm)}(\rho) &\to \frac{1}{\rho}\left[1 \pm i\frac{\ell(\ell+1)}{2\rho} - \cdots\right] \exp\left[\pm i(\rho - \frac{\pi\ell}{2})\right]\,.
\end{aligned} \tag{9.12}$$

On the other hand the behavior of these functions as $\rho \to 0$ are

$$\begin{aligned}
j_\ell(\rho) &\to \frac{\rho^\ell}{(2\ell+1)!!}\left[1 - \frac{\rho^2}{2(2\ell+3)} + \cdots\right] \\
n_\ell(\rho) &\to \frac{(2\ell+1)!!}{(2\ell+1)}\,\frac{1}{\rho^{\ell+1}}\left[1 + \frac{\rho^2}{2(2\ell+1)} + \cdots\right]\,,
\end{aligned} \tag{9.13}$$

where $(2\ell+1)!! = (2\ell+1)(2\ell-1)(2\ell-3)\cdots 1$.

Having established the asymptotic behavior of the two independent solutions of Eq. (9.4), we can write the most general solution for $r < a$ as

$$R_\ell(\rho) = A j_\ell(\rho) + B n_\ell(\rho) \quad \text{for} \quad r < a\,.$$

[1] For more detailed properties of Bessel functions and other functions encountered in this chapter see Abramowitz and Stegun [31]

Using the boundary condition that $R_\ell(\rho)$ should be finite as $\rho \to 0$ and the results of Eq. (9.13), we conclude that $B = 0$, and therefore

$$R_\ell(\rho) = A j_\ell(\rho) \qquad \text{for} \quad r < a\ . \tag{9.14}$$

On the other hand, for $r > a$, the radial Schrödinger equation takes the form

$$\frac{d^2 R_\ell(r)}{dr^2} + \frac{2}{r}\frac{dR_\ell(r)}{dr} + \left(\beta^2 - \frac{\ell(\ell+1)}{r^2}\right) R_\ell(r) = 0 \tag{9.15}$$

with

$$\beta^2 = \frac{2mE}{\hbar^2}\ . \tag{9.16}$$

For bound states, i.e., $E < 0$, we have $\beta^2 = -2m|E|/\hbar^2$. We now define $\rho = \beta r \equiv i\gamma r$ where γ is real, and our general solution is a linear combination of $j_\ell(\rho)$ and $n_\ell(\rho)$, or $h_\ell^{(+)}(\rho)$ and $h_\ell^{(-)}(\rho)$. The latter choice will turn out to be more convenient for imposing the boundary condition for $r \to \infty$. We therefore write the solution of the radial Schrödinger equation for $r > a$ and $E < 0$ as

$$R_\ell(r) = A' h_\ell^{(+)}(\rho) + B' h_\ell^{(-)}(\rho) \qquad \text{for} \quad r > a\ . \tag{9.17}$$

Here ρ is pure imaginary for bound states. Considering the asymptotic behavior of $h_\ell^{(\pm)}(\rho)$ (Eq. (9.12)), and the boundary condition that $R_\ell(r)$ be finite as $r \to \infty$, we find that $B' = 0$. We therefore have for $r > a$ that

$$\begin{aligned} R_\ell(r) &= A'\, h_\ell^{(+)}(\rho) \\ &\to \frac{A'}{i\gamma r}\, e^{-\gamma r - i\pi\ell/2} = \frac{C}{r}\, e^{-\gamma r} \quad \text{for} \quad r \to \infty\ . \end{aligned} \tag{9.18}$$

We now can write the radial wave function for a square well potential as

$$R_\ell(r) = \begin{cases} A\, j_\ell(\alpha r) & \text{for} \quad r < a \\ A' h_\ell^{(+)}(i\gamma r) & \text{for} \quad r > a \end{cases}\ . \tag{9.19}$$

The constants A, A' and the bound state energy E are determined by the requirement that the wave function $R_\ell(r)$ and its derivative be continuous at $r = a$, and the normalization condition

$$\int_0^\infty dr\, r^2\, R_\ell^*(r)\, R_\ell(r) = 1 \tag{9.20}$$

be satisfied.

9.2 The Three-Dimensional Harmonic Oscillator

The harmonic oscillator in three-dimensions is very important because of its simplicity and the fact that the solution is a good approximation to many physically interesting problems [32]. For the present, we are going to use the harmonic oscillator potential to illustrate the solution of the radial Schrödinger equation using the series solution. In particular, we want to illustrate how the discrete eigenvalues arise in a bound state problem as a result of the boundary conditions imposed on the solution of the Schrödinger equation.

The potential for the three dimensional harmonic oscillator is given by

$$V(r) = \frac{1}{2}m\omega^2 r^2 \ . \tag{9.21}$$

The radial Schrödinger equation can be written (see Eq. (8.83)) as

$$\left(-\frac{\hbar^2}{2m}\frac{d^2}{dr^2} + \frac{\hbar^2}{2m}\frac{\ell(\ell+1)}{r^2} + \frac{1}{2}m\omega^2 r^2 - E\right) u_\ell(r) = 0 \ , \tag{9.22}$$

where $u_\ell(r) = r\,R_\ell(r)$ and m is the mass of the particle in the potential well. To simplify the differential equation, we define

$$\beta = \sqrt{\frac{m\omega}{\hbar}} \ , \qquad \xi = \beta r \ , \qquad \text{and} \quad \varepsilon = E/\hbar\omega \ .$$

With these definitions we can write Eq. (9.22) as

$$\left(\frac{d^2}{d\xi^2} - \frac{\ell(\ell+1)}{\xi^2} - \xi^2 + 2\varepsilon\right) u_\ell(\xi) = 0 \ . \tag{9.23}$$

We can solve this differential equation by series solution. However, before we proceed to implement the series solution, let us extract the asymptotic solution for $r \to 0$ and $r \to \infty$. In this way we will find that the final series solution corresponds to a known polynomial when the boundary conditions are applied.

1. For $\xi \to 0$, Eq. (9.23) can be approximated by the equation

$$\left(\frac{d^2}{d\xi^2} - \frac{\ell(\ell+1)}{\xi^2}\right) u_\ell(\xi) = 0 \ , \tag{9.24}$$

since the terms $-\xi^2$ and 2ε are small in comparison to $\ell(\ell+1)/\xi^2$. The solutions of Eq. (9.24) are

$$u_\ell(\xi) = \xi^{\ell+1} \quad \text{or} \quad \xi^{-\ell} \ . \tag{9.25}$$

Since the radial wave function has to be finite at the origin, i.e., $r \to 0$, the second solution in Eq. (9.25) is not acceptable because $u_\ell \to \infty$ as $\xi \to 0$, and the solution would be singular at the origin. From the above analysis we can conclude that the solution to Eq. (9.23) is given by

$$u_\ell(\xi) \to \xi^{\ell+1} \qquad \text{for} \qquad \xi \to 0 \ .$$

Note, this result is valid for all potentials that satisfy the condition that $V(r) \to r^{-2+\delta}$ for $r \to 0$ and $\delta > 0$.

2. For $\xi \to \infty$. In this case we can drop terms like $\ell(\ell+1)/\xi^2$ and 2ε, since they are small in comparison to the term proportional to ξ^2. The asymptotic equation now is given by

$$\left(\frac{d^2}{d\xi^2} - \xi^2\right) u_\ell(\xi) = 0 \ . \tag{9.26}$$

Taking $z = \xi^2$, this equation can be rewritten as

$$\left(4z\frac{d^2}{dz^2} + 2\frac{d}{dz} - z\right) u_\ell(z) = 0 \ , \tag{9.27}$$

or for large z, the differential equation can be approximated by

$$\left(\frac{d^2}{dz^2} - \frac{1}{4}\right) u_\ell(z) = 0 \ . \tag{9.28}$$

This equation has a known solution which is

$$u_\ell = e^{-z/2} = e^{-\xi^2/2} \ . \tag{9.29}$$

From these results we can conclude the $u_\ell(\xi)$ which is a solution of Eq. (9.23) has the following asymptotic behavior

$$u_\ell(\xi) = \begin{cases} \xi^{\ell+1} & \text{for} \quad \xi \to 0 \\ e^{-\xi^2/2} & \text{for} \quad \xi \to \infty \end{cases} \ . \tag{9.30}$$

We therefore can write $u_\ell(\xi)$ as

$$u_\ell(\xi) = \xi^{\ell+1}\, e^{-\xi^2/2}\, f_\ell(\xi) \ . \tag{9.31}$$

In this way we hope $f_\ell(\xi)$ to be a simple function of ξ. Substituting $u_\ell(\xi)$ from Eq. (9.31) into Eq. (9.23), we get an equation for $f_\ell(\xi)$ of the form

$$\frac{d^2 f_\ell(\xi)}{d\xi^2} + 2\left(\frac{\ell+1}{\xi} - \xi\right)\frac{df_\ell(\xi)}{d\xi} + \left[2\varepsilon - (2\ell+3)\right] f_\ell(\xi) = 0 \ . \tag{9.32}$$

If we now define $L(\eta) = f_\ell(\xi)$ where $\eta = \xi^2$, then $L(\eta)$ is a solution to the differential equation

$$\eta \frac{d^2L(\eta)}{d\eta^2} + \left[(\ell + \frac{3}{2}) - \eta\right] \frac{dL(\eta)}{d\eta} + \left[\frac{\varepsilon}{2} - \frac{1}{2}\left(\ell + \frac{3}{2}\right)\right] L(\eta) = 0 \, . \tag{9.33}$$

This is a special case of Laplace's equation

$$\left(z\frac{d^2}{dz^2} + (\beta - z)\frac{d}{dz} - \alpha \right) F = 0 \, . \tag{9.34}$$

This equation has a solution known as a *confluent hypergeometric function* [2] which can be written as a series of the form

$${}_1F_1(\alpha;\beta;z) = \sum_{p=0}^{\infty} \frac{\Gamma(\alpha+p)\,\Gamma(\beta)}{\Gamma(\beta+p)\,\Gamma(\alpha)} \frac{z^p}{p!} \, , \tag{9.35}$$

where $\Gamma(a)$ is a Γ-function [31]. Comparing Eqs. (9.33) and (9.34), we can write the series solution for $L(\eta)$ in the form

$$L(\eta) \;=\; {}_1F_1(a;b;\eta) = \sum_{p=0}^{\infty} \frac{\Gamma(a+p)}{\Gamma(b+p)} \frac{\Gamma(b)}{\Gamma(a)} \frac{\eta^p}{p!} \, , \tag{9.36}$$

where

$$a = \frac{1}{2}\left(\ell + \frac{3}{2}\right) - \frac{\varepsilon}{2} \, , \qquad b = \ell + \frac{3}{2} \, . \tag{9.37}$$

For large η, the confluent hypergeometric function has the asymptotic behavior that

$${}_1F_1(a;b;\eta) \longrightarrow \frac{\Gamma(b)}{\Gamma(a)}\, \eta^{a-b}\, e^{\eta} = \frac{\Gamma(b)}{\Gamma(b)}\, (\beta r)^{2(a-b)}\, e^{\beta^2 r^2} \qquad \text{for} \quad \eta \to \infty \, . \tag{9.38}$$

If we combine this result with that of Eq. (9.31), we observe that $u_\ell(\xi) \propto e^{\beta^2 r^2/2}$, i.e., the wave function goes to infinity as $r \to \infty$ which violates the boundary condition. The only way this situation can be avoided, is for the series solution in Eq. (9.36) to become a finite polynomial. This can be achieved by taking

$$a = \frac{1}{2}\left(\ell + \frac{3}{2}\right) - \frac{\varepsilon}{2} = -n \, , \tag{9.39}$$

where n is a positive integer. Under this condition the series solution becomes

$$L(\eta) = \sum_{p=0}^{\infty} \frac{\Gamma(p-n)}{\Gamma(p+b)} \frac{\Gamma(b)}{\Gamma(-n)} \frac{\eta^p}{p!} \, . \tag{9.40}$$

---

[2]The confluent hypergeometric function is defined as in Abramowitz and Stegun [31] and is available numerically in *Mathematics* [39].

But now we have that [3]

$$\frac{\Gamma(p-n)}{\Gamma(-n)} = \begin{cases} 0 & \text{for} \quad p > n \\ (-1)^p \dfrac{n!}{(n-p)!} & \text{for} \quad p \leq n \end{cases} . \tag{9.41}$$

This result, which turns $L(\eta)$ to a polynomial of degree n, is due to the fact that the gamma function, $\Gamma(n)$, has simple poles for values of n which are negative integers. We now can write $L(\eta)$ as

$$L(\eta) \;=\; {}_1F_1(-n;\ell+3/2;\eta) = \sum_{p=0}^{n} (-1)^p \frac{\Gamma\left(\ell+\frac{3}{2}\right)}{\Gamma\left(\ell+p+\frac{3}{2}\right)} \begin{pmatrix} n \\ p \end{pmatrix} \eta^p . \tag{9.42}$$

This polynomial solution of Eq. (9.33) is known as the *associated Laguerre polynomial.*[4] Combining the results of Eqs. (9.31) and (9.42), we can write the solution of the radial Schrödinger equation for the harmonic oscillator potential as

$$R_{n\ell}(r) = N_{n\ell}(\beta)\, r^{\ell}\, e^{-\beta^2 r^2/2} \sum_{p=0}^{n} (-1)^p \begin{pmatrix} n \\ p \end{pmatrix} \frac{\Gamma\left(\ell+\frac{3}{2}\right)}{\Gamma\left(\ell+p+\frac{3}{2}\right)} \left(\beta^2 r^2\right)^p , \tag{9.43}$$

with

$$\beta = \sqrt{\frac{m\omega}{\hbar}}$$

and the normalization constant $N_{n\ell}(\beta)$ is given by

$$N_{n\ell}^2(\beta) = \frac{2^{\ell-n+2}\,(2\ell+2n+1)!!\,\beta^{2\ell+3}}{\sqrt{\pi}\, n!\,[(2\ell+1)!!]^2} . \tag{9.44}$$

Here we observe that had we not taken the asymptotic behavior of $R_{n\ell}(r)$ for $r \to 0$ and $r \to \infty$ out, we would have not obtained a simple polynomial solution for the remaining wave function. In particular, a series solution for $R_{n\ell}(r)$ would have been an infinite series because it would have included a power series expansion for $e^{-\beta^2 r^2/2}$. Furthermore, it would have been very difficult to impose the boundary condition for $r \to \infty$.

Let us now examine the condition under which $L(\eta)$ became a polynomial, which was the condition to guarantee a finite solution for $r \to \infty$. From Eq. (9.39), we have

$$\varepsilon = 2n + \ell + \frac{3}{2}$$

---

[3]To prove this result, we make use of the reflection property of the Γ-function, i.e.,

$$\Gamma(z)\Gamma(-z) = -\frac{\pi}{z \sin \pi z} ,$$

and the fact that the residue of $\Gamma(z)$ at $z = -n$ is $(-1)^n/n!$ (see problem 1).

[4]In the next section we will derive the series solution for the confluent hypergeometric function and define the associated Laguerre polynomial.

Table 9.1: The lowest eigenvalues and corresponding radial wave function for the Harmonic Oscillator in three-dimension. Also included are spectroscopic names for these states.

n	ℓ	$E_{n\ell}$	$R_{n\ell}/N_{n\ell}$	Name
0	0	$\frac{3}{2}\hbar\omega$	$e^{-\beta^2 r^2/2}$	$0s$
0	1	$\frac{5}{2}\hbar\omega$	$r\, e^{-\beta^2 r^2/2}$	$0p$
1	0	$\frac{7}{2}\hbar\omega$	$\left(1 - \frac{2}{3}\beta^2 r^2\right)\, e^{-\beta^2 r^2/2}$	$1s$
0	2	$\frac{7}{2}\hbar\omega$	$r^2\, e^{-\beta^2 r^2/2}$	$0d$

or

$$E_{n\ell} = \hbar\omega\varepsilon = \hbar\omega\left(2n + \ell + \frac{3}{2}\right) . \tag{9.45}$$

Thus we have found that the condition that the wave function be finite as $r \to \infty$ quantizes the energy levels of the Harmonic Oscillator.

In the Table 9.1 we have the eigenvalues and eigenfunctions of the lowest energy eigenstates with their spectroscopic notation.

9.3 The Coulomb Potential

We started the last chapter by considering the simplest model for the hydrogen atom as a negative charge electron attracted to a positively charged proton by the Coulomb potential. We have already determined the angular part of the wave function to be the spherical harmonics. In this section we solve the radial equation for a hydrogen-like atom which consists of an electron in the Coulomb field of a nucleus of charge Ze, i.e.,

$$V_c(r) = -\frac{Ze^2}{r} . \tag{9.46}$$

In this case the radial equation for $u_\ell(r) = r\, R_\ell(r)$ is given by

$$\left\{\frac{d^2}{dr^2} - \frac{\ell(\ell+1)}{r^2} + \frac{2\mu Ze^2}{\hbar^2 r} + \frac{2\mu E}{\hbar^2}\right\} u_\ell(r) = 0 . \tag{9.47}$$

Since we want to consider bound states only, we have $E < 0$. To simplify our differential equation for $u_\ell(r)$, we define new parameters for the energy and strength of the potential as:

$$\kappa^2 = -\frac{2\mu E}{\hbar^2} \quad \text{and} \quad \beta = \frac{\mu Z e^2}{\hbar^2 \kappa} \ . \tag{9.48}$$

Our differential equation for $u_\ell(r)$ now reduces to

$$\left(\frac{d^2}{dr^2} - \frac{\ell(\ell+1)}{r^2} + \frac{2\beta\kappa}{r} - \kappa^2\right) u_\ell(r) = 0 \ . \tag{9.49}$$

Changing the variable in the differential equation to $\rho = 2\kappa r$, we get

$$\left(\frac{d^2}{d\rho^2} - \frac{\ell(\ell+1)}{\rho^2} + \frac{\beta}{\rho} - \frac{1}{4}\right) u_\ell(\rho) = 0 \ . \tag{9.50}$$

Before we attempt a general solution of this differential equation, let us follow the procedure implemented in the last section and examine the asymptotic solutions for $\rho \to 0$ and $\rho \to \infty$. For $\rho \to 0$, the dominant term is $-\ell(\ell+1)/\rho^2$ and the equation has the asymptotic form

$$\left\{\frac{d^2}{d\rho^2} - \frac{\ell(\ell+1)}{\rho^2}\right\} u_\ell(\rho) = 0 \ . \tag{9.51}$$

This equation has one of two possible solutions:

$$u_\ell \sim \rho^{\ell+1} \ , \quad \text{or} \quad u_\ell \sim \rho^{-\ell} \ .$$

The requirement that $u_\ell(\rho)$ be finite, and in particular zero at $\rho = 0$, demands that we only consider the solution

$$u_\ell(\rho) \sim \rho^{\ell+1} \ . \tag{9.52}$$

For the case of $\rho \to \infty$, our radial Schrödinger equation for $u_\ell(\rho)$ reduces to

$$\left(\frac{d^2}{d\rho^2} - \frac{1}{4}\right) u_\ell(\rho) = 0 \ . \tag{9.53}$$

Here again two solutions are possible,

$$u_\ell(\rho) \sim e^{\rho/2} \ , \quad \text{or} \quad u_\ell(\rho) \sim e^{-\rho/2} \ .$$

The boundary condition requirement that the wave function for a bound state be zero as $r \to \infty$, implies that the only solution that is acceptable is

$$u_\ell(\rho) \sim e^{-\rho/2} \ . \tag{9.54}$$

We now can build the two asymptotic solutions into $u(r)$ by taking

$$u_\ell(\rho) = \rho^{\ell+1} e^{-\rho/2} f_\ell(\rho) \ . \tag{9.55}$$

This procedure, we hope, will imply that $f_\ell(\rho)$ is a simple function like a polynomial. To find the differential equation that $f_\ell(\rho)$ satisfies, we make use of the Schrödinger Eq. (9.50) and the fact that

$$\frac{du_\ell}{d\rho} = (\ell+1)\rho^\ell e^{-\rho/2} f_\ell(\rho) - \frac{1}{2}\rho^{\ell+1} e^{-\rho/2} f_\ell(\rho) + \rho^{\ell+1} e^{-\rho/2}\frac{df_\ell}{d\rho} ,$$

with the second derivative of $u_\ell(\rho)$ given by

$$\begin{aligned} \frac{d^2u_\ell}{d\rho^2} &= \ell(\ell+1)\rho^{\ell-1}e^{-\rho/2}f_\ell(\rho) - (\ell+1)\rho^\ell e^{-\rho/2} f_\ell(\rho) + 2(\ell+1)\rho^\ell e^{-\rho/2}\frac{df_\ell}{d\rho} \\ &+ \frac{1}{4}\rho^{\ell+1}e^{-\rho/2}f_\ell(\rho) - \rho^{\ell+1}e^{-\rho/2}\frac{df_\ell}{d\rho} + \rho^{\ell+1}e^{-\rho/2}\frac{d^2 f_\ell}{d\rho^2} . \end{aligned}$$

This result for the second derivative we substitute in Eq. (9.50) to write a second order differential equation for the function $f_\ell(\rho)$ that is of the form

$$\rho\frac{d^2 f_\ell(\rho)}{d\rho^2} + [2(\ell+1) - \rho]\frac{df_\ell(\rho)}{d\rho} - [(\ell+1) - \beta] f_\ell(\rho) = 0 . \tag{9.56}$$

If we attempt a series solution for this equation,[5] i.e.,

$$f_\ell(\rho) = \sum_{s=0}^{\infty} a_s\, \rho^s , \tag{9.57}$$

then

$$\frac{df}{d\rho} = \sum_{s=0}^{\infty} s\, a_s\, \rho^{s-1} = \sum_{s=0}^{\infty} (s+1)\, a_{s+1}\, \rho^s ,$$

where the second expression results from the fact that the first term in the sum is zero and can be dropped in conjunction with a change in the sum index $s \to s+1$. In a similar manner we have

$$\frac{d^2 f_\ell}{d\rho^2} = \sum_{s=0}^{\infty} s\,(s-1)\, a_s\, \rho^{s-2} = \sum_{s=0}^{\infty} (s+2)(s+1)\, a_{s+2}\, \rho^s .$$

Using these results, our differential equation, Eq. (9.56), becomes

$$\begin{aligned} \sum_{s=0}^{\infty} \Big\{ (s+2)(s+1)a_{s+2}\rho^{s+1} &+ 2(\ell+1)(s+1)a_{s+1}\rho^s - (s+1)a_{s+1}\rho^{s+1} \\ &- [(\ell+1) - \beta]a_s\rho^s \Big\} = 0 . \end{aligned} \tag{9.58}$$

[5]This procedure is identical to that employed in the last chapter to derive the wave function for a one dimensional harmonic oscillator.

Since this sum is zero for any value of ρ, we expect the coefficients of ρ^p,$p = 0, 1, \cdots$ to be zero, i.e.,

$$\begin{aligned}
\rho^0 \quad &: \quad 2(\ell+1)a_1 = (\ell+1-\beta)a_0 \\
&\quad\ \text{therefore} \quad a_1 = \frac{\ell+1-\beta}{2(\ell+1)}a_0 \\
\rho^1 \quad &: \quad 2a_2 + 4(\ell+1)a_2 = a_1 + [(\ell+1)-\beta]a_1 \\
&\quad\ \text{therefore} \quad a_2 = \frac{\ell+2-\beta}{2[2(\ell+1)+1]}a_1 \\
\vdots \quad &\quad \vdots \\
\rho^s \quad &: \quad s(s+1)a_{s+1} + 2(\ell+1)(s+1)a_{s+1} = sa_s + [(\ell+1)-\beta]a_s
\end{aligned}$$

therefore

$$a_{s+1} = \frac{[s+\ell+1-\beta]}{(s+1)[2(\ell+1)+s]}a_s \ . \tag{9.59}$$

From the above recursion relation for the coefficients a_s, we have that

$$\frac{a_{s+1}}{a_s} \sim \frac{1}{s} \quad \text{for large} \quad s \ .$$

Here again, as for the case of the one dimensional harmonic oscillator discussed in the last chapter and the three dimensional harmonic oscillator considered in the last section, if we do not terminate the series, the solution for large ρ will not satisfy the boundary condition that the wave function should go to zero as $\rho \to \infty$. For the series solution with the coefficients given by Eq. (9.59) to terminate, we should for some $s > 0$ have

$$s + \ell + 1 - \beta = 0 \ . \tag{9.60}$$

This can be achieved by taking β to be a positive integer, i.e., $\beta = n \geq 1$. Making use of the definition of β as given in Eq. (9.48), we have

$$\beta = n = \frac{\mu Z e^2}{\hbar^2 \kappa} \ .$$

We can solve this equation for κ, i.e.,

$$\kappa = \frac{\mu Z e^2}{\hbar^2 n} = \frac{Z}{a\,n}$$

or

$$\kappa^2 = -\frac{2\mu E}{\hbar^2} = \frac{\mu^2 Z^2 e^4}{\hbar^4 n^2} = \frac{Z^2}{a^2 n^2} \ ,$$

where $a = \hbar^2/\mu e^2$ is the Bohr radius. Therefore the energy is given by

$$E = -\frac{\mu Z^2 e^4}{2\hbar^2 n^2} \ . \tag{9.61}$$

Defining the *fine structure constant*, α as

$$\alpha = \frac{e^2}{\hbar c} \approx \frac{1}{137} \ , \tag{9.62}$$

the energy is then given in terms of the fine structure constant as

$$E_n = -\frac{\mu c^2}{2}\frac{(Z\alpha)^2}{n^2} \equiv -\frac{R_\infty Z^2}{n^2} \ , \tag{9.63}$$

where the Rydberg, $R_\infty = \mu e^4/2\hbar^2 \approx 13.6$ eV. In Table 9.2 we list the values for the lowest states in this model of the hydrogen atoms. Also included are the spectroscopic name and wave function for each state. Here we observe that for a given n, all states with $\ell \leq n$ have the same energy. This is a result of taking the potential between the nucleus and the electron to be the Coulomb potential only, and is due to a symmetry of the Hamiltonian that is not obvious.[6] In addition, for each ℓ we can have $(2\ell + 1)$ different projections of the angular momentum, i.e., we have a $(2\ell + 1)$ degeneracy. This degeneracy is a result of the rotational symmetry of the Hamiltonian. Although these results are a good approximation for the hydrogen spectrum, it is by no means the full story. We need to include the magnetic interaction, relativistic effects, and the effect of polarization of the vacuum if we are to give a full description of the hydrogen spectrum. Some of these effects will be considered in Chapter 13.

We now return to the wave function for the Coulomb potential in terms of the series solution. For $\beta = n$, the coefficients in the series are related by the recursion relations

$$a_{s+1} = \frac{[s + \ell + 1 - n]}{(s+1)[2(\ell+1) + s]} a_s \ . \tag{9.64}$$

The series solution for $f(\rho)$ can now be written as

$$\begin{aligned} f(\rho) &= a_0 \left\{1 + \frac{a_1}{a_0}\rho + \frac{a_2}{a_0}\rho^2 + \frac{a_3}{a_0}\rho^3 + \cdots\right\} \\ &= a_0 \left\{1 + \frac{a}{b}\rho + \frac{a(a+1)}{b(b+1)}\frac{\rho^2}{2!} + \frac{a(a+1)(a+2)}{b(b+1)(b+2)}\frac{\rho^3}{3!} + \cdots\right\} \ . \end{aligned} \tag{9.65}$$

where $a = \ell + 1 - n$ and $b = 2\ell + 2$. The above infinite series is nothing more than the expansion of the *confluent hypergeometric function* ${}_1F_1$ given in Eq. (9.35) which is:

$${}_1F_1(a; b; \rho) = 1 + \frac{a}{b}\rho + \frac{a(a+1)}{b(b+1)}\frac{\rho^2}{2!} + \frac{a(a+1)(a+2)}{b(b+1)(b+2)}\frac{\rho^3}{3!} + \cdots \ . \tag{9.66}$$

---

[6]The Coulomb spectrum was first calculated by Pauli [33] on the basis that the Hamiltonian has $O(4)$ symmetry. For a detailed discussion of this dynamic symmetry and application to the energy spectrum of Hamiltonians see Schiff [17] Sec. 30.

Table 9.2: The energy and radial wave function for the hydrogen atom ($Z = 1$). The energies are given in terms of Rydbergs R_∞, with $R_\infty = 13.6$ eV. Here $\rho = 2r/na$, where a is the Bohr radius and is given by $a = \hbar^2/\mu e^2$.

name	n	ℓ	E_n	$R_{n\ell}(r)$
$1s$	1	0	R_∞	$2\left(\frac{1}{a}\right)^{3/2} e^{-\rho/2}$
$2s$	2	0	$\frac{1}{4}R_\infty$	$\left(\frac{1}{2a}\right)^{3/2}(2-\rho)\, e^{-\rho/2}$
$2p$	2	1	$\frac{1}{4}R_\infty$	$\frac{1}{\sqrt{3}}\left(\frac{1}{2a}\right)^{3/2} \rho\, e^{-\rho/2}$

For the case when $a = \ell + 1 - n$ is a negative integer, the infinite series becomes a polynomial known as the *associate Laguerre polynomial*,[7] which is defined in this case in terms of the confluent hypergeometric function ${}_1F_1$ as

$$\begin{aligned} L^{2\ell+1}_{n-\ell-1}(\rho) &\equiv \frac{(n+\ell)!}{(n-\ell-1)!\,(2\ell+1)!}\ {}_1F_1(-n+\ell+1, 2\ell+2;\rho) \\ &= \sum_{s=0}^{n-\ell-1} (-1)^s \frac{(n+\ell)!}{(2\ell+1+s)!\,(n-\ell-1-s)!}\,\frac{\eta^s}{s!}\ . \end{aligned} \tag{9.67}$$

Numerical values of this associate Laguerre polynomial can be evaluated with the help of *Mathematica* package [39].

The full solution of the Schrödinger equation for the Coulomb Hamiltonian can now be written as

$$R_{n\ell}(r) = N_{n\ell}\ \rho^\ell e^{-\rho/2}\, L^{2\ell+1}_{n-\ell-1}(\rho) \quad \text{with} \quad \rho = 2\kappa r = \frac{2Z}{an} r\ . \tag{9.68}$$

where $N_{n\ell}$ is the over all normalization of the radial wave function, a is the Bohr radius given by $a = \hbar^2/\mu e^2$, and Z the charge on the nucleus. The requirement that the radial wave function be normalized, i.e.,

$$\int_0^\infty dr\ r^2\ \left[R_{n\ell}(r)\right]^2 = 1$$

[7]This definition of associate Laguerre polynomial is the same as Abramowitz and Stegeun [31] and differs by a factor of $(n+\ell)!$ from that defined by Schiff [17] Sec. 16. Also, Schiff writes this associate Laguerre polynomial as $L^{2\ell+1}_{n+\ell}$.

results in $N_{n\ell}$ being given by [8]

$$N_{n\ell}^2 = \left[\left(\frac{2Z}{na} \right)^3 \frac{(n-\ell-1)!}{2n\,(n+\ell)!} \right] . \tag{9.69}$$

The radial solutions tabulated in Table 9.1 are plotted in Figure 9.1.

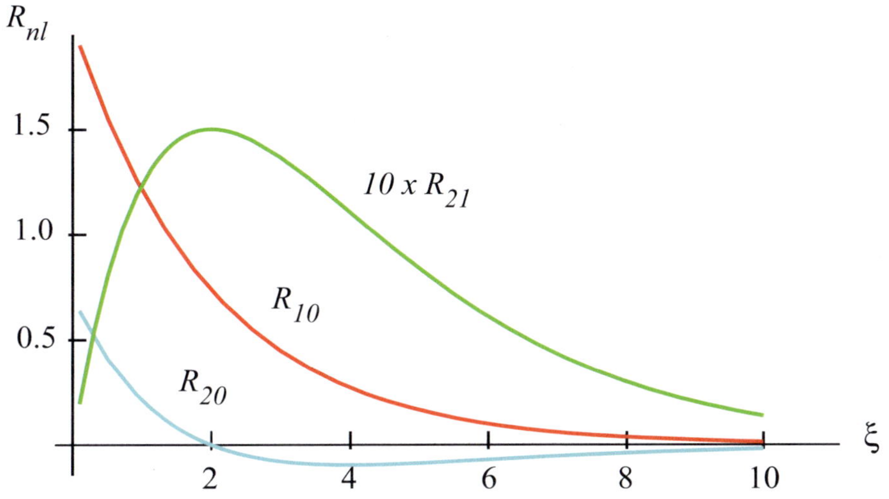

Figure 9.1: The radial wave function for the lowest three states of the Coulomb potential.

The wave functions given in Figure 9.1 cannot be measured in coordinate space, but the momentum space wave function has been measured [34]. To compare our results with experiment, we need to Fourier transform this coordinate space wave function. For the ground state, the momentum space wave function is given by

$$\phi_{100}(\vec{k}) = \frac{1}{(2\pi)^{3/2}} \int d^3r \; e^{i\vec{k}\cdot\vec{r}} \; \psi_{100}(\vec{r}) \; . \tag{9.70}$$

This integral can be performed in spherical polar coordinates. Making use of the fact that

$$\psi_{100}(\vec{r}) = \frac{1}{\sqrt{4\pi}} R_{10}(r)$$

where R_{10} is given in Table 9.1, we get

$$\phi_{100}(\vec{k}) \;=\; \frac{1}{(2\pi)^{3/2}} \frac{2}{a^{3/2}\sqrt{4\pi}} \int d^3r \; e^{i\vec{k}\cdot\vec{r}} \; e^{-r/a}$$

[8]In determining the normalization of the wave function, we have made use of the fact that

$$\int\limits_0^\infty d\rho\, \rho^{2\ell+2}\, e^{-\rho} \left[L_{n-\ell-1}^{2\ell+1}(\rho) \right]^2 = \frac{2n\,(n+\ell)!}{(n-\ell-1)!} \; .$$

$$= \frac{1}{(2\pi)^{3/2}} \frac{2}{\sqrt{4\pi}a^{3/2}} \int_0^\infty drr^2\, e^{-r/a}\, 2\pi \int_{-1}^{+1} dx\, e^{ikrx}\ .$$

In writing the second line we have taken $d^3r = r^2 dr\, d\phi\, dx$, where $x = \cos\theta$ with θ the angle between $\vec{k}$ and $\vec{r}$. The ϕ integration gives a factor of 2π. Performing the x integration we get

$$\phi_{100}(\vec{k}) = \frac{1}{(2\pi)^{3/2}}, \frac{\sqrt{4\pi}}{a^{3/2}} \frac{1}{ik} \int_0^\infty drr\, e^{-r/a}[e^{ikr} - e^{-ikr}]\ .$$

To perform the r integration we define $\xi = 1/a$. This change of variable allows us to write our integral as

$$\begin{aligned}\phi_{100}(\vec{k}) &= \frac{1}{(2\pi)^{3/2}} \frac{2\sqrt{4\pi}}{a^{3/2}} \frac{1}{k} \int_0^\infty drr\, e^{-\xi r} \sin kr \\ &= -\frac{1}{(2\pi)^{3/2}} \frac{2\sqrt{4\pi}}{a^{3/2}} \frac{1}{k} \frac{\partial}{\partial\xi} \int_0^\infty dr\, e^{-\xi r} \sin kr\ .\end{aligned}$$

But we have that

$$\int_0^\infty dr\, e^{-\xi r}\ \sin kr = \frac{k}{\xi^2 + k^2}\ .$$

We can now write the wave function for the ground state of the Coulomb Hamiltonian as

$$\phi_{100}(\vec{k}) = \frac{2\sqrt{2}}{\pi} \frac{a^{3/2}}{(1 + k^2a^2)^2}\ . \tag{9.71}$$

The probability of finding the electron with momentum $\vec{p} = \hbar\vec{k}$ is given by the square of the magnitude of the momentum space wave function, i.e.,

$$|\phi_{100}(\vec{k})|^2 = \frac{8}{\pi^2} \frac{a^3}{(1 + k^2a^2)^4}\ . \tag{9.72}$$

This quantity has been measured in an $(e, 2e)$ experiment on hydrogen. In Figure 9.2 we compare our results with the experimental results, and the agreement could not be better. This illustrates the fact that a direct measurement of the wave function is possible.

9.4 Problems

1. Making use of the fact that the Γ-function can be written as

$$\Gamma(z) = \frac{1}{z} \prod_{l=1}^{\infty} \left(1 + \frac{1}{l}\right)^z \left(1 + \frac{z}{l}\right)^{-1}$$

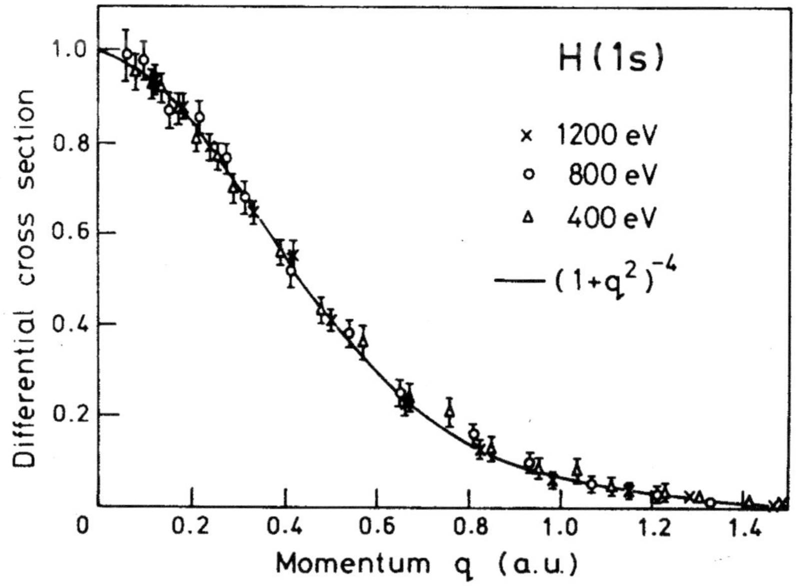

Figure 9.2: Comparison of the momentum wave function for the ground state of Hydrogen with the results of the $(e, 2e)$ experiment on hydrogen by B. Lohmann and E. Weigold [34].

show that the residue of the Γ-function for negative integers, n, is $(-1)^n/n!$, i.e.,

$$\lim_{z \to -n} (z+n)\Gamma(z) = \frac{(-1)^n}{n!} \quad \text{for } n \text{ positive integer} .$$

2. Obtain an approximate analytic expression for the energy levels in a square well potential with depth V_0 and radius a, when $V_0 a^2$ is slightly greater than $\pi^2\hbar^2/8m$. Take the orbital angular momentum to be zero.

3. If the ground state of a particle in a square well is just barely bound, show that the well depth V_0 and radius a are related to the binding energy by the expansion

$$\frac{2\mu V_0 a^2}{\hbar^2} = \frac{\pi^2}{4} + 2\eta + \left(1 + \frac{4}{\pi^2}\right)\eta^2 + \cdots$$

where

$$\eta = \frac{\sqrt{-2\mu E}a}{\hbar} .$$

Here, E is the energy of the ground state and μ is the mass of the particle.

4. Solve the radial Schrödinger equation for $\ell = 0$ and

$$V(r) = -V_0 \, e^{-r/a} .$$

Change variables from r to

$$z = e^{-r/2a}$$

and show that Bessel's equation results. What boundary conditions are to be imposed on the solution $\mathcal{U}_0(r)$ as a function of z? How can these be used to determine the energy levels in this potential?

5. Solve the radial Schrödinger equation for the potential

$$V(r) = \frac{A}{r^2} + Br^2 \ .$$

Consider only the discrete spectrum.

Hint: Compare with the three-dimensional harmonic oscillator in spherical coordinates.

6. To introduce electromagnetic interaction into quantum systems, we make use of the minimal coupling principle which involves the substitution

$$\vec{p} \to \vec{p} - \frac{e}{c}\vec{A} \ ,$$

where $\vec{A}$ is the vector potential and c the velocity of light. This procedure gives, for an electron in an external vector potential $\vec{A}(\vec{r})$, the Schrödinger equation

$$\frac{1}{2m}\left(\frac{\hbar}{i}\nabla - \frac{e}{c}\vec{A}(\vec{r})\right)\cdot\left(\frac{\hbar}{i}\nabla - \frac{e}{c}\vec{A}(\vec{r})\right)\psi(\vec{r}) = E\psi(\vec{r}) \ .$$

Simplify the equation for the case where $\vec{A}(\vec{r})$ describes a uniform magnetic field pointing in the z-direction.

(a) Show that the Hamiltonian can be written as

$$H = -\frac{\hbar^2}{2m}\nabla^2 - \frac{e}{2mc}B_0 L_z + \frac{e^2 B_0^2}{8mc^2}\left(x^2 + y^2\right) \ .$$

(b) Show that in cylindrical coordinates this Hamiltonian is the sum of two Hamiltonians with one corresponding to a two-dimensional harmonic oscillator.

(c) What is the energy spectrum for this Hamiltonian?

CHAPTER 10

Scattering by a Central Potential

Abstract: With the central role played by scattering in any measurement on a quantum system, in this chapter we introduce the basic formulation of scattering theory for a central potential. As a simple application we consider scattering by a square-well, as well as the scattering by a Coulomb potential.

Most scattering experiments consist of a beam of particles incident on a stationary target, with a detector to measure the number of scattered particles in a given direction per unit time. In practice, the beam has many particles in it, but we will neglect any interaction between the particles. Similarly, the target has many particles in a confined region of space and again we will assume the interaction between the target particles can be ignored. Although we should be considering a wave packet incident on the target, it can be shown[1] that we can get the same result for the scattering amplitude assuming the incident beam is described by a plane wave. In this way we reduce the complexity of the algebra in the formulation of the scattering problem. The plane incident wave with momentum $\vec{k}_i$ is then given, up to a normalization, by

$$\psi_{inc}(\vec{r}) \propto e^{i\vec{k}_i\cdot\vec{r}} . \tag{10.1}$$

On the other hand, the scattered particles will be originating at the target and moving in a spherical wave. The amplitude of this scattered wave might depend on the scattering angles. However, because of the cylindrical symmetry about the incident beam, the only angle dependence is the angle between the incident beam direction, $\hat{k}_i$, and the scattered particle's direction of momentum $\hat{k}_f$. Here the subscripts i and f refer to the initial and final states. We now can write the scattered wave in terms of a spherical wave, e^{ikr}/r, as

$$\psi_{scat} \propto f(k_i, \hat{k}_i\cdot\hat{k}_f)\ \frac{e^{ik_f r}}{r} . \tag{10.2}$$

[1]For a full treatment of scattering using wave packets consult Goldberger and Watson [35].

Iraj R. Afnan

For elastic scattering, i.e., when the initial energy of the incident particle is equal to the final energy of the scattered particle, we have that $|\vec{k}_i| = |\vec{k}_f| = k$,[2] and we can write the time independent wave function outside the interaction region as

$$\psi(\vec{r}) \propto e^{i\vec{k}_i\cdot\vec{r}} + f(k,\theta)\,\frac{e^{ikr}}{r}\ , \tag{10.3}$$

where $\cos\theta = \hat{k}_i \cdot \hat{k}_f$ and $f(k,\theta)$ is the amplitude of the scattered wave or the *scattering amplitude.* This wave function is illustrated in Figure 10.1.

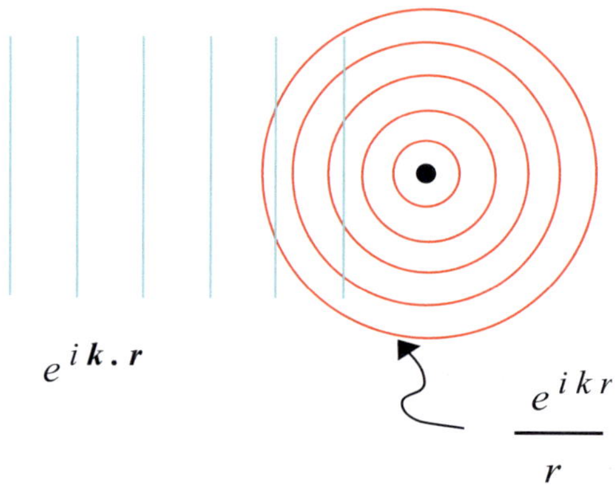

Figure 10.1: Illustration of the asymptotic wave function for a scattering problem. It consists of an incident plane wave and an outgoing spherical wave to represent the scattered particles.

In the next two sections we will relate the probability for scattering in a given direction to the coefficient of the outgoing spherical wave $f(k,\theta)$. We will then relate the amplitude $f(k,\theta)$ to the solution of the Schrödinger equation. In this way we establish the relation between the wave function and the quantities we measure in any scattering experiment. In the final section of this chapter we will consider the problem of scattering by a Coulomb potential, which will require special treatment because the interaction in this case has an infinite range.

10.1 The Cross Section

In any scattering experiment, we measure the number of particles scattered at a given angle θ. If we divide the number of scattered particles per unit time by the flux of incident particles, we get the probability for a particle scattering at a given angle. This is referred

[2]This definition of elastic scattering assumes we are in the center of mass where $\vec{k}_i$ and $\vec{k}_f$ are the relative initial and final momenta (see Sec. 10.2).

to as the differential cross section for that angle. In this section we derive the relation between the cross section and the scattering amplitude $f(k,\theta)$.

The time dependent Schrödinger equation for the potential $V(r)$ is given by

$$i\hbar\frac{\partial\Psi(\vec{r},t)}{\partial t} = \left(-\frac{\hbar^2}{2\mu}\nabla^2 + V(r)\right)\Psi(\vec{r},t)\ , \tag{10.4}$$

where μ is the reduced mass (see Eq. (8.10)) of the incident and target particles. The complex conjugate solution, $\Psi^*(\vec{r},t)$ satisfies the equation

$$-\,i\hbar\frac{\partial\Psi^*(\vec{r},t)}{\partial t} = \left(-\frac{\hbar^2}{2\mu}\nabla^2 + V(r)\right)\Psi^*(\vec{r},t)\ . \tag{10.5}$$

Multiplying Eq. (10.4) by Ψ^* from the left, and Eq. (10.5) by Ψ from the right and subtracting, we get

$$\begin{aligned} i\hbar\,\frac{\partial}{\partial t}\,(\Psi^*\Psi) &= -\frac{\hbar^2}{2\mu}\,\Big(\Psi^*\nabla^2\Psi - (\nabla^2\Psi^*)\Psi\Big) \\ &= -\nabla\cdot\frac{\hbar^2}{2\mu}\,(\,\Psi^*\nabla\Psi - (\nabla\Psi^*)\Psi\,)\ . \end{aligned} \tag{10.6}$$

If we now define the density, $\rho = \Psi^*\Psi$, then Eq. (10.6) can be written as

$$\frac{\partial\rho}{\partial t} + \nabla\cdot\vec{j} = 0\ , \tag{10.7}$$

where the current $\vec{j}$ is given by

$$\begin{aligned} \vec{j} &= \frac{\hbar}{2\mu i}\,(\Psi^*\nabla\Psi - (\nabla\Psi^*)\Psi) \\ &= \frac{\hbar}{2\mu i}\,(\psi^*\nabla\psi - (\nabla\psi^*)\psi)\ . \end{aligned} \tag{10.8}$$

Here we have dropped the time dependence in the wave function since we are not dealing with wave packets. In any scattering experiment the scattered particle is measured at a distance r which is large compared to the range of the interaction. This means that the detector is in the asymptotic region where the wave function is given by

$$\psi(\vec{r}) = e^{i\vec{k}_i\cdot\vec{r}} + f(k,\theta)\,\frac{e^{ikr}}{r}\ . \tag{10.9}$$

Using Eq. (10.9) in Eq, (10.8), we get, after some algebra that involves writing ∇ in spherical polar coordinates,[3] an expression of the form

$$\begin{aligned}\vec{j} &= \frac{\hbar\vec{k}_i}{\mu} + \frac{\hbar k}{\mu}\frac{\hat{e}_r}{r^2}|f(k,\theta)|^2 \\ &\quad + e^{ikr(1-\cos\theta)}[\cdots] + e^{-ikr(1-\cos\theta)}[\cdots] \ . \end{aligned} \tag{10.10}$$

Since the detector is never put in the forward direction, i.e., $\theta = 0$, and it subtends a finite solid angle, we need to integrate over the angular range of the detector, i.e.,

$$\int d\theta\, d\phi\, \sin\theta\, g(\theta,\phi)\, e^{\pm ikr(1-\cos\theta)} \ ,$$

where $g(\theta,\phi)$ is a smooth function that depends on the detector's properties. For $kr \gg 1$, we are integrating a highly oscillatory function, and the integral is zero according to the Riemann-Labegue lemma. We thus have for the current

$$\vec{j} = \frac{\hbar\vec{k}_i}{\mu} + \frac{\hbar\, k}{\mu}\frac{\hat{e}_r}{r^2}|f(k,\theta)|^2 \ . \tag{10.11}$$

The first term represents the incident beam of particles, and is present even if there was no scattering target. The radial flux of scattered particles is then given by

$$\vec{j}\cdot\hat{e}_r = \frac{\hbar k}{\mu}\frac{|f(k,\theta)|}{r^2} \ . \tag{10.12}$$

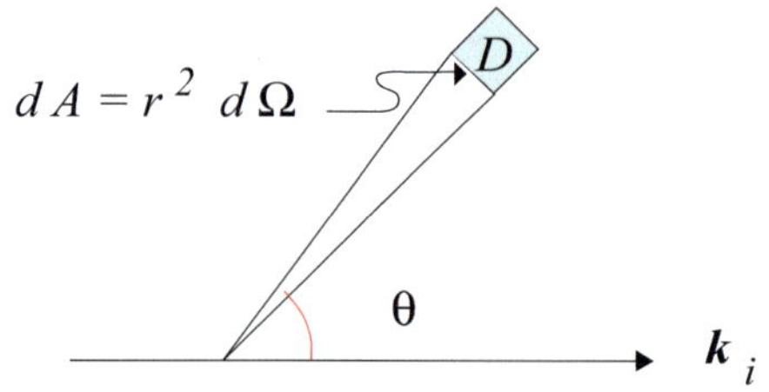

Figure 10.2: Illustration of the scattering angle θ, and the solid angle $d\Omega$, subtended by the detector D.

The number of particles crossing the area that subtends a solid angle $d\Omega$ is given by (see Figure 10.2)

$$\vec{j}\cdot\hat{e}_r\, dA = \frac{\hbar k}{\mu}|f(k,\theta)|^2\, d\Omega \ . \tag{10.13}$$

---

[3]In spherical polar coordinates we have

$$\nabla = \hat{e}_r\frac{\partial}{\partial r} + \hat{e}_\theta\frac{1}{r}\frac{\partial}{\partial\theta} + \hat{e}_\phi\frac{1}{r\sin\theta}\frac{\partial}{\partial\phi} \ .$$

The differential cross section is the flux of scattered particles per unit area divided by the incident flux, $\hbar k/\mu$. Therefore we have that

$$d\sigma = |f(k,\theta)|^2 \, d\Omega$$

and we have

$$\frac{d\sigma}{d\Omega} = |f(k,\theta)|^2 \; . \tag{10.14}$$

This in fact is the result we would have expected from the construction of Eq. (10.3) for the asymptotic wave function. We now have to relate the scattering amplitude $f(k,\theta)$ to the solution of the Schrödinger equation. One important observation we can make at this stage is that the cross section measured experimentally depends on the form of the wave function for large r, i.e., outside the interaction region.

10.2 Kinematics

Consider the scattering of two particles where the potential between the particles is a function of the relative distance between the particles, i.e., $V(|\vec{r}_1 - \vec{r}_2|)$. The Hamiltonian for such a system is given by

$$H = \frac{p_1^2}{2m_1} + \frac{p_2^2}{2m_2} + V(|\vec{r}_1 - \vec{r}_2|) \; , \tag{10.15}$$

where $p_1^2 = -\hbar^2\nabla_1^2$ and $p_2^2 = -\hbar^2\nabla_2^2$. We introduce relative and center of mass coordinates and momenta (see Eqs. (8.6) and (8.7)), i.e.,

$$\begin{aligned} \vec{r} &= \vec{r}_1 - \vec{r}_2 \qquad \text{and} \qquad \vec{R} = \frac{m_1\vec{r}_1 + m_2\vec{r}_2}{m_1 + m_2} \\ \vec{p} &= \frac{m_2\vec{p}_1 - m_1\vec{p}_2}{m_1 + m_2} \qquad \text{and} \qquad \vec{P} = \vec{p}_1 + \vec{p}_2 \; . \end{aligned} \tag{10.16}$$

In terms of these new variables, the Hamiltonian takes the form

$$H = \frac{P^2}{2M} + \frac{p^2}{2\mu} + V(r) \; , \tag{10.17}$$

where

$$M = m_1 + m_2 \qquad \text{and} \qquad \mu = \frac{m_1 m_2}{m_1 + m_2} \; . \tag{10.18}$$

Here μ is the reduced mass and M is the total mass. In the center of mass we have $\vec{P} = 0$, and the two-particle Hamiltonian reduces to the one-particle Hamiltonian with a reduced mass μ, i.e.,

$$H = \frac{p^2}{2\mu} + V(r) \; . \tag{10.19}$$

We thus have reduced the two-particle problem to a one particle problem of mass μ scattering from a potential $V(r)$. The Schrödinger equation in the two-particle center of mass is

$$\left(-\frac{\hbar^2}{2\mu}\nabla^2 + V(r)\right)\psi(\vec{r}) = E\psi(\vec{r}) \,. \tag{10.20}$$

10.3 The Square Well Potential

Before we proceed to a general discussion of the solution of Eq. (10.20), let us consider the solution of Eq. (10.20) for the case of a simple square well of radius a, i.e.,

$$V(r) = \begin{cases} -V_0 & \text{for} \quad r < a \\ 0 & \text{for} \quad r > a \end{cases} \,. \tag{10.21}$$

In the last chapter, we considered the bound state (i.e., $E < 0$) problem for this potential. We now have to consider the case of $E > 0$, i.e., the scattering problem. For $r < a$, the solution of the radial equation is the same as for the bound state (see Eq. (9.19)), which is

$$R_\ell(r) = A\, j_\ell(\alpha r) \quad \text{with} \quad \alpha^2 = \frac{2\mu}{\hbar^2}(E + V_0) \qquad \text{for} \qquad r < a \,. \tag{10.22}$$

For $r > a$, we have a linear combination of the two solutions of the radial equation, i.e.,

$$R_\ell(r) = B\, j_\ell(kr) + C\, n_\ell(kr) \,, \tag{10.23}$$

or

$$R_\ell(r) = B'\, h_\ell^{(+)}(kr) + C'\, h_\ell^{(-)}(kr) \,, \tag{10.24}$$

where $k^2 = 2\mu E/\hbar^2$. Since we are not considering bound states, the wave function $R_\ell(r)$ does not go to zero as $r \to \infty$, otherwise, the current at the detector would be zero. This means neither B nor C (B' or C') is zero. In this case either combination is valid and we will use both at different times.

To get a feeling for the form of the scattering wave function, let us take $\ell = 0$. In this case we have that

$$R_0(r) = \frac{u_0(r)}{r} \,, \tag{10.25}$$

with

$$u_0(r) = \begin{cases} A \sin \alpha r & \text{for} \quad r < a \\ B \sin kr + C \cos kr & \text{for} \quad r > a \end{cases} \,. \tag{10.26}$$

An alternative way of writing the solution for $r > a$ is

$$u_0(r) = A' \sin(kr + \delta) \qquad \text{for} \qquad r > a \,. \tag{10.27}$$

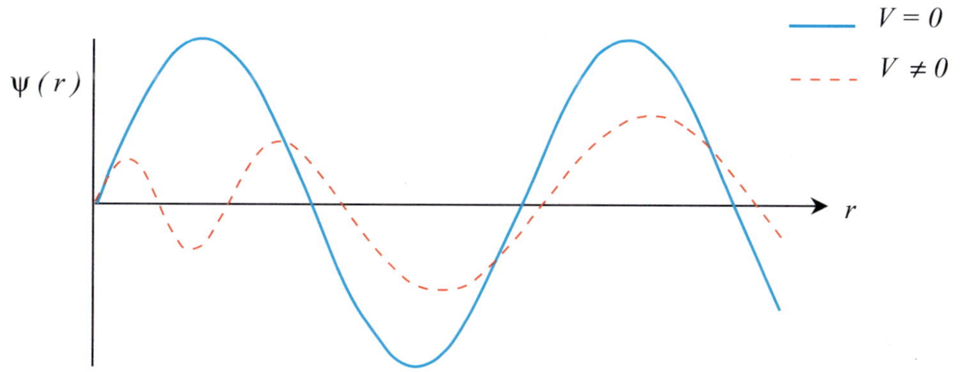

Figure 10.3: A comparison of the radial scattering wave function in the presence and absence of a potential V.

Note that both forms for the solution for $r > a$ have two constants to be adjusted by the boundary condition which requires that the wave function and its derivative be continuous at $r = a$. Before we determine these constants, let us examine the form of the wave function for the case of $V_0 = 0$, and $V_0 \neq 0$ at large distances, i.e., for $r > a$. In Figure 10.3, we sketch both wave functions. We observe that for $V_0 \neq 0$ the wave length for $r < a$ is smaller than is the case for $r > a$. However, for $V_0 = 0$, the wave length is the same for all r. Thus the presence of the potential, shifts the wave function for $r > a$ relative to the wave function for the case $V_0 = 0$ by an amount δ. In other words, the constant δ introduced in Eq. (10.27) depends on the parameters of the potential, in this case V_0 and a. The inverse might also be possible, i.e., if we measure δ we might be able to determine the parameters of the potential. The δ introduced in Eq. (10.27) is called the scattering *phase shift.*

To guarantee that the wave function and its derivative are continuous at $r = a$, we take

$$\begin{aligned} A \sin \alpha a &= A' \sin(ka + \delta) \\ \alpha A \cos \alpha a &= kA' \cos(ka + \delta) \ . \end{aligned}$$

Therefore, to determine δ, we take the ratio of the above two equations. This gives us the result that

$$\begin{aligned} \alpha \cot \alpha a &= k \cot(ka + \delta) \\ &= k \frac{\cot ka \cot \delta - 1}{\cot \delta + \cot ka} \ . \end{aligned} \tag{10.28}$$

Solving this equation for $\cot \delta$ we get

$$k \cot \delta = \frac{\alpha \cot \alpha a + k \tan ka}{1 - \frac{\alpha}{k} \cot \alpha a \tan ka} \ . \tag{10.29}$$

Having determined the phase shift δ we can determine one of the other two constants (A or A'). This leaves one overall multiplicative constant to be determined. For bound states, this constant was determined by the requirement that the wave function be normalized. However, the scattering wave function is not normalizable in the same manner as the bound state, because the normalization integral is mathematically not well defined.[4] This is a consequence of the fact that the wave function does not go to zero as $r \to \infty$.

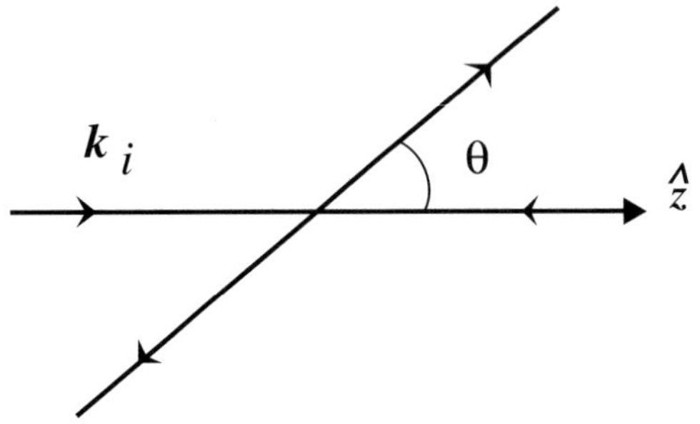

Figure 10.4: The scattering plane in the two-body center of mass. The incident beam is along the positive z-axis.

10.4 The Scattering Amplitude

Having considered the simple problem of S-wave (i.e., $\ell = 0$) scattering by a square well, we now turn to the more general problem of two-particle scattering. In the two-body center of mass, we have two particles with opposite momenta, initially along the z-axis (see Figure 10.4). Since the scattering takes place in a plane, we can eliminate the ϕ dependence in this problem. Furthermore, the initial momentum defines a direction in space, thus braking the isotropy of the space. This implies that the angular momentum of the system is not fixed as was the case for the bound state problem, but depends on the momentum in the initial state. In fact, the angular momentum, classically, is perpendicular to the scattering plane. In general, the wave function is a linear combination of many such angular momenta. As we have chosen our z-axis to be along the direction of the incident momentum and therefore in the scattering plane, we expect the projection of our angular momentum along the z-axis to be zero. This means that the wave function for a given angular momentum is of the form

$$R_\ell(r)\, Y_{\ell 0}(\theta, \phi) \ . \tag{10.30}$$

But

$$Y_{\ell 0}(\theta, \phi) = \sqrt{\frac{2\ell + 1}{4\pi}}\, P_\ell(\cos\theta) \ , \tag{10.31}$$

[4]We will show in a later chapter that the scattering wave function has a δ-function normalization.

where $\cos\theta = \hat{r}\cdot\hat{k}_i$. We now can write the general form of the wave function as

$$\psi(\vec{r}) = \frac{1}{(2\pi)^{3/2}} \sum_{\ell} i^{\ell}(2\ell+1)\,\psi_\ell(r)\,P_\ell(\cos\theta)\ . \tag{10.32}$$

The choice of the factor of $(2\pi)^{-3/2}\, i^{\ell}\,(2\ell+1)$ is for later convenience in the normalization. At this stage we would like to point out that this choice for the normalization is not unique.

Since $P_\ell(\cos\theta)$ is related to the spherical harmonics, it is an eigenstate of the angular momentum operator square, L^2. Making use of the orthogonality of the P_ℓ, we can show that the radial Schrödinger equation for $\psi_\ell(r)$ is given by

$$\frac{d^2\psi_\ell(r)}{dr^2} + \frac{2}{r}\frac{d\psi_\ell(r)}{dr} - \frac{\ell(\ell+1)}{r^2}\psi_\ell(r) + [k^2 - U(r)]\psi_\ell(r) = 0\ , \tag{10.33}$$

where

$$k^2 = \frac{2\mu E}{\hbar^2} \qquad \text{and} \qquad U(r) = \frac{2\mu}{\hbar^2}V(r)\ . \tag{10.34}$$

We now assume that the potential satisfies the condition that; $U(r) \to 0$ faster than r^{-1}. In this case the radial equation for large r is given by

$$\frac{d^2\psi_\ell(r)}{dr^2} + \frac{2}{r}\frac{d\psi_\ell(r)}{dr} + \left(k^2 - \frac{\ell(\ell+1)}{r^2}\right)\psi_\ell(r) = 0 \tag{10.35}$$

which is the spherical Bessel's equation (see Eq. (9.15)).

> **Note**: Our assumption that $U(r) \to 0$ faster than r^{-1} excludes the Coulomb potential. From this point on we restrict ourselves to the class of potentials that satisfy the above condition. We will examine the Coulomb potential as a special case at the end of this chapter.

Before we write the general solution to Eq. (10.33), let us rewrite the solution to the square well potential in terms of the spherical Bessel function. We have from Eq. (10.25) and (10.27) that, for $\ell = 0$,

$$\begin{aligned} \psi_0(r) &= \frac{A'}{r}\sin(kr+\delta) = \frac{A'}{r}\left[\sin kr\,\cos\delta + \cos kr\,\sin\delta\right] \\ &= kA'\left[\cos\delta\, j_0(kr) + \sin\delta\, n_0(kr)\right]\ . \end{aligned} \tag{10.36}$$

This gives us the idea of writing the general solution to Eq. (10.35), for any angular momentum, ℓ, as

$$\psi_\ell(r) \to A_\ell\left[\cos\delta_\ell\, j_\ell(kr) + \sin\delta_\ell\, n_\ell(kr)\right] \quad \text{for} \quad r\to\infty\ . \tag{10.37}$$

Since for $r \to \infty$ we have that (see Eq. (9.12))

$$\begin{aligned} j_\ell(\rho) &\longrightarrow \frac{1}{\rho}\sin\left(\rho - \frac{\pi\ell}{2}\right) \\ n_\ell(\rho) &\longrightarrow \frac{1}{\rho}\cos\left(\rho - \frac{\pi\ell}{2}\right) , \end{aligned} \tag{10.38}$$

then the two terms on the right hand side of Eq. (10.37) can be combined to give an asymptotic solution of the form

$$\psi_\ell(r) \to \frac{A_\ell}{kr}\sin\left(kr - \frac{\pi\ell}{2} + \delta_\ell\right) \quad \text{for} \quad r \to \infty . \tag{10.39}$$

Here A_ℓ and δ_ℓ are the two constants to be determined by the normalization of the wave function and the continuity of the logarithmic derivative.[5] We now would like to make use of Eqs. (10.39) and (10.32) to write the total asymptotic wave function in the form given by Eq. (10.3). In this way we hope to relate the scattering amplitude $f(k,\theta)$ to the phase shifts δ_ℓ. To achieve this result we first write $\sin(kr - \pi\ell/2 + \delta_\ell)$ in terms of exponentials as

$$\begin{aligned} \psi_\ell(r) &\longrightarrow \frac{A_\ell}{kr}\frac{1}{2i}\left[e^{i\left(kr-\frac{\pi\ell}{2}+\delta_\ell\right)} - e^{-i\left(kr-\frac{\pi\ell}{2}+\delta_\ell\right)}\right] \\ &= \frac{A_\ell}{2ikr}e^{-i\delta_\ell}\left[-e^{-i\left(kr-\frac{\pi\ell}{2}\right)} + e^{2i\delta_\ell}\, e^{i\left(kr-\frac{\pi\ell}{2}\right)}\right] \\ &= \frac{A_\ell}{2ikr}e^{-i\delta_\ell}\left[2i\sin\left(kr - \frac{\pi\ell}{2}\right) + \left(e^{2i\delta_\ell} - 1\right)\, e^{i\left(kr-\frac{\pi\ell}{2}\right)}\right] . \end{aligned} \tag{10.40}$$

Second, using Eq. (10.38) to relate the first term on the right hand side of Eq. (10.40) to the spherical Bessel function, we can now write the asymptotic radial wave function for a given ℓ as

$$\psi_\ell(r) \to A_\ell e^{-i\delta_\ell}\left(j_\ell(kr) + (-i)^\ell\frac{1}{2ik}\left(e^{2i\delta_\ell} - 1\right)\frac{e^{ikr}}{r}\right) . \tag{10.41}$$

Using this result in Eq. (10.32), and the fact that $A_\ell = e^{i\delta_\ell}$, we can write the total scattering wave function for $r \to \infty$ in the form of Eq. (10.3), i.e.,

$$\psi(\vec{r}) \to \frac{1}{(2\pi)^{3/2}}\left(e^{i\vec{k}\cdot\vec{r}} + f(k,\theta)\,\frac{e^{ikr}}{r}\right) , \tag{10.42}$$

[5] The logarithmic derivative is given by

$$\frac{d}{dr}\log\psi_\ell(r) = \frac{1}{\psi_\ell(r)}\frac{d\psi_\ell(r)}{dr}$$

which is identical to the condition given in Eq. (10.28) for the square well with $\ell = 0$.

where

$$e^{i\vec{k}\cdot\vec{r}} = \sum_\ell i^\ell \, (2\ell+1) \, j_\ell(kr) \, P_\ell(\cos\theta) \; , \tag{10.43}$$

and

$$f(k,\theta) = \frac{1}{k} \sum_\ell (2\ell+1) \left(\frac{e^{2i\delta_\ell} - 1}{2i} \right) P_\ell(\cos\theta) \; , \tag{10.44}$$

where $\cos\theta = \hat{k}\cdot\hat{r}$. Now since the direction of the final momentum $\vec{k}_f$ is the radial direction, i.e., $\hat{r} = \hat{k}_f$, we have $\cos\theta = \hat{k}_i \cdot \hat{k}_f$, with $\vec{k}_i$ being the initial momentum of the particles in the beam.

In Eq. (10.44), we have established the relation between the scattering amplitude and the phase shifts. In this way we have completed the relation between the experimentally measured cross section and the wave function which is a solution to the Schrödinger equation. Finally, we can write Eq. (10.44) as

$$f(k,\theta) = \frac{1}{k} \sum_\ell (2\ell+1) \, f_\ell(k) \, P_\ell(\cos\theta) \; , \tag{10.45}$$

where $f_\ell(k)$, the *partial wave amplitude*, is given by

$$f_\ell(k) = \frac{1}{2i} \left(e^{2i\delta_\ell} - 1 \right) = e^{i\delta_\ell} \sin\delta_\ell \; . \tag{10.46}$$

In general, this partial wave amplitude is a complex number, i.e., it has a magnitude and a phase. In Eq. (10.45) we have a *partial wave expansion* for the scattering amplitude.[6] With the help of Eq. (10.14) we can write the differential cross section in terms of the partial wave amplitude as

$$\frac{d\sigma}{d\Omega} = \frac{1}{k^2} \left| \sum_\ell (2\ell+1) \, f_\ell(k) \, P_\ell(\cos\theta) \right|^2 . \tag{10.47}$$

This gives the probability for an incident particle, with momentum k and energy $E = \frac{k^2}{2\mu}$, to be scattered in the direction defined by the angle θ. To get the total cross section, i.e., the probability of scattering in any direction, we have to integrate the differential cross section over the 4π solid angle, i.e.,

$$\begin{aligned} \sigma_T &= \int \left(\frac{d\sigma}{d\Omega} \right) d\Omega \\ &= \frac{1}{k^2} \sum_\ell \sum_{\ell'} (2\ell+1)(2\ell'+1) \, f_\ell(k) f^*_{\ell'}(k) \int d\Omega \, P_\ell(\cos\theta) P_{\ell'}(\cos\theta) \; . \end{aligned}$$

[6] In Chapter 14 we will detail a more general derivation of this result.

Using the orthogonality of the Legendre polynomials (i.e., Eq. (8.60)), we get

$$\sigma_T = \frac{4\pi}{k^2} \sum_\ell (2\ell+1)\,|f_\ell(k)|^2 = \frac{4\pi}{k^2} \sum_\ell (2\ell+1)\,\sin^2\delta_\ell\ . \tag{10.48}$$

Comparing Eqs. (10.47) and (10.48), we observe that the differential cross section has more information about the scattering amplitude than the total cross section. To illustrate this, consider the case when only two partial waves are important, the $\ell = 0$ and 1. In this case we have for the differential cross section

$$\frac{d\sigma}{d\Omega} = \frac{1}{k^2}\,\Big(|f_0|^2 + 9|f_1|^2\cos^2\theta + 3(f_0 f_1^* + f_0^* f_1)\cos\theta\Big)\ , \tag{10.49}$$

while for the total cross section we have

$$\sigma_T = \frac{4\pi}{k^2}\,\Big(|f_0|^2 + 3|f_1|^2\Big)\ . \tag{10.50}$$

Thus we see that the differential cross section will give us information on the relative phase of the $\ell = 0$ and $\ell = 1$ amplitudes. Here we note that in the event of S-wave scattering only, the differential cross section is angle independent, i.e., the differential cross section is isotropic.

10.5 The Optical Theorem

This theorem is a special case of a more general theorem that sets a nonlinear constraint on the scattering amplitude called *unitarity*. In particular it gives a relation between the scattering amplitude in the forward direction, $f(k,\theta = 0)$, and the total cross section, σ_T. It results from the condition that probability should be conserved in the scattering process.

From Eq. (10.45), we have that the forward scattering amplitude is given by

$$\begin{aligned} f(k,\theta=0) &= \frac{1}{k}\sum_\ell (2\ell+1)\,f_\ell(k)\,P_\ell(1) \\ &= \frac{1}{k}\sum_\ell (2\ell+1)\,f_\ell(k) \end{aligned} \tag{10.51}$$

since $P_\ell(1) = 1$. Taking the imaginary part of this equation we get

$$\mathcal{I}m\,[f(k,0)] = \frac{1}{k}\sum_\ell (2\ell+1)\,\mathcal{I}m\,[f_\ell(k)]\ . \tag{10.52}$$

But we have from Eq. (10.46) that for real phase shifts

$$\mathcal{I}m\,[f_\ell(k)] = \sin^2\delta_\ell\ . \tag{10.53}$$

Therefore, we can write

$$\mathcal{I}m\,[f(k,0)] = \frac{1}{k}\sum_{\ell}(2\ell+1)\,\sin^2\delta_\ell\;. \tag{10.54}$$

Making use of Eq. (10.48), we can write

$$\mathcal{I}m\,[f(k,0)] = \frac{k}{4\pi}\,\sigma_T\;. \tag{10.55}$$

This result is commonly known as the *Optical Theorem*, and relates the forward scattering amplitude to the total cross section. Considering the fact that the total cross section is proportional to the scattering amplitude squared, then Eq. (10.55) imposes a non-linear constraint on the scattering amplitude, $f(k,\theta)$.

To illustrate the relation between unitarity and the optical theorem, we recall from Eq. (10.46) that the partial wave scattering amplitude is given by

$$\begin{aligned} f_\ell(k) &= \frac{1}{2i}\left(e^{2i\delta_\ell}-1\right) \\ &\equiv \frac{1}{2i}\left(S_\ell(k)-1\right)\;, \end{aligned} \tag{10.56}$$

where $S_\ell(k)$ is the partial wave S-matrix element. Solving Eq. (10.56) for $S_\ell(k)$, we get

$$S_\ell(k) = 1 + 2i f_\ell(k)\;. \tag{10.57}$$

Unitarity is the result of the fact that the S-matrix is unitary, i.e.,

$$S^\dagger\,S = I\;, \tag{10.58}$$

which is obviously the case for $S_\ell(k)$ if the the phase shifts δ_ℓ are real. Also, the unitarity of $S_\ell(k)$ gives us the result

$$S_\ell^\dagger(k)\,S_\ell(k) = 1\;, \tag{10.59}$$

or

$$\mathcal{I}m\,f_\ell(k) = |f_\ell(k)|^2\;. \tag{10.60}$$

This non-linear relation for the partial wave scattering amplitude is identical to the optical theorem. We will see in a later chapter on formal scattering theory, Chapter 14, that this result can be derived from the Schrödinger equation directly.

10.6 The Phase Shifts for Two-Body Scattering

So far we have determined the cross section in terms of the scattering amplitude or the phase shifts. However, the phase shifts are constants used in writing the asymptotic form of the wave function. In the case of a square well potential these phase shifts were determined by matching the logarithmic derivative of the radial wave function at the well radius $r = a$. In general, we can follow the same procedure and integrate the differential equation from the origin to a large radial distance r_0, which is larger than the range of the potential. We then match the logarithmic derivative, as calculated from integrating the radial Schrödinger equation, with the logarithmic derivative from the asymptotic solution at $r = r_0$. Thus from integrating the Schrödinger equation we have

$$\gamma = \frac{d}{dr} \log \psi_\ell(r)\Big|_{r=r_0-\epsilon} , \tag{10.61}$$

where ϵ is infinitesimal. On the other hand, for $r > r_0$, we have the asymptotic solution

$$\psi_\ell(r) = \cos\delta_\ell \left(j_\ell(kr) + \tan\delta_\ell \, n_\ell(kr) \right)$$

and its derivative

$$\frac{d\psi_\ell(r)}{dr} = k \cos\delta_\ell \left(j'_\ell(kr) + \tan\delta_\ell \, n'_\ell(kr) \right)$$

where j'_ℓ and n'_ℓ are the derivative of the spherical Bessel and Neumann functions. We now can write the logarithmic derivative for $r > r_0$ as

$$\begin{aligned} \frac{d}{dr} \log \psi_\ell(r)\Big|_{r=r_0+\epsilon} &= k \frac{j'_\ell(kr_0) + \tan\delta_\ell \, n'_\ell(kr_0)}{j_\ell(kr_0) + \tan\delta_\ell \, n_\ell(kr_0)} \\ &= \gamma \, . \end{aligned} \tag{10.62}$$

This equation can be solved for the phase shift, or $\tan\delta_\ell$, in terms of γ with the result that

$$\tan\delta_\ell = \frac{k \, j'_\ell(kr_0) - \gamma \, j_\ell(kr_0)}{\gamma \, n_\ell(kr_0) - k \, n'_\ell(kr_0)} \, . \tag{10.63}$$

Here, we observe that given γ, we can determine the phase shift δ_ℓ.

> **Note**: Here γ is a function of r_0, and r_0 should be chosen large enough so that δ_ℓ or $\tan\delta_\ell$ is independent of r_0.

From the above results we can study the behavior of δ_ℓ at low energies, i.e., $k \to 0$. For $kr_0 \ll \ell$, we can write the spherical Bessel and Neumann functions and their derivative as

$$\begin{aligned} j_\ell(kr_0) &\to \frac{(kr_0)^\ell}{(2\ell+1)!!} \quad &\text{and} \quad n_\ell(kr_0) &\to \frac{(2\ell-1)!!}{(kr_0)^{\ell+1}} \\ j'_\ell(kr_0) &\to -\frac{\ell(kr_0)^{\ell-1}}{(2\ell+1)!!} \quad &\text{and} \quad n'_\ell(kr_0) &\to \frac{(\ell+1)(2\ell-1)!!}{(kr_0)^{\ell+2}} \, . \end{aligned} \tag{10.64}$$

We then can write the phase shift for $kr_0 \ll \ell$ as

$$\tan\delta_\ell \to \frac{(kr_0)^{2\ell+1}}{(2\ell+1)!!\,(2\ell-1)!!}\,\frac{\ell-\gamma r_0}{\ell+1+\gamma r_0}\,, \tag{10.65}$$

or

$$\tan\delta_\ell \propto k^{2\ell+1} \qquad \text{for} \qquad kr_0 \ll \ell\,. \tag{10.66}$$

From this result we may deduce that for small wave number k, i.e., low energy, $\sin\delta_\ell \propto k^{2\ell+1}$. If we now write the cross section as

$$\sigma_T = \frac{4\pi}{k^2}\sum_\ell (2\ell+1)\,\sin^2\delta_\ell \equiv \sum_\ell \sigma_\ell\,, \tag{10.67}$$

then the partial wave cross section, σ_ℓ, is given by

$$\begin{aligned}\sigma_\ell &= \frac{4\pi}{k^2}\,(2\ell+1)\,\sin^2\delta_\ell \\ &\to (\text{const.})\,k^{4\ell} \qquad \text{for} \qquad kr_0 \ll \ell\,.\end{aligned} \tag{10.68}$$

From this result we may conclude that for $k \to 0$ we have

$$\sigma_\ell = \begin{cases} \text{const.} & \text{for} \quad \ell = 0 \\ 0 & \text{for} \quad \ell \neq 0 \end{cases}\,. \tag{10.69}$$

Thus at low energies, we expect the $\ell = 0$ partial wave to dominate the cross section.

10.7 Coulomb Scattering

The analysis in this chapter has so far been restricted to finite range potentials. This excludes the Coulomb potential which is considered to be infinite in range. Because of the central role played by the Coulomb potential in both atomic and molecular physics, and the fact that all accelerators produce beams of charged particles which are scattered by targets often made of charged particles (e.g., nuclei), a discussion of scattering theory that excludes the Coulomb problem is incomplete. The aim of this section is to derive the amplitude for the scattering of two charged particles, and from that, extract the Rutherford cross section.

Consider for the present, the scattering of two charged particles with charges Ze and $Z'e'$, with the Coulomb potential between the particles given by

$$V(\vec{r}) = \frac{ZZ'ee'}{r}\,. \tag{10.70}$$

The Schrödinger equation for this potential is then given by

$$\left(-\frac{\hbar^2}{2\mu}\nabla^2 + \frac{ZZ'ee'}{r}\right)\psi(\vec{r}) = E\,\psi(\vec{r})\,, \tag{10.71}$$

where μ is the reduced mass of the two particles. We can rewrite this equation, after multiplication by $-\frac{2\mu}{\hbar^2}$, as

$$\left(\nabla^2 + k^2 - \frac{2\gamma k}{r}\right)\psi(\vec{r}) = 0 \; , \tag{10.72}$$

where

$$k^2 = \frac{2\mu E}{\hbar^2} \qquad \text{and} \qquad \gamma = \frac{\mu Z Z' e e'}{\hbar^2 k} \; . \tag{10.73}$$

We are going to consider the solution of this equation for $E > 0$, i.e., we would like to derive the scattering wave function $\psi(\vec{r})$, and from its asymptotic behavior extract the scattering amplitude.

For any scattering experiment, we can define the direction of the incident beam to be the z-axis. The presence of this preferred direction in space breaks the symmetry (that space is isotropic) and it gives us a problem which has cylindrical symmetry. As a result of this new symmetry we expect the wave function and the scattering amplitude to be independent of the angle ϕ, as illustrated in Eq. (10.32). Furthermore, the wave function asymptotically should have an incident beam given up to a normalization by

$$e^{ikz} \; ,$$

and a scattered beam that is proportional to a spherical out going wave, i.e.,

$$\frac{e^{ikr}}{r} \; .$$

This suggests that the solution to Eq. (10.72) can be written as

$$\psi(\vec{r}) = e^{ikz}\, g(r - z) \; ,$$

and excludes the possibility of having a function of the form

$$\psi(\vec{r}) = e^{ikz}\, g(r + z) \; ,$$

since the latter leads to an incoming spherical wave given by

$$\frac{e^{-ikr}}{r} \; .$$

The fact that the boundary condition on this scattering problem requires the total wave function to be a function of z and $r - z$, suggests that the optimum choice for a coordinate system for solving the Schrödinger equation is the parabolic coordinate given by [36]

$$\begin{aligned} \xi &= r - z = r(1 - \cos\theta) \\ \eta &= r + z = r(1 + \cos\theta) \\ \phi &= \phi \; . \end{aligned} \tag{10.74}$$

In this new coordinate system, our total scattering wave function is a product of an incident plane wave and a function of ξ, i.e.,

$$\psi(\vec{r}) = e^{ik(\eta-\xi)/2}\, g(\xi)\ . \tag{10.75}$$

The Laplacian, ∇^2, in this coordinate system is given by

$$\nabla^2 = \frac{4}{\xi+\eta}\left\{\frac{\partial}{\partial\xi}\left(\xi\frac{\partial}{\partial\xi}\right) + \frac{\partial}{\partial\eta}\left(\eta\frac{\partial}{\partial\eta}\right)\right\} + \frac{1}{\xi\eta}\frac{\partial^2}{\partial\phi^2}\ . \tag{10.76}$$

We now can write the Schrödinger equation in this coordinate system as a partial differential equation in two variables, since we have no ϕ dependence. Furthermore, for the wave function with the structure given in Eq. (10.75), we have

$$\nabla^2\psi(\vec{r}) = \frac{4\,e^{ik(\eta-xi)/2}}{\eta+\xi}\left[\xi\frac{d^2g}{d\xi^2} + (1-ik\xi)\,\frac{dg}{d\xi} - \frac{k^2}{4}\,(\eta+\xi)\,g\right]\ . \tag{10.77}$$

This allows us to rewrite Eq. (10.72) for $g(\xi)$ as

$$\left(\xi\frac{d^2}{d\xi^2} + (1-ik\xi)\,\frac{d}{d\xi} - \gamma k\right)g(\xi) = 0\ , \tag{10.78}$$

which is the *Confluent Hypergeometric* equation,[7] with the solution given by

$$g(\xi) = {}_1F_1(-i\gamma; 1; ik\xi)\ , \tag{10.79}$$

where the Confluent Hypergeometric function ${}_1F_1(\alpha;\beta;z)$ is given in Eq. (9.35) as an infinite series. We thus can write the total wave function for the scattering of two charged particles as

$$\begin{aligned}\psi(\vec{r}) &= e^{ikz}\,g(r-z)\\ &= e^{ikz}\,{}_1F_1(-i\gamma; 1; ik(r-z))\ .\end{aligned} \tag{10.80}$$

To get the asymptotic form of this wave function, and thus determine the scattering amplitude, we need to know the behavior of ${}_1F_1(\alpha;\beta;\rho)$ for large ρ. We have that [8]

$${}_1F_1(\alpha;\beta;\rho) \to \Gamma(\beta)\left\{\frac{(-\rho)^{-\alpha}}{\Gamma(\beta-\alpha)} + \frac{\rho^{\alpha-\beta}\,e^{\rho}}{\Gamma(\alpha)}\right\} \qquad \text{as}\quad |\rho|\to\infty\ . \tag{10.81}$$

---

[7]To get Eq. (10.78) into the form of the confluent hypergeometric equation given in Eq. (9.34) we need to change variables to $z = ik\xi$.

[8]See Abramowitz and Stegun [31] Eq. 13.5.1

With this result for the asymptotic behavior of the Confluent Hypergeometric equation, we can write the total scattering wave function for $r \to \infty$ as

$$\begin{aligned} \psi(\vec{r}) &= e^{ikz}\ {}_1F_1(-i\gamma;1;ik\xi) \\ &\to \frac{e^{\pi\gamma/2}}{\Gamma(1+i\gamma)} \left\{ e^{i(kz+\gamma\log k(r-z))} + \frac{\Gamma(1+i\gamma)}{i\Gamma(-i\gamma)} \frac{e^{-i\gamma\log\sin^2\theta/2}}{2k\sin^2\theta/2} \frac{e^{i(kr-\gamma\log 2kr}}{r} \right\} \\ &\equiv \frac{e^{\pi\gamma/2}}{\Gamma(1+i\gamma)} \left[e^{i(kz+\gamma\log k(r-z))} + f(k,\theta)\,\frac{e^{i(kr-\gamma\log 2kr)}}{r} \right] , \end{aligned} \tag{10.82}$$

where the scattering amplitude $f(k,\theta)$ is given by [9]

$$\begin{aligned} f(k,\theta) &= -\gamma\frac{\Gamma(1+i\gamma)}{\Gamma(1-i\gamma)}\frac{e^{-i\gamma\log\sin^2\theta/2}}{2k\sin^2\theta/2} \\ &= -\frac{\gamma}{2k\sin^2\theta/2}\, e^{i(\eta_0-\gamma\log\sin^2\theta/2)} , \end{aligned} \tag{10.83}$$

where

$$\eta_0 = \arg\Gamma(1+i\gamma) \ . \tag{10.84}$$

The corresponding differential cross section is then given by

$$\frac{d\sigma}{d\Omega} = \frac{\gamma^2}{4k^2\sin^4\theta/2} \ . \tag{10.85}$$

This cross section is commonly known as the Rutherford cross section. Here we observe:

1. That the differential cross section for the scattering of two charged particles, $\frac{d\sigma}{d\Omega} \to \infty$ as $\theta \to 0$. This in fact is the case both for electron-atom scattering, and proton-nucleus scattering.

2. The total cross section is also infinite. This is a result of the fact that the Coulomb potential has an infinite range and the force between two changed particles is never zero.

[9] In writing this result we have made use of the fact that

$$(1-\cos\theta) = 2\sin^2\theta/2 \ ,$$

and

$$-z\,\Gamma(-z) = \Gamma(1-z) \ .$$

3. The asymptotic wave function as presented in Eq. (10.82) consists of an incident plane wave and an outgoing spherical scattering wave. However, if we compare this result with the equation we derived for finite range potentials, i.e., Eq. (10.42), we observe that both the plane wave and the scattering wave are modified by a factor proportional to γ, which is basically the product of the charges on the two particles. This distortion of the incident plane wave and the scattered spherical wave is the result of the fact that the Coulomb potential has an infinite range.

10.8 Problems

1. Calculate the S-wave phase shift (i.e., $\ell = 0$) for neutron proton scattering at center of mass energies of 5 and 10 MeV given that the potential between a neutron and proton is a square well of radius $r_0 = 2.51$ fm and the depth is $V_0 = 17.8$ MeV. Calculate the corresponding cross section.

   Note: 1 fm $= 10^{-13}$ cm, and $\hbar^2/2\mu = 41.47$ MeV fm^2.

2. Calculate the cross section for scattering off a hard sphere of radius R at very low energies, i.e., in the limit where $kR \ll 1$.

   Hint: A hard sphere can be represented by the potential

$$V(r) = \begin{cases} +\infty & r \leq R \\ 0 & r > R \end{cases} .$$

3. Show that for P-wave ($\ell - 1$) scattering by an attractive square well potential of radius a and depth V_0, the phase shifts satisfy the equation

$$\alpha^2 \left[ka \cot(ka + \delta_1) - 1\right] = k^2 \left(\alpha a \cot \alpha a - 1\right) ,$$

   where $k^2 = \frac{2\mu}{\hbar^2}E$ and $\alpha^2 = \frac{2\mu}{\hbar^2}(E + V_0)$. For neutron proton scattering we can take $V_0 = 36.2$ MeV and $a = 2.02$ fm. Taking $\frac{\hbar^2}{2\mu} = 41.47$ MeV fm^2, calculate the phase shift at $E = 10$ MeV.

4. Using the orthogonality of the Legendre polynomials, show that if the scattering wave function is written as

$$\psi(\vec{r}) = \frac{1}{(2\pi)^{3/2}} \sum_{\ell} i^{\ell}(2\ell + 1)\psi_{\ell}(r)\, P_{\ell}(\cos\theta) ,$$

   then $\psi_{\ell}(r)$ satisfies the radial Schrödinger equation.

5. Show that for complex phase shifts the total cross section as calculated from the Optical Theorem is larger than the total elastic cross section obtained from integrating the differential cross section.

   Hint: To prove the above, you can restrict your argument to one partial wave, e.g., $\ell = 0$.

6. The amplitude of S-wave scattering at at low energies is given by

$$f_0 = -\frac{ak}{1 + iak - \frac{1}{2}ar_e k^2}$$

   where a and r_e are real constants, and k is the momentum.

   (a) Show that this amplitude satisfies unitarity.

   (b) Show that the corresponding cross section goes to a constant as the energy goes to zero, i.e., $k \longrightarrow 0$.

   (c) Considering your result for part (b), can you give the constant a a physical meaning?

7. Prove that a simple statement equivalent to the Optical Theorem for S-waves is

$$\mathcal{I}m\left(\frac{1}{f_0}\right) = -1 \ .$$

8. The P-wave amplitude for positive pion proton (i.e., $\pi^+ p$) scattering is given by

$$f_1(k) = \Lambda k^3 \left\{k^4 - i\Lambda k^3 + \left(2\beta^2 - \frac{3}{2}\Lambda\beta\right)k^2 + \left(\beta^4 - \frac{1}{2}\Lambda\beta^3\right)\right\}^{-1}$$

   where $\beta = 5.3344\ \text{fm}^{-1}$ and $\Lambda = 10.337\ \text{fm}^{-1}$.

   (a) Calculate the phase shift $\delta_1(k)$ as a function of k for $0 < k < 2.5\ \text{fm}^{-1}$. Plot the phase shifts.

   (b) Calculate the total cross section for P-wave $\pi^+ p$ scattering as a function of energy.

   (c) What happens to the cross section when the phase shift is $\pi/2$?

CHAPTER 11

Matrix Formulation of Quantum Mechanics

Abstract: Having solved the Schrödinger equation for several potentials, we turn our attention in this chapter to the matrix formulation of quantum mechanics and the different representations we can use to simplify the solution of problems for both bound and scattering problems.

So far in our study of quantum mechanics we have come across a number of quantities that can be measured, e.g., momentum, energy and angular momentum. For each of these quantities or *observables*, we can introduce an operator. Thus in Table 11.1 shown below we have a number of observables with their corresponding operators. The question is, what properties should these and other operators that correspond to observables, satisfy. In this chapter we will develop the properties of these operators, the states of the system under observation, and their interrelation. In the process we develop a theoretical framework of quantum mechanics that is more general than just the solution of the Schrödinger equation. In the following we first examine the properties of operators corresponding to observables and the eigenvalues and eigenfunctions of these operators. This is followed with an introduction of Dirac notation [15][1] which is used to detail the unitary transformations relating the different representations of the operators and their eigenfunctions. Finally we consider the three representations most often used. These are the coordinate, the momentum and the angular momentum representations. The latter will allow us to introduce a spin 1/2 basis for the representation of the angular momentum operator.

[1]The classic book by the Dirac *The Principles of Quantum Mechanics* [15] presents a concise logical mathematical presentation of quantum mechanics that was initially written at the height of the birth of quantum mechanics and is considered as essential reading for aspiring theoretical physicists.

Iraj R. Afnan

Table 11.1: Commonly encountered observables and the corresponding operators

	Physical Observable	Operator
Coordinate	$\vec{r}$ $x,\ y,\ z$	$\tilde{\vec{r}}$ $\tilde{x},\ \tilde{y},\ \tilde{z}$
Momentum	$\vec{p}$ $p_x,\ p_y,\ p_z$	$-i\hbar\nabla$ $-i\hbar\frac{\partial}{\partial x},\ -i\hbar\frac{\partial}{\partial y},\ -i\hbar\frac{\partial}{\partial z}$
Angular Momentum	$\vec{L} = \vec{r}\times\vec{p}$ $L_x = yp_z - zp_y$ $L_y = zp_x - xp_z$ $L_z = xp_y - yp_x$	$\tilde{\vec{L}} = -i\hbar\vec{r}\times\nabla$ $\tilde{L}_x = -i\hbar\left(y\frac{\partial}{\partial z} - z\frac{\partial}{\partial y}\right)$ $\tilde{L}_y = -i\hbar\left(z\frac{\partial}{\partial x} - x\frac{\partial}{\partial z}\right)$ $\tilde{L}_z = -i\hbar\left(x\frac{\partial}{\partial y} - y\frac{\partial}{\partial x}\right)$
Energy	$H = \frac{p^2}{2m} + V(\vec{r})$	$\tilde{H} = -\frac{\hbar^2}{2m}\nabla^2 + V(\tilde{\vec{r}})$

11.1 Operators and Observables

Let us consider the operator $\tilde{F}$ which corresponds to a quantity F that can be measured. If we have a system in a state ψ_α, then the average value of the observable F is

$$\langle\tilde{F}\rangle = \int \psi_\alpha^*\,\tilde{F}\,\psi_\alpha \equiv (\psi_\alpha,\,\tilde{F}\,\psi_\alpha)\ . \tag{11.1}$$

For example, if we need to know the average momentum of an electron in the ground state of the hydrogen atom, i.e., the 0s state, then we need to calculate

$$\langle\,\tilde{\vec{p}}\,\rangle = \int d^3r\,\psi_{100}^*(\vec{r})\,[-i\hbar\nabla]\,\psi_{100}(\vec{r}) \equiv (\psi_{100},\,-i\hbar\nabla\,\psi_{100})\ . \tag{11.2}$$

In general the action of the operator $\tilde{F}$ on the state ψ_α gives a new state ψ'_α, i.e.,

$$\psi'_\alpha = \tilde{F}\,\psi_\alpha\ . \tag{11.3}$$

To satisfy the condition of the addition of probability amplitudes, i.e., superposition of states, $\tilde{F}$ should satisfy the two conditions

$$\tilde{F}\,(\psi_1+\psi_2) = \tilde{F}\,\psi_1 + \tilde{F}\,\psi_2 \tag{11.4}$$

and

$$\tilde{F}\,a\,\psi_\alpha = a\,\tilde{F}\,\psi_\alpha\ , \tag{11.5}$$

where a is any complex number. These two conditions define a *linear operator*. We will find that most operators that correspond to physically measurable quantities are linear operators. One exception is the time reversal operator which is anti-linear, i.e.,

$$\tilde{F}\, a\, \psi_\alpha = a^*\, \tilde{F}\, \psi_\alpha \ , \tag{11.6}$$

where a^* is the complex conjugate of a.

For the operator $\tilde{F}$ to correspond to a measurable quantity, the average value of $\tilde{F}$ should be real, i.e.,

$$\langle \tilde{F} \rangle = \langle \tilde{F} \rangle^*$$

or

$$\left(\psi_\alpha,\, \tilde{F}\, \psi_\alpha\right) = \left(\psi_\alpha,\, \tilde{F}\, \psi_\alpha\right)^* = \left(\tilde{F}\, \psi_\alpha,\, \psi_\alpha\right) \ . \tag{11.7}$$

In writing the above result we have made use of the definition of the bracket in terms of an integral over the wave function as defined in Eq. (11.1). If we define a matrix element of $\tilde{F}$, $F_{\alpha\beta}$, as

$$F_{\alpha\beta} \equiv \left(\psi_\alpha,\, \tilde{F}\, \psi_\beta\right) \ , \tag{11.8}$$

then the generalization of Eq. (11.7) is [2]

$$F_{\alpha\beta} = F^*_{\beta\alpha}$$

or

$$\begin{aligned} \left(\psi_\alpha,\, \tilde{F}\, \psi_\beta\right) &= \left(\psi_\beta,\, \tilde{F}\, \psi_\alpha\right)^* = \left(\tilde{F}\, \psi_\alpha,\, \psi_\beta\right) \\ &\equiv \left(\psi_\alpha,\, \tilde{F}^\dagger\, \psi_\beta\right) \ . \end{aligned} \tag{11.9}$$

This means the operator $\tilde{F}$ should be self-adjoint or *Hermitian*, i.e.,

$$\tilde{F} = \tilde{F}^\dagger = \left(\tilde{F}^T\right)^* \ , \tag{11.10}$$

where $\tilde{F}^T$ is the transpose of $\tilde{F}$. Here, if we recall the fact that Hermitian matrices have real eigenvalues, then taking $F_{\alpha\beta}$ as a matrix with $F_{\alpha\beta} = F^*_{\beta\alpha}$ guarantees that the eigenvalues of $\tilde{F}$ are real. We therefore can make the general statement:

In quantum mechanics all operators are linear and Hermitian if they are to correspond to physical observables.

[2] This corresponds to the matrix $F_{\alpha\beta}$ being Hermitian.

Example: Consider the momentum operator

$$\tilde{\vec{p}} = -i\hbar\vec{\nabla}\ .$$

To show that $\tilde{\vec{p}}$ is a linear operator, we have to show that it satisfies the two conditions for a linear operator, i.e., Eqs. (11.4) and (11.5) are satisfied. The first condition is satisfied because

$$\begin{aligned}\tilde{\vec{p}}\,(\psi_1+\psi_2) &= -i\hbar\vec{\nabla}\,(\psi_1+\psi_2) = -i\hbar\vec{\nabla}\psi_1 - i\hbar\vec{\nabla}\psi_2\\ &= \tilde{\vec{p}}\,\psi_1 + \tilde{\vec{p}}\,\psi_2\ ,\end{aligned}$$

while the second condition is valid since

$$\tilde{\vec{p}}\,a\,\psi_\alpha = -i\hbar\vec{\nabla}\,a\,\psi_\alpha = a\,(-i\hbar\vec{\nabla})\ \psi_\alpha = a\,\tilde{\vec{p}}\,\psi_\alpha\ ,$$

for any complex constant a. We therefore have established that $\tilde{\vec{p}}$ is a linear operator. To show that $\tilde{\vec{p}}$ is a Hermitian operator, we have to show that

$$\left(\psi_\alpha,\,\tilde{\vec{p}}\psi_\beta\right) = \left(\psi_\beta,\,\tilde{\vec{p}}\psi_\alpha\right)^*\ .$$

But we have that

$$\left(\psi_\alpha,\,\tilde{\vec{p}}\psi_\beta\right) = -i\hbar\int d^3r\,\psi_\alpha^*(\vec{r})\,\vec{\nabla}\,\psi_\beta(\vec{r})\ .$$

Integrating by parts, assuming the wave functions $\psi_\alpha^*(\vec{r})$ and $\psi_\beta(\vec{r})$ go to zero asymptotically, we get

$$\begin{aligned}\left(\psi_\alpha,\,\tilde{\vec{p}}\psi_\beta\right) &= i\hbar\int d^3r\,\left[\vec{\nabla}\psi_\alpha(\vec{r})\right]^*\,\psi_\beta(\vec{r})\\ &= \int d^3r\,\left[-i\hbar\vec{\nabla}\psi_\alpha(\vec{r})\right]^*\,\psi_\beta(\vec{r})\\ &= \left\{\int d^3r\,\psi_\beta^*(\vec{r})\,\left[-i\hbar\vec{\nabla}\psi_\alpha(\vec{r})\right]\right\}^*\\ &= \left(\psi_\beta,\,\tilde{\vec{p}}\psi_\alpha\right)^*\ .\end{aligned}$$

Therefore, the momentum operator $\tilde{\vec{p}} = -i\hbar\vec{\nabla}$ is a linear and Hermitian operator.

Consider the case when the operator $\tilde{F}$ acts on a state ψ_α and gives the state ψ'_α which is proportional to ψ_α, i.e.,

$$\tilde{F}\psi_\alpha = \psi'_\alpha = F_\alpha\psi_\alpha\ , \tag{11.11}$$

where F_α is the constant of proportionality. If we compare this result to the equivalent matrix equation, i.e., if $\tilde{F}$ was a matrix rather than an operator, then we would refer to F_α as the eigenvalue and ψ_α as the eigenstate or eigenfunction. For example, for the three-dimensional harmonic oscillator, we have

$$\tilde{H}\,\psi_{n\ell m} = E_{n\ell}\,\psi_{n\ell m}\ ,$$

where $E_{n\ell}$ is the eigenvalue and $\psi_{n\ell m}$ is the eigenfunction of $\tilde{H}$.

To show that *if* $\tilde{F}$ is Hermitian, then F_α is real, we write Eq. (11.11) as

$$\tilde{F}\,\psi_\alpha = F_\alpha\,\psi_\alpha\ , \tag{11.12}$$

and then the complex conjugate equation with α replaced by β is given by

$$\tilde{F}^*\,\psi^*_\beta = F^*_\beta\,\psi^*_\beta\ . \tag{11.13}$$

We now multiply Eq. (11.12) by ψ^*_β from the left and Eq. (11.13) by ψ_α from the right, and then integrate the two equations. If we now subtract one equation from the other we get

$$\begin{aligned}\left(F_\alpha - F^*_\beta\right)\int d^3r\,\psi^*_\beta(\vec{r})\,\psi_\alpha(\vec{r}) &= \int d^3r\,\psi^*_\beta(\vec{r})\,\tilde{F}\,\psi_\alpha(\vec{r}) - \int d^3r\,(\tilde{F}^*\,\psi^*_\beta(\vec{r}))\,\psi_\alpha(\vec{r})\\ &= F_{\beta\alpha} - F^*_{\alpha\beta}\ .\end{aligned}$$

Since $\tilde{F}$ is Hermitian, $F_{\alpha\beta} = F^*_{\beta\alpha}$, the above equation reduces to

$$(F_\alpha - F^*_\beta)\int d^3r\,\psi^*_\beta(\vec{r})\,\psi_\alpha(\vec{r}) = 0\ . \tag{11.14}$$

For $\alpha = \beta$ since $\int d^3r\,|\psi_\alpha|^2 \neq 0$, except for the uninteresting case of $\psi_\alpha(\vec{r}) = 0$, we therefore have

$$F_\alpha = F^*_\alpha\ . \tag{11.15}$$

We can now state that:

The eigenvalues of Hermitian operators are real.

On the other hand for $F_\alpha \neq F_\beta$ we have that [3]

$$\int d^3r\,\psi^*_\beta(\vec{r})\,\psi_\alpha(\vec{r}) = (\psi_\beta,\,\psi_\alpha) = 0 \qquad \text{for} \quad \alpha \neq \beta\ . \tag{11.16}$$

which implies that:

[3] In the event that $\alpha \neq \beta$ and $F_\alpha = F_\beta$, then ψ_α and ψ_β are not orthogonal, and we need to use the Schmidt orthogonalization procedure.

The eigenstates of a Hermitian operator are orthogonal.

If the states ψ_α correspond to a bound state, the wave function $\psi_\alpha(\vec{r}) \to 0$ as $|\vec{r}| \to \infty$, and the wave function is normalizable. In that case we have

$$(\psi_\alpha, \psi_\beta) = \delta_{\alpha\beta} , \tag{11.17}$$

and we can state that:

The eigenstates of a Hermitian operator are orthonormal.

Another property of the eigenstates of Hermitian operators is that they form a complete set of states, i.e., if we have any state Ψ that is normalizable, then we can write

$$\Psi(x) = \sum_\alpha a_\alpha \psi_\alpha(x) . \tag{11.18}$$

Using the orthogonality of the ψ_α's, i.e., Eq. (11.17), we get

$$(\psi_\beta, \Psi) = \sum_\alpha a_\alpha (\psi_\beta, \psi_\alpha) = \sum_\alpha a_\alpha \delta_{\alpha\beta} = a_\beta .$$

Therefore,

$$a_\beta = (\psi_\beta, \Psi) . \tag{11.19}$$

The statement of completeness of the eigenstates ψ_α is represented by

$$\sum_\alpha \psi_\alpha(x) \psi_\alpha^*(x') = \delta(x - x') . \tag{11.20}$$

To prove this result we expand the δ-function $\delta(x - x')$, in terms of our basis states as

$$\delta(x - x') = \sum_\alpha a_\alpha(x') \psi_\alpha(x) .$$

To determine the coefficient a_α we make use of the orthonormality condition, i.e., Eq. (11.17), to get

$$a_\alpha(x') = \int dx\, \psi_\alpha^*(x) \delta(x' - x) = \psi_\alpha^*(x') .$$

This proves the result of Eq. (11.20), which is a statement of completeness of the eigenstates ψ_α.

Summary:

1. For every observable there is a linear Hermitian operator.
2. The eigenvalues of this Hermitian operator are real.
3. The eigenstates of the Hermitian operator are orthonormal, i.e.,

$$(\psi_\alpha, \psi_\beta) = \delta_{\alpha\beta}$$

4. The eigenstates of the Hermitian operator form a complete set, i.e.,

$$\sum_\alpha \psi_\alpha(x)\, \psi_\alpha^*(x') = \delta(x - x') \ .$$

In the above discussion, we have ignored two facts:

1. In some problems we have more than one eigenstate with the same energy, e.g., if we consider the Hamiltonian for the Coulomb problem

$$\tilde{H} = -\frac{\hbar^2}{2m}\nabla^2 - \frac{Ze^2}{r}$$

then the eigenstates of $\tilde{H}$ are $\psi_{n\ell m}$ and

$$\tilde{H}\, \psi_{n\ell m} = E_n\, \psi_{n\ell m} \ ,$$

i.e., the energy of the system is independent of ℓ and m. A similar situation arises for all central potentials in that the eigenvalues are independent of m. In these cases it is not obvious that states with different m are orthogonal. We have guaranteed the orthogonality of the eigenstates of $\tilde{H}$ by taking $\psi_{n\ell m} \propto Y_{\ell m}(\theta, \phi)$, i.e., that the state $\psi_{n\ell m}$ be also an eigenstate of the total angular momentum square, $\tilde{L}^2$, and its z-component, $\tilde{L}_z$.

2. In most problems in quantum mechanics the eigenvalues of the Hamiltonian include a discreet as well as a continuous spectrum. The continuous part of the spectrum corresponds to the scattering states. For such states, $\psi_\alpha(r)$ does not go to zero as $r \to \infty$ and the eigenstates are not normalizable, e.g., for the free Hamiltonian,

$$\tilde{H} = -\frac{\hbar^2}{2m}\nabla^2 \ ,$$

the eigenstates are

$$\phi_{\vec{p}}(\vec{r}\,) = \frac{1}{(2\pi\hbar)^{3/2}}\, e^{i\vec{k}\cdot\vec{r}}$$

where $\vec{p} = \hbar\vec{k}$ and the normalization taken to be

$$(\phi_{\vec{p}}, \phi_{\vec{p}'}) = \delta(\vec{p} - \vec{p}\,') \ ,$$

which is infinite for $\vec{p} = \vec{p}\,'$. One either accepts this $\delta-$function normalization, or puts the system in a box and at the end of the calculation takes the limit as the volume of the box goes to ∞.

With this extended basis that includes scattering as well as bound states, Eqs. (11.18) and (11.20) become

$$\Psi(\vec{r}) = \sum_\alpha a_\alpha\, \psi_\alpha(\vec{r}) + \int d^3k\, a(\vec{k}\,)\, \psi_{\vec{k}}(\vec{r}\,)\ , \tag{11.21}$$

and

$$\sum_\alpha \psi_\alpha(\vec{r}\,)\, \psi^*_\alpha(\vec{r}\,') + \int d^3k\, \psi_{\vec{k}}(\vec{r}\,)\, \psi^*_{\vec{k}}(\vec{r}\,') = \delta(\vec{r} - \vec{r}\,')\ . \tag{11.22}$$

Let us now show that if two observables can be measured simultaneously, then the operators corresponding to these observables commute. If two observables can be measured at the same time, then we have one eigenstate that is an eigenstate of both operators corresponding to the two observables, i.e.,

$$\tilde{F}\psi_\alpha = F_\alpha \psi_\alpha \qquad \text{and} \qquad \tilde{M}\psi_\alpha = M_\alpha \psi_\alpha\ , \tag{11.23}$$

where $\tilde{F}$ and $\tilde{M}$ are the two operators corresponding to the two observables, and ψ_α is the eigenstate of both operators. Multiplying the first of the equations in Eq. (11.23) by $\tilde{M}$, and the second equation by $\tilde{F}$ we get

$$\begin{aligned} \tilde{M}\tilde{F}\psi_\alpha &= \tilde{M}F_\alpha\psi_\alpha = F_\alpha\tilde{M}\psi_\alpha = F_\alpha M_\alpha \psi_\alpha \\ \tilde{F}\tilde{M}\psi_\alpha &= \tilde{F}M_\alpha\psi_\alpha = M_\alpha\tilde{F}\psi_\alpha = M_\alpha F_\alpha \psi_\alpha\ . \end{aligned}$$

Subtracting one equation from the other, we get

$$\left(\tilde{M}\tilde{F} - \tilde{F}\tilde{M}\right)\psi_\alpha = \left(F_\alpha M_\alpha - M_\alpha F_\alpha\right)\psi_\alpha = 0\ .$$

For any state Ψ we have

$$\Psi = \sum_\alpha a_\alpha\, \psi_\alpha$$

and then

$$\left(\tilde{M}\tilde{F} - \tilde{F}\tilde{M}\right)\Psi = \sum_\alpha a_\alpha \left(\tilde{M}\tilde{F} - \tilde{F}\tilde{M}\right)\psi_\alpha = 0\ .$$

Since this is true for any Ψ, it follows that

$$\left(\tilde{M}\tilde{F} - \tilde{F}\tilde{M}\right) = \left[\tilde{M},\, \tilde{F}\right] = 0\ . \tag{11.24}$$

Thus the operators corresponding to two observables that can be measured simultaneously commute.

The inverse of the above result can be easily shown; i.e., if $\left[\tilde{M},\, \tilde{F}\right] = 0$, then

$$\tilde{F}\psi_\alpha = F_\alpha \psi_\alpha \qquad \text{and} \qquad \tilde{M}\psi_\alpha = M_\alpha \psi_\alpha \ . \tag{11.25}$$

To prove that the statement in Eq. (11.25) is valid, given Eq. (11.24), we assume that

$$\tilde{M}\psi_\alpha = M_\alpha \psi_\alpha \ .$$

We then multiply by $\tilde{F}$ from the left to get

$$\tilde{F}\tilde{M}\psi_\alpha = \tilde{F}M_\alpha\psi_\alpha = M_\alpha \tilde{F}\psi_\alpha \ .$$

Using the fact that $\tilde{F}$ and $\tilde{M}$ commute, i.e., $\tilde{F}\tilde{M} = \tilde{M}\tilde{F}$, we can write the above equation as

$$\tilde{M}\left(\tilde{F}\psi_\alpha\right) = M_\alpha \left(\tilde{F}\psi_\alpha\right) \ ,$$

i.e., $\tilde{F}\psi_\alpha$ is an eigenstate of $\tilde{M}$ with eigenvalue M_α. This implies that

$$\tilde{F}\psi_\alpha \propto \psi_\alpha \ .$$

In particular, we can define the constant of proportionality to be F_α such that

$$\tilde{F}\psi_\alpha = F_\alpha \psi_\alpha \ .$$

This proves the result that if two operators commute, then we have a state that is an eigenstate of both operators, and this corresponds to the fact that the observables corresponding to the two operators can be measured simultaneously.

Finally, we will prove that if

$$\left[\tilde{K},\, \tilde{F}\right] = i\tilde{M} \ , \tag{11.26}$$

then

$$\langle(\Delta F)^2\rangle\,\langle(\Delta K)^2\rangle \geq \frac{1}{4}\langle M\rangle^2 \tag{11.27}$$

with

$$\begin{aligned} \Delta F &= \tilde{F} - \langle\tilde{F}\rangle \\ \Delta K &= \tilde{K} - \langle\tilde{K}\rangle \\ \langle\tilde{K}\rangle &\equiv \left(\Phi,\, \tilde{K}\,\Phi\right) \ . \end{aligned} \tag{11.28}$$

This basically says that if two operators do not commute, then there is an uncertainty relation between the corresponding measurements, i.e., we cannot measure both observables to any desired degree of accuracy.

Proof: Take a state

$$\Psi = \phi + \lambda\psi \qquad \lambda \text{ real} \tag{11.29}$$

then

$$\begin{aligned} 0 \leq (\Psi,\, \Psi) &= (\phi + \lambda\psi,\, \phi + \lambda\psi) \\ &= (\phi,\, \phi) + \lambda\left\{(\phi,\, \psi) + (\psi,\, \phi)\right\} + \lambda^2\, (\psi,\, \psi)\ . \end{aligned} \tag{11.30}$$

The right hand side of this equation can be considered as a quadratic in λ. For the case when this quadratic is equal to zero, we have λ given by

$$\lambda = \frac{1}{2\,(\psi,\, \psi)}\left[-\{(\phi,\, \psi) + (\psi,\, \phi)\}\pm\Big\{[(\phi,\, \psi) + (\psi,\, \phi)]^2 - 4\,(\psi,\, \psi)\,(\phi,\, \phi)\Big\}^{1/2}\right]. \tag{11.31}$$

The requirement that the r.h.s. of Eq. (11.30) be 'greater than' zero for any real λ can only be satisfied if the quantity in the square root bracket in Eq. (11.31) is negative, i.e.,

$$(\psi,\, \psi)\,(\phi,\, \phi) \geq \frac{1}{4}|\,(\phi,\, \psi) + (\psi,\, \phi)\,|^2\ . \tag{11.32}$$

The equality sign in Eq. (11.32) corresponds to the equal sign in Eq. (11.30). To prove Eq. (11.27), we take

$$\phi = \left(\tilde{F} - \langle\tilde{F}\rangle\right)\Phi = \Delta F\,\Phi \quad \text{and} \quad \psi = i\left(\tilde{K} - \langle\tilde{K}\rangle\right)\Phi = i\,\Delta K\,\Phi\ . \tag{11.33}$$

Then

$$\begin{aligned} (\phi,\, \phi) &= \langle(\Delta F)^2\rangle = \left(\Phi,\, \left(\tilde{F} - \langle\tilde{F}\rangle\right)\left(\tilde{F} - \langle\tilde{F}\rangle\right)\Phi\right) \\ (\psi,\, \psi) &= \langle(\Delta K)^2\rangle = \left(\Phi,\, \left(\tilde{K} - \langle\tilde{K}\rangle\right)\left(\tilde{K} - \langle\tilde{K}\rangle\right)\Phi\right) \end{aligned} \tag{11.34}$$

while

$$\begin{aligned} (\phi,\, \psi) &= i\left(\Phi,\, \left[\tilde{F}\tilde{K} - \tilde{F}\langle\tilde{K}\rangle - \langle\tilde{F}\rangle\tilde{K} - \langle\tilde{F}\rangle\,\langle\tilde{K}\rangle\right]\Phi\right) \\ (\psi,\, \phi) &= -i\left(\Phi,\, \left[\tilde{K}\tilde{F} - \tilde{K}\langle\tilde{F}\rangle - \langle\tilde{K}\rangle\tilde{F} - \langle\tilde{K}\rangle\,\langle\tilde{F}\rangle\right]\Phi\right) \end{aligned} \tag{11.35}$$

and then

$$(\phi,\, \psi) + (\psi,\, \phi) = i\left(\Phi,\, \left[\tilde{F}\tilde{K} - \tilde{K}\tilde{F}\right]\Phi\right) = \left(\Phi,\, \tilde{M}\,\Phi\right) = \langle\tilde{M}\rangle\ . \tag{11.36}$$

Using the results of Eqs. (11.34) and (11.36) in Eq. (11.32), we get the result in Eq. (11.27).

11.2 Dirac Notation

In the last section we showed that:

1. For each observable there is a linear Hermitian operator $\tilde{F}$.
2. For each such operator there is a set of eigenstates ψ_α, and eigenvalues F_α such that
$$\tilde{F}\psi_\alpha = F_\alpha \psi_\alpha \ .$$
3. The eigenvalues F_α are real and the eigenstates form a complete and orthonormal set, i.e.,
$$(\psi_\alpha,\, \psi_\beta) = \delta_{\alpha\beta} \ .$$
Any state Ψ can now be written as
$$\Psi = \sum_\alpha a_\alpha \, \psi_\alpha \qquad \text{with} \qquad a_\alpha = (\psi_\alpha,\, \Psi) \ ,$$
or
$$\sum_\alpha \psi_\alpha \, \psi_\alpha^* = \tilde{I} \ ,$$
where $\tilde{I}$ is the unit operator.

Let us compare the above properties of ψ_α with the basis vectors used in three-dimension (3-D). In 3-D we have that our basis vectors are $\hat{e}_1$, $\hat{e}_2$, and $\hat{e}_3$. These basis vectors satisfy the condition that

$$\hat{e}_i \cdot \hat{e}_j = \delta_{ij} \qquad ,$$

i.e., they are orthonormal. Any vector $\vec{V}$ can be written as

$$\vec{V} = \sum_{i=1}^{3} v_i \hat{e}_i \qquad \text{with} \qquad v_i = \hat{e}_i \cdot \vec{V} \ .$$

Comparing the properties of $\hat{e}_i$ with those of ψ_α, we may deduce:

1. The expression $(\psi_\alpha,\, \psi_\beta)$ is equivalent to the scalar product, i.e.,
$$(\psi_\alpha,\, \psi_\beta) = \int d^3r \, \psi_\alpha^*(\vec{r}) \, \psi_\beta(\vec{r})$$
is equivalent to the 3-D vector product
$$\vec{V} \cdot \vec{U} = \sum_i V_i U_i \ .$$

However, unlike the case in three-dimensional space where the basis states are unit vectors $\hat{e}_i$ $(i = 1, 2, 3)$, in quantum mechanics we need both the space of basis states ψ_α $(\alpha = 1, 2, \cdots)$, and the adjoint space ψ^*_α $(\alpha = 1, 2, \cdots)$. To distinguish between the two spaces, Dirac [15] introduced the notation

$$\psi_\alpha \quad \Longrightarrow \quad |\alpha\rangle \qquad \text{called ket} \tag{11.37}$$

and

$$\psi^*_\alpha \quad \Longrightarrow \quad \langle\alpha| \qquad \text{called bra} \tag{11.38}$$

so that the scalar product is given by the bra|ket, i.e.,

$$(\psi_\alpha, \psi_\beta) \quad \Longrightarrow \quad \langle\alpha|\beta\rangle = \delta_{\alpha\beta} \ , \tag{11.39}$$

which is a statement of the orthonormality of the basis states.

2. In most problems in 3-D we introduce a set of basis vectors $\hat{e}_1$, $\hat{e}_2$, and $\hat{e}_3$ in terms of which we can write any vector, e.g., $\hat{e}_1$, $\hat{e}_2$, and $\hat{e}_3$ can be the unit vectors in the x, y, and z direction, or they can be the unit vectors in the $\hat{r}$, $\hat{\theta}$, and $\hat{\phi}$ direction if we are working in spherical polar coordinates. In a similar manner in quantum mechanics we take a Hermitian operator and take its eigenstates to form a basis in terms of which we can write any state. Thus we can take the eigenstates of $\tilde{F}$ to define our basis states, i.e.,

$$\tilde{F}\psi_\alpha = F_\alpha\psi_\alpha \qquad \Longrightarrow \qquad \tilde{F}|\alpha\rangle = F_\alpha|\alpha\rangle \ , \tag{11.40}$$

and now we can write the state $\Psi = |\cdots\rangle$ in terms of our basis as

$$\Psi = \sum_\alpha a_\alpha \, \psi_\alpha \quad \Longrightarrow \quad |\cdots\rangle = \sum_\alpha |\alpha\rangle \, a_\alpha \ , \tag{11.41}$$

with

$$a_\alpha = (\psi_\alpha, \ \Psi) \quad \Longrightarrow \quad a_\alpha = \langle\alpha|\cdots\rangle \ . \tag{11.42}$$

In this notation a statement of completeness of our basis states is given by

$$\sum_\alpha \psi_\alpha \, \psi^*_\alpha = \tilde{I} \quad \Longrightarrow \quad \sum_\alpha |\alpha\rangle\langle\alpha| = 1 \ . \tag{11.43}$$

To illustrate the simplicity of the Dirac notation, consider the state $|a\rangle$ and the set of basis vectors $|\alpha\rangle$ $(\alpha = 1, 2, \cdots)$. Making use of the completeness and orthonormality of the basis states, we can write

$$|a\rangle = \left(\sum_\alpha |\alpha\rangle\langle\alpha|\right) |a\rangle = \sum_\alpha |\alpha\rangle\langle\alpha|a\rangle \ . \tag{11.44}$$

This is basically the expansion of the state $|a\rangle$ in terms of the basis $|\alpha\rangle$. In a similar manner we have

$$\langle a| = \langle a| \left(\sum_\alpha |\alpha\rangle\langle\alpha| \right) = \sum_\alpha \langle a|\alpha\rangle\langle\alpha| \ . \tag{11.45}$$

Taking into consideration the fact that $\langle a| = (|a\rangle)^\dagger$, it is clear that

$$\langle a|\alpha\rangle = (\langle\alpha|a\rangle)^\dagger \ . \tag{11.46}$$

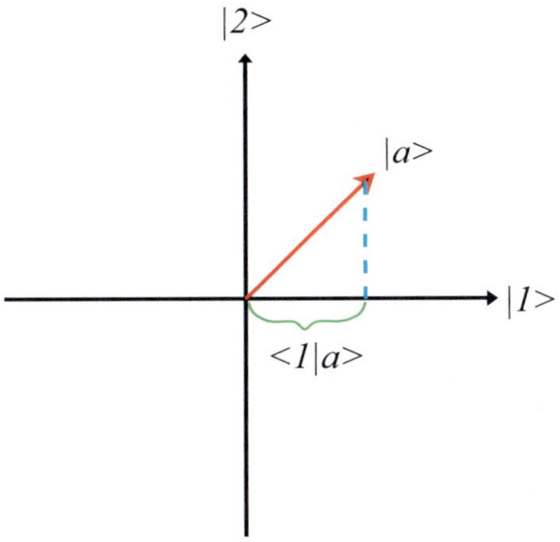

Figure 11.1: The projection of the vector $|a\rangle$ on to the unit vector $|1\rangle$.

Let us now turn our attention to the physical interpretation of the scalar product $\langle\alpha|a\rangle$. If the state $|a\rangle$ is a vector in the space where the basis vectors are $|\alpha\rangle$ $(\alpha = 1, 2, \cdots)$, then $\langle\alpha|a\rangle$ is the projection of the state vector $|a\rangle$ along the $|\alpha\rangle$ axis (see Figure 11.2). We now can define a projection operator which when acting on a state vector $|a\rangle$ will give us the projection of $|a\rangle$ along that axis. Such a projection operator is

$$\tilde{P}_\alpha = |\alpha\rangle\langle\alpha| \ . \tag{11.47}$$

When $\tilde{P}_\alpha$ acts on $|a\rangle$ we get

$$\tilde{P}_\alpha |a\rangle = |\alpha\rangle\langle\alpha|a\rangle \ ,$$

which is the component of $|a\rangle$ along $|\alpha\rangle$ times the unit vector along $|\alpha\rangle$. Thus $\tilde{P}_\alpha|a\rangle$ is a vector along the $|\alpha\rangle$-axis with a magnitude equal to the projection of $|a\rangle$ along $|\alpha\rangle$. Some of the properties of the projection operator are

$$\begin{aligned} \tilde{P}_\alpha\tilde{P}_\beta &= |\alpha\rangle\langle\alpha|\beta\rangle\langle\beta| = |\alpha\rangle\, \delta_{\alpha\beta}\, \langle\beta| \\ &= \tilde{P}_\alpha\, \delta_{\alpha\beta} \ , \end{aligned} \tag{11.48}$$

and

$$\sum_\alpha \tilde{P}_\alpha = \sum_\alpha |\alpha\rangle\langle\alpha| = \tilde{I} \ . \tag{11.49}$$

So far, we have assumed that if we make a measurement for which there is an operator $\tilde{F}$ when the system is in the state $|\alpha\rangle = \psi_\alpha$ then the resultant of the measurement is F_α, i.e.,

$$\tilde{F}|\alpha\rangle = F_\alpha|\alpha\rangle \ .$$

Suppose we prepare the system in a state $|a\rangle$ which is not an eigenstate of the operator $\tilde{F}$ whose corresponding observable we want to measure. For example, consider the measurement of the position of the electron in a hydrogen atom. The electron in the hydrogen atom is in an eigenstate of the Hamiltonian (i.e., energy), but is not in an eigenstate of position. We first write the state $|a\rangle$ in terms of the eigenstate of $\tilde{F}$, i.e.,

$$|a\rangle = \sum_\alpha |\alpha\rangle\langle\alpha|a\rangle$$

and then we operate on $|a\rangle$ with the operator $\tilde{F}$ corresponding to the measurement we want to perform, i.e.,

$$\tilde{F}|a\rangle = \sum_\alpha \tilde{F}|\alpha\rangle\langle\alpha|a\rangle = \sum_\alpha F_\alpha|\alpha\rangle\langle\alpha|a\rangle \ .$$

Note: The operation of $\tilde{F}$ on the state $|a\rangle$ gives a new state $|a'\rangle$, so that

$$|a'\rangle = \sum_\alpha F_\alpha|\alpha\rangle\langle\alpha|a\rangle = \sum_\alpha |\alpha\rangle F_\alpha\langle\alpha|a\rangle \ .$$

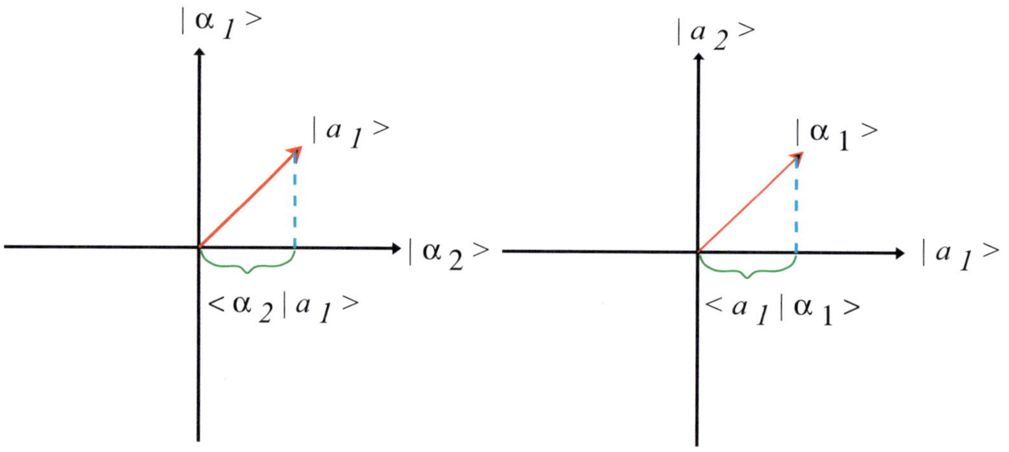

Figure 11.2: The projection of the state $|a_1\rangle$ on to the basis $|\alpha_1\rangle$ left, and the projection of the state $|\alpha_1\rangle$ on the basis $|a_1\rangle$ on the right.

Here $|a'\rangle \neq (\text{const.})|a\rangle$. To determine the average value of the operator $\tilde{F}$, we calculate

$$\begin{aligned} \langle a|\tilde{F}|a\rangle &= \langle a|a'\rangle = \sum_\alpha \langle a|\alpha\rangle F_\alpha\langle\alpha|a\rangle \\ &= \sum_\alpha |\langle\alpha|a\rangle|^2 \, F_\alpha \ . \end{aligned} \tag{11.50}$$

This implies that $|\langle\alpha|a\rangle|^2$ is the probability of finding the system described by the state $|a\rangle$ in the quantum state that is an eigenstate of the operator $\tilde{F}$. Therefore

$$\langle\alpha|a\rangle = \text{probability amplitude of obtaining } F_\alpha \text{ for the measurement of the observable } \tilde{F} \text{ on a system in a state } |a\rangle.$$

If $|a\rangle$ is an eigenstate of the operator $\tilde{A}$, i.e.,

$$\tilde{A}|a\rangle = A_a|a\rangle$$

then

$$\langle a|\alpha\rangle = \text{probability amplitude of obtaining } A_a \text{ for the measurement of the observable } \tilde{A} \text{ on a system in state } |\alpha\rangle.$$

Note that

$$\langle a|\alpha\rangle = (\langle\alpha|a\rangle)^\dagger \ .$$

11.3 Representation of Operators

Having established the fact that we can find a set of basis states $|\alpha\rangle$ $(\alpha = 1, 2, \cdots)$, such that

$$\sum_\alpha |\alpha\rangle\langle\alpha| = \tilde{I} \ ,$$

we now can write operators in a format that allows us to extract numbers from quantum theory that can be compared with experiment. For example, consider the operator $\tilde{A}$ in whose eigenvalues and eigenstates we are interested. We can write the operator $\tilde{A}$ in terms of the basis states $|\alpha\rangle$ $(\alpha = 1, 2, \cdots)$ as

$$\tilde{A} = \tilde{I}\tilde{A}\tilde{I} = \sum_\alpha \sum_{\alpha'} |\alpha\rangle\langle\alpha|\tilde{A}|\alpha'\rangle\langle\alpha'| \ . \tag{11.51}$$

It is clear that if we know all the matrix elements of the operator $\tilde{A}$, i.e., $\langle\alpha|\tilde{A}|\alpha'\rangle$, then we know the operator $\tilde{A}$.

In most problems in quantum mechanics when we perform a measurement, the system is in an eigenstate of the operator corresponding to the quantity being measured. For example, when we measure the energy of an electron in the hydrogen atom, the electron is in an eigenstate of the Hamiltonian, and is described by the wave function which is a solution of the Schrödinger equation. That means given the operator $\tilde{A}$ we need to find the state $|a\rangle$ such that

$$\tilde{A}|a\rangle = A_a|a\rangle \ , \tag{11.52}$$

or

$$\sum_{\alpha} \tilde{A}|\alpha\rangle\langle\alpha|a\rangle = A_a|a\rangle$$

and therefore

$$\sum_{\alpha} \langle\alpha'|\tilde{A}|\alpha\rangle\langle\alpha|a\rangle = A_a\langle\alpha'|a\rangle \ . \tag{11.53}$$

Thus given the operator $\tilde{A}$, we can determine the matrix elements $\langle\alpha'|\tilde{A}|\alpha\rangle$ and then solve the above equation for $\langle\alpha|a\rangle$. In Eq. (11.53) we have a standard eigenvalue problem, and the equation can be written as

$$\sum_{\alpha} \left\{\langle\alpha'|\tilde{A}|\alpha\rangle - A_a\delta_{\alpha'\alpha}\right\} \langle\alpha|a\rangle = 0 \ . \tag{11.54}$$

For this equation to have solutions, we require that the determinant of coefficient be zero, i.e.,

$$\det\left\{\langle\alpha'|\tilde{A}|\alpha\rangle - A_a\delta_{\alpha'\alpha}\right\} = 0 \ . \tag{11.55}$$

This determines the eigenvalues. Since $\tilde{A} = \tilde{A}^\dagger$, i.e., $\tilde{A}$ is Hermitian then

$$\left(\langle\alpha'|\tilde{A}|\alpha\rangle\right)^\dagger = \langle\alpha|\tilde{A}^\dagger|\alpha'\rangle = \langle\alpha|\tilde{A}|\alpha'\rangle \ , \tag{11.56}$$

and the matrix $\langle\alpha'|\tilde{A}|\alpha\rangle$ is Hermitian. If we make our basis states $|\alpha\rangle$ finite in dimension, i.e., $\alpha = 1, 2, \cdots, N$, then the solution of Eqs. (11.54) and (11.55) reduces to that of solving N homogeneous algebraic equations. Furthermore, since the matrix $\langle\alpha'|\tilde{A}|\alpha\rangle$ is Hermitian, the eigenvalues are real. Having determined the eigenvalues from Eq. (11.55), we can determine the eigenvectors from Eq. (11.54). From these eigenvectors we can construct the matrix U defined as

$$U_{a\alpha} = \langle a|\alpha\rangle \ . \tag{11.57}$$

Then since the eigenvectors are orthogonal

$$\left[\tilde{U}\tilde{U}^\dagger\right]_{aa'} = \sum_{\alpha} \langle a|\alpha\rangle\langle\alpha|a'\rangle = \delta_{aa'} \tag{11.58}$$

and the matrix $\tilde{U}$ is unitary. We also have

$$\sum_{\alpha\alpha'} \langle a|\alpha\rangle\langle\alpha|\tilde{A}|\alpha'\rangle\langle\alpha'|a'\rangle = \langle a|\tilde{A}|a'\rangle = A_a\,\delta_{aa'}$$

or

$$\left[\tilde{U}\tilde{A}\tilde{U}^\dagger\right]_{aa'} = A_a\,\delta_{aa'} \ , \tag{11.59}$$

i.e., $\tilde{U}\tilde{A}\tilde{U}^\dagger$ is diagonal with its diagonal elements the eigenvalues of $\tilde{A}$.

Suppose we have two operators $\tilde{A}$ and $\tilde{B}$ which are linear and Hermitian. We then can construct two different sets of basis states given by

$$\tilde{A}|a\rangle = A_a|a\rangle \qquad \text{and} \qquad \tilde{B}|b\rangle = B_b|b\rangle \ . \tag{11.60}$$

If we have a general state vector $|\cdots\rangle$ which is written in terms of the basis of the eigenstates of the operator $\tilde{A}$, i.e., we have $\langle a|\cdots\rangle$, how can we write this general state vector $|\cdots\rangle$ in terms of the eigenstates of the operator $\tilde{B}$? In other words, what is the relationship between $\langle b|\cdots\rangle$ and $\langle a|\cdots\rangle$? Using the completeness and orthonormality of the two basis, we have

$$\langle b|\cdots\rangle = \sum_a \langle b|a\rangle\langle a|\cdots\rangle = \sum_a U_{ba}\langle a|\cdots\rangle \ , \tag{11.61}$$

while

$$\langle a|\cdots\rangle = \sum_b \langle a|b\rangle\langle b|\cdots\rangle = \sum_b U^*_{ba}\langle b|\cdots\rangle \ . \tag{11.62}$$

Thus the matrix $\tilde{U}$ can be used to relate $\langle a|\cdots\rangle$ and $\langle b|\cdots\rangle$.

In a similar way, if we have an operator $\tilde{F}$ in terms of one set of basis states, we can write it in terms of another set of basis states. This is achieved by making use of the completeness of the states which allows us to write

$$\langle a|\tilde{F}|a'\rangle = \sum_{bb'} \langle a|b\rangle\langle b|\tilde{F}|b'\rangle\langle b'|a'\rangle \tag{11.63}$$

with the inverse relation given by

$$\langle b|\tilde{F}|b'\rangle = \sum_{aa'} \langle b|a\rangle\langle a|\tilde{F}|a'\rangle\langle a'|b'\rangle \ . \tag{11.64}$$

The above two equations can be written in matrix form as

$$\tilde{F}^a = \tilde{U}^\dagger\tilde{F}^b\tilde{U} \qquad \text{and} \qquad \tilde{F}^b = \tilde{U}\tilde{F}^a\tilde{U}^\dagger \ . \tag{11.65}$$

It is clear from these relations between the matrix elements of $\tilde{F}$ in the two basis states, that $\tilde{U}$ must satisfy the condition

$$\tilde{U}\tilde{U}^\dagger = \tilde{U}^\dagger\tilde{U} = \tilde{I} \ , \tag{11.66}$$

i.e.,

$$\tilde{U}^\dagger = \tilde{U}^{-1} \ . \tag{11.67}$$

In other words, the matrix $\tilde{U}$ is unitary. From this it follows that the scalar product is independent of the basis, i.e., if

$$|\alpha_u\rangle = \tilde{U}|\alpha\rangle \qquad \text{and} \qquad |\beta_u\rangle = \tilde{U}|\beta\rangle \tag{11.68}$$

then

$$\langle\alpha_u|\beta_u\rangle = \langle\alpha|\tilde{U}^\dagger\tilde{U}|\beta\rangle = \langle\alpha|\beta\rangle \ , \tag{11.69}$$

i.e., the scalar product is the same in the transformed basis as in the original basis.

Definitions:

1. The basis in terms of which a state vector and operator are written is referred to as the *representation.*

2. The change of basis through a unitary matrix or operator is referred to as a *change of representation.*

11.3.1 The Coordinate Representation

So far the representation we have been using is the coordinate representation (or r-representation), i.e, the basis states are the eigenstates of the position operator

$$\tilde{\vec{r}}|\vec{r}\rangle = \vec{r}\,|\vec{r}\rangle \tag{11.70}$$

with the orthonormality and completeness of the states $|\vec{r}\rangle$ given by

$$\langle\vec{r}\,|\vec{r}\,'\rangle = \delta\left(\vec{r} - \vec{r}\,'\right) \quad \text{and} \quad \int d^3r\,|\vec{r}\,\rangle\langle\vec{r}\,| = \tilde{I} \ . \tag{11.71}$$

Here we have replaced the sum by an integral since $\vec{r}$ takes on continuous values.

To illustrate some operators in this representation let us consider the momentum operator in one dimension. We have that the position and momentum operator satisfy the commutation relation

$$\tilde{x}\tilde{p} - \tilde{p}\tilde{x} = i\hbar \ . \tag{11.72}$$

If we take the matrix element of this equation in coordinate representation, we get

$$\langle x|\left(\tilde{x}\tilde{p} - \tilde{p}\tilde{x}\right)|x'\rangle = i\hbar\langle x|x'\rangle = i\hbar\delta(x - x') \tag{11.73}$$

but we have that

$$\tilde{x}|x\rangle = x|x\rangle \tag{11.74}$$

so that Eq. (11.73) can be written as

$$(x - x')\langle x|\tilde{p}|x'\rangle = i\hbar\delta(x - x') = -i\hbar(x - x')\delta'(x - x') \ , \tag{11.75}$$

where $\delta'(x) = \frac{d}{dx}\delta(x)$. The second equality can be justified by recalling that the Dirac δ-function always appears in an integral, and an integration by parts allows us to write [4]

$$f'(x)\,\delta(x-a) = -f(x)\,\delta'(x-a)\ . \tag{11.76}$$

We now can write Eq. (11.75) as

$$\langle x|\tilde{p}|x'\rangle = -i\hbar\frac{d}{dx}\delta(x-x') = -i\hbar\frac{d}{dx}\langle x|x'\rangle = i\hbar\frac{d}{dx'}\langle x|x'\rangle$$

or

$$\tilde{p}|x'\rangle = i\hbar\frac{d}{dx'}\,|x'\rangle\ . \tag{11.77}$$

For the momentum operator acting on a general state $|\cdots\rangle$, we have, using the completeness of our basis states,

$$\begin{aligned} \tilde{p}|\cdots\rangle &= \int \tilde{p}\,|x\rangle\,dx\,\langle x|\cdots\rangle = i\hbar\int \frac{d}{dx}\,|x\rangle\,dx\,\langle x|\cdots\rangle \\ &= -i\hbar\int |x\rangle\,dx\,\frac{d}{dx}\,\langle x|\cdots\rangle\ , \end{aligned} \tag{11.78}$$

where the final expression is a result of an integration by parts and the assumption that the state $\langle x|\cdots\rangle$ goes to zero for $x \to \pm\infty$. In general, we have for a function of the momentum operator that [5]

---

[4]For the integral of a δ-function we have

$$\begin{aligned} \int_{-\infty}^{+\infty} dx\,\delta(x-x') &= \int_{-\infty}^{+\infty} dx\,\left[\frac{d}{dx}(x-x')\right]\,\delta(x-x') \\ &= -\int_{-\infty}^{+\infty} dx\,(x-x')\,\delta'(x-x') \end{aligned}$$

[5]The function of an operator $f(\tilde{p})$ is only defined in terms of a power series in the operator $\tilde{p}$, i.e.,

$$f(\tilde{p}) = a_0 + a_1\tilde{p} + a_2\tilde{p}^2 + \cdots\ .$$

This is analogous to the definition of a function of a matrix, e.g.,

$$e^{-\tilde{p}} = 1 - \tilde{p} + \frac{1}{2!}\tilde{p}^2 - \frac{1}{3!}\tilde{p}^3 + \cdots\ ,$$

where $\tilde{p}^2 = \tilde{p}\tilde{p}$ and $\tilde{p}^3 = \tilde{p}\tilde{p}\tilde{p}$, $\cdots$. In the event that $f(\tilde{p})$ acts on an eigenstate of $\tilde{p}$, i.e., $\tilde{p}|p_n\rangle = p_n|p_n\rangle$ then

$$f(\tilde{p})\,|p_n\rangle = f(p_n)\,|p_n\rangle\ .$$

$$f(\tilde{p})\,|\cdots\rangle = \int dx\,|x\rangle\, f\left(-i\hbar\frac{d}{dx}\right)\langle x|\cdots\rangle\ . \tag{11.79}$$

In writing this result we have made use of the fact that

$$\tilde{p} \longrightarrow \left(-i\hbar\frac{d}{dx}\right) \quad \text{and} \quad \tilde{p}^2 \longrightarrow \left(-i\hbar\frac{d}{dx}\right)^2\ .$$

If we now have a function of the momentum and position operators, i.e., $\tilde{A}(\tilde{p},\tilde{x})$, and we want to solve the equation

$$\tilde{A}\,(\tilde{p},\tilde{x})\ |a\rangle = A_a\,|a\rangle\ , \tag{11.80}$$

we first have to write $|a\rangle$ in terms of our basis states, i.e.,

$$|a\rangle = \int dx\,|x\rangle\,\langle x|a\rangle$$

so that

$$\tilde{A}\,|a\rangle = \int dx\,|x\rangle\, A\left(-i\hbar\frac{d}{dx}, x\right)\langle x|a\rangle\ . \tag{11.81}$$

We now use the orthogonality of our basis states (i.e., $\langle x|x'\rangle = \delta(x-x')$) to write Eq. (11.80) in the coordinate representation as

$$A\left(-i\hbar\frac{d}{dx}, x\right)\langle x|a\rangle = A_a\,\langle x|a\rangle\ . \tag{11.82}$$

In a similar manner, we have the coordinate representation in three-dimensions to correspond to

$$\tilde{\vec{r}} \longrightarrow \vec{r}\ , \qquad \tilde{\vec{p}} \longrightarrow -i\hbar\vec{\nabla}\ , \quad \text{and} \quad |a\rangle \longrightarrow \langle\vec{r}\,|a\rangle\ . \tag{11.83}$$

As an illustration of Eq. (11.82) consider the case when the operator $\tilde{A}$ is the Hamiltonian which is a function of the momentum and position operators. In this case our operator equation takes the form

$$\tilde{H}\left(\tilde{\vec{p}},\tilde{\vec{r}}\right)\,|E_\alpha\rangle = E_\alpha\,|E_\alpha\rangle\ . \tag{11.84}$$

In coordinate representation this equation reduces to the standard Schrödinger equation as we are familiar with, i.e., in Dirac notation it takes the form

$$H\left(-i\hbar\vec{\nabla},\vec{r}\right)\langle\vec{r}\,|E_\alpha\rangle = E_\alpha\,\langle\vec{r}\,|E_\alpha\rangle\ , \tag{11.85}$$

where the wave function in coordinate space is defined as $\langle\vec{r}\,|E_\alpha\rangle$.

11.3.2 The Momentum Representation

If we take for our basis the eigenstates of the momentum operator $\tilde{\vec{p}}$, then we have

$$\tilde{\vec{p}}\,|\vec{p}\rangle = \vec{p}\,|\vec{p}\rangle \ , \quad \langle\vec{p}\,|\vec{p}\,'\rangle = \delta\left(\vec{p}-\vec{p}\,'\right) \quad \text{and} \quad \int |\vec{p}\rangle\, d^3p\, \langle\vec{p}\,| = \tilde{I} \ . \tag{11.86}$$

Proceeding in a similar manner as in the coordinate representation, we can write any function of the momentum operator acting on a general state as

$$f(\tilde{\vec{p}})\,|\cdots\rangle = \int d^3p\,|\vec{p}\rangle\, f(\vec{p})\,\langle\vec{p}\,|\cdots\rangle \ . \tag{11.87}$$

To determine the coordinate operator in momentum representation, we consider again the matrix element of the commutation relation of the coordinate and the momentum in one-dimension, as given in Eq. (11.72), in momentum representation, i.e.,

$$\langle p|\left(\tilde{x}\tilde{p}-\tilde{p}\tilde{x}\right)|p'\rangle = i\hbar\,\langle p|p'\rangle = i\hbar\,\delta(p-p') \ .$$

Using Eq. (11.86), we have

$$(p-p')\,\langle p|\tilde{x}|p'\rangle = -i\hbar\,\delta(p-p') = i\hbar\,(p-p')\,\delta'(p-p') \ .$$

Therefore, we have for the position operator, the matrix element

$$\langle p|\tilde{x}|p'\rangle = i\hbar\,\frac{d}{dp}\,\langle p|p'\rangle = -i\hbar\,\frac{d}{dp'}\,\langle p|p'\rangle \ .$$

This allows us to write

$$\tilde{x}\,|p'\rangle = -i\hbar\,\frac{d}{dp'}\,|p'\rangle \ . \tag{11.88}$$

Now any function of the position operator, when acting on an eigenstate of the momentum operator, gives

$$f(\tilde{x})\,|p\rangle = f\left(-i\hbar\frac{d}{dp}\right)\,|p\rangle \ ,$$

and for the most general state $|\cdots\rangle$, we have

$$f(\tilde{x})\,|\cdots\rangle = \int dp\,|p\rangle\, f\left(i\hbar\frac{d}{dp}\right)\,\langle p|\cdots\rangle \ . \tag{11.89}$$

We are now in a position to write our standard eigenvalue problem for the operator $\tilde{A}(\tilde{\vec{p}},\tilde{\vec{r}})$ in momentum space. Given that

$$\tilde{A}(\tilde{\vec{p}},\tilde{\vec{r}})\,|a\rangle = A_a\,|a\rangle \ ,$$

we can write this equation in momentum representation as

$$A(\vec{p}, i\hbar\vec{\nabla}_p)\,\langle\vec{p}\,|a\rangle = A_a\,\langle\vec{p}\,|a\rangle \ . \tag{11.90}$$

Example 1: As a first example, let us consider the case when the operator $\tilde{A}(\tilde{\vec{p}}, \tilde{\vec{r}})$ is just the position operator, i.e.,

$$\tilde{A}(\tilde{\vec{p}}, \tilde{\vec{r}}) = \tilde{\vec{r}} \, . \tag{11.91}$$

Then Eq. (11.90) takes the form

$$i\hbar \vec{\nabla}_p \langle \vec{p} | \vec{r} \rangle = \vec{r} \langle \vec{p} | \vec{r} \rangle \, .$$

The solution to this first order differential equation is given by

$$\langle \vec{p} | \vec{r} \rangle = \frac{1}{(2\pi\hbar)^{3/2}} \, e^{-i\vec{k}\cdot\vec{r}} \quad \text{with} \quad \vec{k} = \frac{\vec{p}}{\hbar} \, . \tag{11.92}$$

This is the eigenstate of the position operator in momentum representation. The eigenstate of the momentum operator in coordinate representation is $\langle \vec{r} | \vec{p} \rangle$, and is given by

$$\langle \vec{r} | \vec{p} \rangle = (\langle \vec{p} | \vec{r} \rangle)^\dagger = \frac{1}{(2\pi\hbar)^{3/2}} \, e^{i\vec{k}\cdot\vec{r}} \, . \tag{11.93}$$

Example 2: For our second example, we consider the case when $\tilde{A}$ is the Hamiltonian operator, i.e.,

$$\tilde{H}(\tilde{\vec{p}}, \tilde{\vec{r}}) = \frac{1}{2m} \, \tilde{p}^2 + \tilde{V}(\tilde{\vec{r}}, \tilde{\vec{p}}) \, , \tag{11.94}$$

We have taken the potential to be as general as possible and thus a function of both the position and the momentum operator. In coordinate representation, $\tilde{H}$ takes the form

$$\begin{aligned} \langle \vec{r} | \tilde{H} | \vec{r}\,' \rangle &= \frac{1}{2m} \langle \vec{r} | \tilde{p}^2 | \vec{r}\,' \rangle + \langle \vec{r} | \tilde{V} | \vec{r}\,' \rangle \\ &= -\frac{\hbar^2}{2m} \, \nabla^2_{r'} \, \delta(\vec{r} - \vec{r}\,') + \langle \vec{r} | \tilde{V} | \vec{r}\,' \rangle \, . \end{aligned} \tag{11.95}$$

The δ-function in the first term tells us that the operator $\tilde{p}^2$ is diagonal in the coordinate representation.

The time independent Schrödinger equation (see Eq. (4.7a)) now takes the form of

$$\tilde{H} \, | E \rangle = E \, | E \rangle$$

or, in coordinate representation it reduces to

$$\langle \vec{r} | \tilde{H} | E \rangle = E \, \langle \vec{r} | E \rangle \, .$$

Introducing a complete set of eigenstates of the position operator $\tilde{\vec{r}}$, we get

$$\int d^3r' \, \langle \vec{r}\,|\tilde{H}|\vec{r}\,'\rangle\langle \vec{r}\,'|E\rangle = E\,\langle \vec{r}\,|E\rangle$$

Using the expression for the Hamiltonian in coordinate representation as given in Eq. (11.95), we get

$$\int d^3r' \left(-\frac{\hbar^2}{2m}\,\nabla^2_{r'}\right)\,\delta(\vec{r}-\vec{r}\,')\,\langle \vec{r}\,'|E\rangle + \int d^3r' \,\langle \vec{r}\,|\tilde{V}|\vec{r}\,'\rangle\langle \vec{r}\,'|E\rangle = E\,\langle \vec{r}\,|E\rangle\ .$$

Integrating the first term on the left hand side of the above equation by parts, twice, we get

$$-\frac{\hbar^2}{2m}\,\nabla^2_r\,\langle \vec{r}\,|E\rangle + \int d^3r' \,\langle \vec{r}\,|\tilde{V}|\vec{r}\,'\rangle\langle \vec{r}\,'|E\rangle = E\,\langle \vec{r}\,|E\rangle\ . \tag{11.96}$$

For most of the potentials we have considered, $\tilde{V}$ is a function of the position operator only, and in that case

$$\langle \vec{r}\,|\tilde{V}(\vec{r}\,)|\vec{r}\,'\rangle = V(\vec{r}\,)\,\delta(\vec{r}-\vec{r}\,')\ , \tag{11.97}$$

and Eq. (11.96) reduces to the ordinary Schrödinger equation, i.e.,

$$\left\{-\frac{\hbar^2}{2m}\nabla^2_r + V(\vec{r}\,)\right\}\,\langle \vec{r}\,|E\rangle = E\,\langle \vec{r}\,|E\rangle\ . \tag{11.98}$$

In other words, $\langle \vec{r}\,|E\rangle$ is the usual wave function encountered before.

In momentum representation, the Schrödinger equation takes the form

$$\langle \vec{p}\,|\tilde{H}|E\rangle = E\,\langle \vec{p}\,|E\rangle\ ,$$

or on introducing a complete set of eigenstates of the momentum operator, we get

$$\int d^3p' \,\langle \vec{p}\,|\tilde{H}|\vec{p}\,'\rangle\langle \vec{p}\,'|E\rangle = E\,\langle \vec{p}\,|E\rangle\ .$$

But we have for the Hamiltonian in momentum space that

$$\langle \vec{p}\,|\tilde{H}|\vec{p}\,'\rangle = \frac{p^2}{2m}\,\delta(\vec{p}-\vec{p}\,') + \langle \vec{p}\,|\tilde{V}|\vec{p}\,'\rangle\ . \tag{11.99}$$

This allows us to write the Schrödinger equation in momentum representation as

$$\left(E - \frac{p^2}{2m}\right)\,\langle \vec{p}\,|E\rangle = \int d^3p' \,\langle \vec{p}\,|\tilde{V}|\vec{p}\,'\rangle\langle \vec{p}\,'|E\rangle\ . \tag{11.100}$$

We will see when we come to formal scattering theory, that in some cases it is advantageous to work in momentum space to get the scattering amplitude. In these cases we will need to start with the Schrödinger equation in momentum representation. In cases when the potential is not diagonal in coordinate representation, the Schrödinger equation in coordinate representation is an integrodifferential equation. On the other hand, in momentum representation we have an integral equation that is easier to solve. Thus for each problem we have a representation in which the Schrödinger equation is simplest to solve, and we should take advantage of this convenience.

11.3.3 Angular Momentum Representation

In Chapter 8, we showed that if $\vec{\tilde{L}}$ is the angular momentum operator then

$$\begin{aligned} \left[\tilde{L}^2, \tilde{L}_i\right] &= 0 \quad \text{for} \quad i = 1,2,3 \\ \left[\tilde{L}_i, \tilde{L}_j\right] &= i\hbar \sum_k \varepsilon_{ijk}\, \tilde{L}_k \end{aligned} \tag{11.101}$$

where the totally antisymmetric tensor ε_{ijk} is defined (see Eq. (8.23)) as

$$\varepsilon_{ijk} = \begin{cases} +1 & \text{If } (i,j,k) \text{ is an even permutation of } (1,2,3). \\ -1 & \text{If } (i.j.k) \text{ is an odd permutation of } (1,2,3). \\ 0 & \text{If any two of } (i,j,k) \text{ are equal.} \end{cases} \quad . \tag{11.102}$$

Since L^2 and L_3 commute, we can construct simultaneous eigenstates of L^2 and L_3, i.e.,

$$\begin{aligned} \tilde{L}^2\, Y_{\ell m}(\theta,\phi) &= \hbar^2 \ell(\ell+1)\, Y_{\ell m}(\theta,\phi) \\ \tilde{L}_3\, Y_{\ell m}(\theta,\phi) &= \hbar m\, Y_{\ell m}(\theta,\phi) \end{aligned} \tag{11.103}$$

where $Y_{\ell m}$ turns out to be the usual spherical harmonics as defined in Chapter 8.

In the present section we would like to generalize this result using the matrix formulation. Consider the vector operator $\vec{\tilde{J}}$ with components $\tilde{J}_1$, $\tilde{J}_2$ and $\tilde{J}_3$ which satisfy a set of commutation relations identical to the angular momentum operator, i.e.,

$$\left[\tilde{J}_i, \tilde{J}_j\right] = i\hbar\varepsilon_{ijk}\, \tilde{J}_k \ . \tag{11.104}$$

From this point on, we use of the convention that there is an explicit sum over the repeated index, which in the case of Eq. (11.104) is the index k. Taking

$$\tilde{J}^2 = \sum_{i=1}^{3} \tilde{J}_i^2 \ , \tag{11.105}$$

we can show, using Eq. (11.104), that

$$\left[\tilde{J}^2, \tilde{J}_k\right] = \sum_{i=1}^{3} \left[\tilde{J}_i^2, \tilde{J}_k\right] = 0 \ . \tag{11.106}$$

We now define two new operators

$$\tilde{J}_+ = \tilde{J}_1 + i\tilde{J}_2 \qquad \tilde{J}_- = \tilde{J}_1 - i\tilde{J}_2 \ , \tag{11.107}$$

and then using the commutation relation, Eq. (11.104), we can show that

$$\left[\tilde{J}_3, \tilde{J}_\pm\right] = \pm\hbar\tilde{J}_\pm \qquad \left[\tilde{J}_+, \tilde{J}_-\right] = 2\hbar\tilde{J}_3 \ . \tag{11.108}$$

The commutation relations in Eqs. (11.104) and (11.108) are a special case of a Lie Algebra. Since $\tilde{J}^2$ and $\tilde{J}_3$ are two commuting operators, we have states that are simultaneous eigenstates of both $\tilde{J}^2$ and $\tilde{J}_3$, i.e.,

$$\tilde{J}^2 \left|\beta, m\right\rangle = \hbar^2\beta|\beta, m\rangle \tag{11.109a}$$

$$\tilde{J}_3 \left|\beta, m\right\rangle = \hbar m \left|\beta, m\right\rangle \ . \tag{11.109b}$$

The problem now is to determine the eigenvalues of $\tilde{J}^2$ and $\tilde{J}_3$, i.e., β and m. Since $\tilde{J}_3$ is a component of $\tilde{\vec{J}}$, we expect that for a given value of β, the eigenvalue m has a maximum value, m_{max}, and a minimum value, m_{min}. Multiplying Eq. (11.109b) by $\tilde{J}_+$ results in

$$\tilde{J}_+ \tilde{J}_3 \left|\beta, m\right\rangle = \hbar m \, \tilde{J}_+ \left|\beta, m\right\rangle \ .$$

Now using of the commutation relation in Eq. (11.108) in the form

$$\tilde{J}_+\tilde{J}_3 = \tilde{J}_3\tilde{J}_+ - \hbar\tilde{J}_+$$

allows us to write

$$\tilde{J}_3\tilde{J}_+ \left|\beta, m\right\rangle = \hbar(m+1)\tilde{J}_+ \left|\beta, m\right\rangle \ . \tag{11.110}$$

This means that $\tilde{J}_+ \left|\beta, m\right\rangle$ is an eigenstate of $\tilde{J}_3$ with eigenvalue $\hbar(m+1)$ except when $m = m_{max}$ in which case

$$\tilde{J}_+ \left|\beta, m_{max}\right\rangle = 0 \ . \tag{11.111}$$

In a similar manner, we multiply Eq. (11.109b) by $\tilde{J}_-$ to have

$$\tilde{J}_-\tilde{J}_3 \left|\beta, m\right\rangle = \hbar m \, \tilde{J}_- \left|\beta, m\right\rangle \ ,$$

and using the commutation relation given in Eq. (11.108), we get

$$\tilde{J}_3\tilde{J}_- \left|\beta, m\right\rangle = \hbar(m-1) \, \tilde{J}_- \left|\beta, m\right\rangle \ . \tag{11.112}$$

In other words, $\tilde{J}_- \left|\beta, m\right\rangle$ is an eigenstate of $\tilde{J}_3$ with an eigenvalue $\hbar(m-1)$, except when $m = m_{min}$ in which case

$$\tilde{J}_- \left|\beta, m_{min}\right\rangle = 0 \ . \tag{11.113}$$

We thus have established that $\tilde{J}_+$ ($\tilde{J}_-$) increases (decreases) m by one unit of $\hbar$. In this way, given the one state $|\beta, m\rangle$, we can generate all the states with different m and a fixed value of β by the action of $\tilde{J}_+$ or $\tilde{J}_-$. Using the definition of $\tilde{J}_+$ and $\tilde{J}_-$, we can write

$$\begin{aligned} \tilde{J}_-\tilde{J}_+ &= \left(\tilde{J}_1 - i\tilde{J}_2\right)\left(\tilde{J}_1 + i\tilde{J}_2\right) \\ &= \tilde{J}^2 - \tilde{J}_3^2 - \hbar\tilde{J}_3 \ , \end{aligned} \tag{11.114}$$

and then

$$\tilde{J}_- \tilde{J}_+ |\beta, m_{max}\rangle = \hbar^2 \left(\beta - m_{max}^2 - m_{max}\right) |\beta, m_{max}\rangle = 0 \ .$$

Assuming the state $|\beta, m_{max}\rangle \neq 0$, then the quantity in brackets should be zero and this can be satisfied if

$$\beta = m_{max}\,(m_{max} + 1) \ . \tag{11.115}$$

On the other hand, since

$$\tilde{J}_+ \tilde{J}_- = \tilde{J}^2 - \tilde{J}_3^3 + \hbar \tilde{J}_3$$

the action of this operator on the state $|\beta, m_{min}\rangle$ gives

$$\tilde{J}_+ \tilde{J}_- |\beta, m_{min}\rangle = \hbar^2 \left(\beta - m_{min}^2 + m_{min}\right) |\beta, m_{min}\rangle = 0 \ .$$

From this we have that

$$\beta = m_{min}^2 - m_{min} \ . \tag{11.116}$$

Comparing Eqs. (11.115) and (11.116), we get

$$m_{max}\,(m_{max} + 1) - m_{min}^2 + m_{min} = 0 \ .$$

This can be written as

$$(m_{max} + m_{min})\,(m_{max} - m_{min} + 1) = 0$$

and has two possible solutions: $m_{max} = -m_{min}$ and $m_{max} - m_{min} = -1$. Since $m_{max} \geq m_{min}$, the only possible solution is

$$m_{max} = -m_{min} \ . \tag{11.117}$$

Since there are as many values of m greater than zero as there are less than zero $m_{max} - m_{min} = 2j$. On the other hand since m changes by intervals of one then $j = 0, \frac{1}{2}, 1, \cdots$, i.e.,

$$m_{max} - m_{min} = 2j \quad \text{with} \quad j = 0,\ \frac{1}{2},\ 1,\ \frac{3}{2}, \cdots \ . \tag{11.118}$$

In other words, m has the range of values

$$-\,j \leq m \leq j \qquad \text{with } (2j+1) \text{ possible values.} \tag{11.119}$$

This means that $m_{max} = j$, and the eigenvalue corresponding to the operator $\tilde{J}^2$, i.e., β, is given using Eq. (11.115) as

$$\beta = j(j+1) \ , \tag{11.120}$$

and we can write our eigenvalue problem for the operators $\tilde{J}^2$ and $\tilde{J}_3$ as

$$\begin{aligned} \tilde{J}^2 \,|j, m\rangle &= \hbar^2 j(j+1)\,|j, m\rangle \\ \tilde{J}_3 \,|j, m\rangle &= \hbar m \,|j, m\rangle \ . \end{aligned} \tag{11.121}$$

Here we observe that we have solved the above eigenvalue problem to the extent of determining the eigenvalues, the number of eigenstates and how they are related, without solving any differential equations. However, we have had to make use of the commutation relation of the operators associated with this problem. This procedure can be applied to a number of problems, and we will use this method to solve the harmonic oscillator problem in Chapter 13 when we consider the occupation number representation.

Although we have demonstrated that the operators $\tilde{J}_+$ ($\tilde{J}_-$) will increase (decrease) the value of m, we need to determine the normalization of the new state generated by these two operators. To determine this normalization let us first consider the action of $\tilde{J}_-$ on the state with maximum m, i.e.,

$$|j, j-1\rangle = C_{j,j-1}\, \tilde{J}_- \,|j, j\rangle \ . \tag{11.122}$$

We then can write the general state as

$$|j, m\rangle = C_{j,m} \left(\tilde{J}_-\right)^{j-m} |j, j\rangle \ , \tag{11.123}$$

where $C_{j,m}$ is a normalization constant to be determined. This can be rewritten as

$$\begin{aligned} |j, m\rangle &= C_{j,m}\, \tilde{J}_- \left\{ \left(\tilde{J}_-\right)^{j-m-1} |j, j\rangle \right\} \\ &= \frac{C_{j,m}}{C_{j,m+1}}\, \tilde{J}_- \,|j, m+1\rangle \ , \end{aligned} \tag{11.124}$$

and therefore

$$\langle j, m| = \left(\frac{C_{j,m}}{C_{j,m+1}}\right)^* \langle j, m+1|\, \tilde{J}_+ \ , \tag{11.125}$$

since $\left(\tilde{J}_-\right)^* = \tilde{J}_+$. The requirement that $|j, m\rangle$ be normalized is then

$$\begin{aligned} \langle j, m|j, m\rangle = 1 &= \left|\frac{C_{j,m}}{C_{j,m+1}}\right|^2 \langle j, m+1|\tilde{J}_+\, \tilde{J}_-|j, m+1\rangle \\ &= \left|\frac{C_{j,m}}{C_{j,m+1}}\right|^2 \langle j, m+1| \left(\tilde{J}^2 - \tilde{J}_3^2 + \hbar \tilde{J}_3\right) |j, m+1\rangle \\ &= \left|\frac{C_{j,m}}{C_{j,m+1}}\right|^2 \hbar^2 \left(j(j+1) - m(m+1)\right) \ , \end{aligned}$$

and as a result

$$\left|\frac{C_{j,m+1}}{C_{j,m}}\right| = \hbar\sqrt{j(j+1) - m(m+1)} \ .$$

We now can write

$$\tilde{J}_- \,|j, m\rangle = \hbar\sqrt{j(j+1) - m(m-1)}\, |j, m-1\rangle \ . \tag{11.126}$$

Similarly, we can show that

$$\tilde{J}_+ \, |j,m\rangle = \hbar\sqrt{j(j+1) - m(m+1)} \; |j,m+1\rangle \; . \tag{11.127}$$

The above results are valid for $j = \ell$ (where ℓ is an integer) and in this case the states $|\ell, m\rangle$ are the spherical harmonics, which in coordinate representation are given by

$$\langle \theta, \phi \,|\, \ell, m\rangle = Y_{\ell m}(\theta, \phi) \; . \tag{11.128}$$

For the case of $j = \frac{1}{2}, \frac{3}{2}, \cdots$ the eigenstates of $\tilde{J}^2$ and $\tilde{J}_3$ can be considered as a spinor representation. We will discuss the case of $j = \frac{1}{2}$ in the next section. The above basis of eigenstates of the two commuting operators $\tilde{J}^2$ and $\tilde{J}_3$, with $\tilde{J}^2 = \tilde{J}_1^2 + \tilde{J}_2^2 + \tilde{J}_3^2$ and the $\tilde{J}_i$ satisfying the commutation relation given in Eq. (11.104), form a basis in terms of which we can write any eigenstate of a linear Hermitian operator.

11.4 Spin $\frac{1}{2}$ Particles

As an example of the usefulness of the angular momentum, let us consider the case of $j = 1/2$, and as a first step calculate the matrices

$$\langle j, m|\tilde{J}_i|j, m'\rangle \; .$$

To determine these matrices, we recall that we have determined the result of the action of $\tilde{J}_\pm$ on the state $|j,m\rangle$, and we have that

$$\tilde{J}_1 = \frac{1}{2}\left(\tilde{J}_+ + \tilde{J}_-\right) \qquad \tilde{J}_2 = \frac{1}{2i}\left(\tilde{J}_+ - \tilde{J}_-\right) \; .$$

We then have, using the orthonormality of the states $|j,m\rangle$, that

$$\begin{aligned} \langle j, m|\tilde{J}_1|j, m'\rangle \;&=\; \frac{\hbar}{2}\Big\{ \sqrt{j(j+1) - m(m-1)} \; \delta_{m,m'+1} \\ &\qquad + \sqrt{j(j+1) - m(m+1)} \; \delta_{m,m'-1} \Big\} \; . \end{aligned} \tag{11.129}$$

For $j = 1/2$, $m = \pm 1/2$ and the above expression is the element of a 2×2 matrix σ_1, i.e.,

$$\langle j, m|\tilde{J}_1|j, m'\rangle = \frac{\hbar}{2} \, [\sigma_1]_{mm'} \; ,$$

where σ_1 is given as

$$\sigma_1 = \begin{pmatrix} 0 & 1 \\ 1 & 0 \end{pmatrix} \; . \tag{11.130}$$

In a similar manner, the matrix elements of $\tilde{J}_2$ are given by

$$\begin{aligned} \langle j,m|\tilde{J}_2|j,m'\rangle &= \frac{\hbar}{2i}\Big\{\sqrt{j(j+1)-m(m-1)}\,\delta_{m,m'+1} \\ &\qquad -\sqrt{j(j+1)-m(m+1)}\,\delta_{m,m'-1}\Big\}\,, \end{aligned} \tag{11.131}$$

and for $j = 1/2$ this reduces to the matrix element of σ_2, *i.e*

$$\langle j,m|\tilde{J}_2|j,m'\rangle = \frac{\hbar}{2}\,[\sigma_2]_{mm'}$$

where the matrix σ_2 is given as

$$\sigma_2 = \begin{pmatrix} 0 & -i \\ i & 0 \end{pmatrix}\,. \tag{11.132}$$

Finally, we have for the matrix elements of $\tilde{J}_3$ for $j = 1/2$ as the elements of the matrix σ_3, i.e.,

$$\langle j,m|\tilde{J}_3|j,m'\rangle = \frac{\hbar}{2}\,[\sigma_3]_{mm'}$$

where the matrix σ_2 is

$$\sigma_3 = \begin{pmatrix} 1 & 0 \\ 0 & -1 \end{pmatrix}\,. \tag{11.133}$$

Here, σ_1, σ_2 and σ_3 are known as the Pauli spin matrices, and in conjunction with the unit 2×2 matrix form a complete basis for writing any 2×2 matrix (see Problem 3). We also have

$$\langle j,m|\tilde{J}^2|j,m'\rangle = \frac{\hbar^2}{4}\,[\,\sigma_1\sigma_1+\sigma_2\sigma_2+\sigma_3\sigma_3\,]_{mm'}$$

with

$$[\,\sigma_1\sigma_1+\sigma_2\sigma_2+\sigma_3\sigma_3\,] = 3\begin{pmatrix} 1 & 0 \\ 0 & 1 \end{pmatrix}\,. \tag{11.134}$$

The last step follows from the fact that $\sigma_i^2 = I$. From the commutation relation of the $\tilde{J}_i$, Eq. (11.104), we can show that the Pauli spin matrices σ_i satisfy the commutation relation

$$[\sigma_i,\sigma_j] = 2i\,\varepsilon_{ijk}\,\sigma_k\,. \tag{11.135}$$

They also satisfy the anti-commutation relation

$$\{\sigma_i,\sigma_j\} \equiv \sigma_i\sigma_j+\sigma_j\sigma_i = 0 \quad \text{for } i\neq j\,.$$

or

$$\{\sigma_i,\sigma_j\} = 2\,\delta_{ij} \tag{11.136}$$

In this case, the states $|j, m\rangle$ for $j = \frac{1}{2}$ turn out to be two component spinors. To show this we consider the eigenvalue equation for the operator (or matrix) $S_3 = \tilde{J}_3 = \frac{\hbar}{2}\sigma_3$, i.e.,

$$S_3 \begin{pmatrix} u \\ v \end{pmatrix} = \pm\frac{\hbar}{2} \begin{pmatrix} u \\ v \end{pmatrix}$$

or

$$\begin{pmatrix} 1 & 0 \\ 0 & -1 \end{pmatrix} \begin{pmatrix} u \\ v \end{pmatrix} = \pm \begin{pmatrix} u \\ v \end{pmatrix} .$$

After matrix multiplication on the left hand side, this reduces to

$$\begin{pmatrix} u \\ -v \end{pmatrix} = \pm \begin{pmatrix} u \\ v \end{pmatrix} .$$

Therefore, the eigenstate corresponding to the eigenvalue of $+\frac{\hbar}{2}$ is $\begin{pmatrix} u \\ 0 \end{pmatrix}$, while the eigenstate corresponding to the eigenvalue $-\frac{\hbar}{2}$ is $\begin{pmatrix} 0 \\ v \end{pmatrix}$. After normalization, the two eigenstates of S_3 are

$$\alpha \equiv \begin{pmatrix} 1 \\ 0 \end{pmatrix} \qquad \text{and} \qquad \beta \equiv \begin{pmatrix} 0 \\ 1 \end{pmatrix} . \tag{11.137}$$

These column matrices of length two are referred to as *two-component Pauli spinors.*[6]

In general, the eigenstates of S_i are a linear combination of α and β, e.g., if we consider the case of $i = 3$ then, as observed above, α and β are the eigenstates of S_3. To determine the eigenstates of S_1 and S_2, we consider the operator

$$S_1 \cos\phi + S_2 \sin\phi ,$$

and then for $\phi = 0$ we will have the eigenstates of S_1, while for $\phi = \frac{\pi}{2}$ we will have the eigenstates of S_2. The eigenvalue equation in this case is given by

$$(S_1 \cos\phi + S_2 \sin\phi) \begin{pmatrix} u \\ v \end{pmatrix} = \frac{\hbar}{2}\lambda \begin{pmatrix} u \\ v \end{pmatrix} .$$

[6]Although spinors are different from the standard wave functions we have encountered in coordinate or momentum space, we can recast the Pauli spinor to look like an eigenstate of S_3 with eigenvalues of $\pm\frac{\hbar}{2}$, i.e.,

$$S_3 \, |\tfrac{1}{2}m\rangle = m\hbar \, |\tfrac{1}{2}m\rangle \quad \text{with} \quad m = \pm\frac{1}{2} .$$

We can write the state $|\frac{1}{2}m\rangle$ in a 'coordinate' representation, i.e., $\langle\xi|\frac{1}{2}m\rangle$, where now the coordinate can take on two values $\xi = \pm\frac{1}{2}$. In this case

$$\langle\xi|\tfrac{1}{2}m\rangle = \delta_{\xi m} .$$

Making use of the explicit matrix representation of S_1 and S_2, we can can write

$$\frac{\hbar}{2}\begin{pmatrix} 0 & e^{-i\phi} \\ e^{i\phi} & 0 \end{pmatrix}\begin{pmatrix} u \\ v \end{pmatrix} = \frac{\hbar}{2}\lambda\begin{pmatrix} u \\ v \end{pmatrix} .$$

This gives us two algebraic equations of the form

$$e^{-i\phi}\, v = \lambda\, u \qquad \text{and} \qquad e^{i\phi}\, u = \lambda\, v$$

which have a solution for

$$\lambda = \pm 1 .$$

For $\lambda = 1$, the eigenstates are

$$\frac{1}{\sqrt{2}}\begin{pmatrix} e^{-i\phi/2} \\ e^{i\phi/2} \end{pmatrix} = \frac{1}{\sqrt{2}}\left\{ e^{-i\phi/2}\,\alpha + e^{i\phi/2}\,\beta \right\} , \tag{11.138}$$

while for $\lambda = -1$, we have

$$\frac{1}{\sqrt{2}}\begin{pmatrix} e^{-i\phi/2} \\ -e^{i\phi/2} \end{pmatrix} = \frac{1}{\sqrt{2}}\left\{ e^{-i\phi/2}\,\alpha - e^{i\phi/2}\,\beta \right\} . \tag{11.139}$$

Thus for $\phi = 0$ the eigenstates of S_1 corresponding to $\lambda = \pm 1$ are

$$\frac{1}{\sqrt{2}}\,(\alpha \pm \beta) = \frac{1}{\sqrt{2}}\begin{pmatrix} 1 \\ \pm 1 \end{pmatrix} , \tag{11.140}$$

while for $\phi = \frac{\pi}{2}$ the eigenstates of S_2 are

$$\frac{e^{i\pi/4}}{\sqrt{2}}\,(\alpha \pm i\beta) = \frac{e^{i\pi/4}}{\sqrt{2}}\begin{pmatrix} 1 \\ \pm i \end{pmatrix} . \tag{11.141}$$

Here we note that for $\phi \to \phi + 2\pi$, the solution changes sign and in fact the solution is invariant under the transformation $\phi \to \phi + 4\pi$. This is a property of the spin 1/2 wave function which we will discuss in more detail when considering the relation of symmetry to conservation laws in the next chapter.

There are many particles in nature with spin 1/2, e.g., the electron, proton, quarks, $\cdots$. Associated with the spin there is a magnetic moment, which in the case of the electron is given by

$$\vec{M} = -\frac{eg}{2mc}\,\vec{S} = -\frac{e\hbar g}{4mc}\,\vec{\sigma} . \tag{11.142}$$

Here, e, and m are the charge and mass of the electron, while c is the velocity of light and g is the gyromagnetic ratio, which is 2 according to the Dirac equation [37], and more specifically

$$g = 2.0023192 .$$

For an electron in a constant magnetic field $\vec{B}$, the Hamiltonian for the interaction of the magnetic moment and magnetic field is given by[7]

$$H = -\vec{M}\cdot\vec{B} = \frac{eg\hbar}{4mc}\,\vec{\sigma}\cdot\vec{B}\ . \tag{11.143}$$

Since the Hamiltonian H is a 2×2 matrix in spin space, then the wave function is given by a two component spinor of the form

$$\psi(t) = \begin{pmatrix} u(t) \\ v(t) \end{pmatrix} ,$$

and the time dependent Schrödinger equation becomes

$$i\hbar\frac{d\psi}{dt} = \frac{eg\hbar}{4mc}\,\vec{\sigma}\cdot\vec{B}\,\psi(t)\ .$$

Since the Hamiltonian has no space or time dependence, we expect the wave function ψ to be independent of the position of the electron, and its time dependence to be of the form

$$\psi(t) = \begin{pmatrix} u(t) \\ v(t) \end{pmatrix} = e^{-i\omega t}\begin{pmatrix} \alpha_+ \\ \alpha_- \end{pmatrix} .$$

We now can write the time independent Schrödinger equation as

$$\hbar\omega\begin{pmatrix} \alpha_+ \\ \alpha_- \end{pmatrix} = \frac{eg\hbar}{4mc}\,\vec{\sigma}\cdot\vec{B}\begin{pmatrix} \alpha_+ \\ \alpha_- \end{pmatrix} .$$

Taking the direction of the magnetic field to be along the 3−axis, we get

$$\hbar\omega\begin{pmatrix} \alpha_+ \\ \alpha_- \end{pmatrix} = \frac{eg\hbar B}{4mc}\begin{pmatrix} 1 & 0 \\ 0 & -1 \end{pmatrix}\begin{pmatrix} \alpha_+ \\ \alpha_- \end{pmatrix} = \frac{eg\hbar B}{4mc}\begin{pmatrix} \alpha_+ \\ -\alpha_- \end{pmatrix} .$$

The solutions to this eigenvalue problem are

$$\omega = \frac{egB}{4mc} \qquad \text{with} \qquad \begin{pmatrix} \alpha_+ \\ \alpha_- \end{pmatrix} = \begin{pmatrix} 1 \\ 0 \end{pmatrix} = \alpha$$

and

$$\omega = -\frac{egB}{4mc} \qquad \text{with} \qquad \begin{pmatrix} \alpha_+ \\ \alpha_- \end{pmatrix} = \begin{pmatrix} 0 \\ 1 \end{pmatrix} = \beta\ .$$

[7]The fact that the magnetic moment depends on c, the velocity of light, suggests that relativity has something to do with magnetic moments of a spin $\frac{1}{2}$ particle. In fact, Dirac's[37] equation which combined the special theory of relativity and quantum mechanics predicted the spin, and the magnetic moment of the electron as well as the fact that the electron has a negative energy partner, the positron, to be discovered some four years later. This was the first theoretical prediction of anti-matter.

We now can write the general solution as

$$\begin{aligned}\psi(t) &= ae^{-i\omega t}\begin{pmatrix}1\\0\end{pmatrix}+be^{i\omega t}\begin{pmatrix}0\\1\end{pmatrix}\\ &= \begin{pmatrix}ae^{-i\omega t}\\ be^{i\omega t}\end{pmatrix} \quad \text{with} \quad \omega=\frac{egB}{4mc}\ .\end{aligned}$$

Let us consider the case when the spin in the initial state (i.e., t=0) is an eigenstate of S_1, i.e.,

$$\psi(0)=\begin{pmatrix}a\\b\end{pmatrix}=\frac{1}{\sqrt{2}}\begin{pmatrix}1\\1\end{pmatrix}$$

or

$$S_1\,\psi(0)=\frac{\hbar}{2}\,\psi(0)$$

The solution at a later time t is then given by

$$\psi(t)=\frac{1}{\sqrt{2}}\begin{pmatrix}e^{-i\omega t}\\ e^{i\omega t}\end{pmatrix}\ .$$

The average value of the spin along the 1-axis at a time t is by definition

$$\begin{aligned}\langle S_1\rangle &= \langle\psi(t)|S_1|\psi(t)\rangle\\ &= \frac{\hbar}{2}\frac{1}{\sqrt{2}}\begin{pmatrix}e^{i\omega t} & e^{-i\omega t}\end{pmatrix}\begin{pmatrix}0 & 1\\ 1 & 0\end{pmatrix}\frac{1}{\sqrt{2}}\begin{pmatrix}e^{-i\omega t}\\ e^{i\omega t}\end{pmatrix}\\ &= \frac{\hbar}{2}\cos 2\omega t\ .\end{aligned}$$

On the other hand, the average value of the spin along the 2-axis is

$$\begin{aligned}\langle S_2\rangle &= \langle\psi(t)|S_2|\psi(t)\rangle\\ &= \frac{\hbar}{2}\frac{1}{\sqrt{2}}\begin{pmatrix}e^{i\omega t} & e^{-i\omega t}\end{pmatrix}\begin{pmatrix}0 & -i\\ i & 0\end{pmatrix}\frac{1}{\sqrt{2}}\begin{pmatrix}e^{-i\omega t}\\ e^{i\omega t}\end{pmatrix}\\ &= \frac{\hbar}{2}\sin 2\omega t\ .\end{aligned}$$

Combining the above two results we observe that as a function of time the spin is precessing about the 3-axis with a frequency of precession of

$$2w=\frac{egB}{2mc}\approx\frac{eB}{mc}\ .$$

11.5 Problems

1. The vector $\vec{r}$ in real three-dimensional space is subject to the transformation

$$\vec{r}' = \tilde{A}\,\vec{r} = \vec{a} \times \vec{r}$$

where $\vec{a}$ is a given fixed vector. Show that $\tilde{A}$ is a linear operator which satisfies the equation

$$\tilde{A}^3 + a^2\tilde{A} = 0 \qquad \text{with} \qquad a = |\vec{a}| \ .$$

2. Construct the matrix of a linear transformation in real two-dimensional space which doubles the length of every vector drawn from the origin, and rotate it through a positive angle of 45°. Show that this matrix satisfies the equation $\tilde{A}^4 = -16$.

3. Show that every 2×2 matrix can be written in the form

$$a_0 I + a_1\sigma_1 + a_2\sigma_2 + a_3\sigma_3$$

where the a_i are complex numbers, and the σ_i are the Pauli matrices given by

$$\sigma_1 = \begin{pmatrix} 0 & 1 \\ 1 & 0 \end{pmatrix} \qquad \sigma_2 = \begin{pmatrix} 0 & -i \\ i & 0 \end{pmatrix} \qquad \sigma_3 = \begin{pmatrix} 1 & 0 \\ 0 & -1 \end{pmatrix} \ .$$

In the above, I is the unit matrix.

4. Given the matrix

$$\tilde{A} = \begin{pmatrix} 2 & \sqrt{2/3} & 0 \\ \sqrt{2/3} & 2 & \sqrt{1/3} \\ 0 & \sqrt{1/3} & 2 \end{pmatrix} \ .$$

   (a) Show that the equation $(x, \tilde{A}x) = 1$ represents an ellipsoid in real three-dimensional space, and find the length of the principle axes of the ellipsoid.

   (b) What are the direction cosines of the principle axis?

   (c) Construct the unitary matrix $\tilde{U}$ which diagonalizes $\tilde{A}$.

   (d) Show by direct matrix multiplication that $\tilde{U}\tilde{A}\tilde{U}^{-1}$ is a diagonal matrix.

5. The characteristic equation for a matrix $\tilde{A}$ has the simple form

$$\det\{a_{ij} - \alpha\delta_{ij}\} = 0$$

only if the basis of the representation is orthonormal. Write down the eigenvalue equation

$$\tilde{A}\,x = \lambda\,x$$

using an arbitrary basis $|y_i|$, and show that the characteristic equation assumes the form

$$\det\{a_{ij} - \lambda\,\Delta_{ij}\} = 0$$

where

$$a_{ij} = \left(y_i,\, \tilde{A}\, y_j\right) \qquad \text{and} \qquad \Delta_{ij} = (y_i,\, y_j)\ .$$

6. Show that the set of all polynomials of a degree smaller than N in a real variable u can be regarded as a linear vector space of N dimensions. Let the scalar product in this space be defined as

$$(x,\, y) = \int_{-1}^{1} du\, x^*(u)\, y(u)\ .$$

Prove that the operator

$$\tilde{A} = -\frac{d}{du}\,(1-u^2)\,\frac{d}{du}$$

is Hermitian.

7. Construct the characteristic equation for the operator A of the previous problem in the special case of $N = 3$. Show that it has the eigenvalues 0, 2, 6. What are the corresponding eigenvectors?

8. Show that for an operator $\tilde{A}$,

$$\operatorname{tr}\left\{\tilde{A}\right\} = \sum_{\alpha}\langle\alpha|\tilde{A}|\alpha\rangle$$

is independent of the choice of basis $|\alpha\rangle$.

9. Let $|u\rangle$ and $|v\rangle$ be two vectors of finite norm. Show that

$$\begin{aligned} \operatorname{tr}\{\,|u\rangle\langle v|\,\} &= \langle v|u\rangle \\ \operatorname{tr}\{\,|v\rangle\langle u|\,\} &= \langle u|v\rangle \end{aligned}$$

10. Let $\tilde{H}$ be a positive definite Hermitian operator.

(a) Show that for any $|u\rangle$ and $|v\rangle$

$$|\langle u|\tilde{H}|v\rangle|^2 \leq \langle u|\tilde{H}|u\rangle\,\langle v|\tilde{H}|v\rangle\ ,$$

and that the equality $\langle u|\tilde{H}|u\rangle = 0$ necessarily implies that $\tilde{H}|u\rangle = 0$.

(b) Show that $\operatorname{tr}\tilde{H} \geq 0$, and that the equality implies that $\tilde{H} = 0$.

11. Solve the eigenvalue problem

$$\vec{\sigma} \cdot \hat{n} \, \chi_\lambda = \lambda \, \chi_\lambda$$

where $\hat{n}$ is a unit vector given by

$$\hat{n} = (\sin\theta \, \cos\phi, \, \sin\theta \, \sin\phi, \, \cos\theta)$$

and $\vec{\sigma}$ is the Pauli spin matrix.

(a) Find the states χ_μ.

(b) Write the states χ_μ in terms of the two component spinors for spin up, α, and spin down, β.

(c) construct the projection operator

$$\tilde{P}_\lambda = |\chi_\lambda\rangle \, \langle\chi_\lambda| \, .$$

(d) If $\hat{n} = (1, 0, 0)$, what is the projection operator P_μ?

12. Show that it is impossible to construct a non-vanishing 2×2 matrix which anticommutes with all three Pauli matrices.

13. Prove that for any two vectors $\vec{A}$ and $\vec{B}$, we have

$$\left(\vec{\sigma} \cdot \vec{A}\right) \, \left(\vec{\sigma} \cdot \vec{B}\right) = \vec{A} \cdot \vec{B} + i\vec{\sigma} \cdot \left(\vec{A} \times \vec{B}\right)$$

where $\vec{\sigma}$ is the Pauli spin matrix.

CHAPTER 12

Symmetry and Conservation

Abstract: With symmetry playing a central role in quantum mechanics, in this chapter we commence with the simplest of symmetries of translation in space and time. Here we consider the group structure of the symmetry operators. This is followed by a rotational symmetry that plays a central role in atomic and nuclear physics and is a major component of this chapter. Finally we examine the time and space reversion and isospin that are central to the study of nuclear and particle physics.

We have seen that for a central potential, the Hamiltonian commutes with the angular momentum. This means we can write the wave function for the system to be an eigenstate of both the energy and angular momentum. This will allow us to write the wave function in terms of a radial part $R_\ell(r)$, and an angle dependent part, given by the spherical harmonics $Y_{\ell m}(\theta, \phi)$, i.e.,

$$\psi(\vec{r}) = R_\ell(r)\, Y_{\ell m}(\hat{r}) \quad .$$

This in turn reduces the Schrödinger equation to a second order linear differential equation in the radial variable r. What we have accomplished, is to use the symmetry of the system to divide the problem into two parts: (a) The geometry given by the spherical harmonics $Y_{\ell m}(\theta, \phi)$, (b) The dynamics as given by $R_\ell(r)$.[1]

What we want to discuss in this chapter is how we can make use of the symmetry of the problem to separate the geometry from the dynamics. Here it is important to remember that the geometry has all the symmetries and thus the conservation laws resulting from these symmetries. We will examine four symmetries in detail:

[1]In this chapter we will not consider symmetries that give rise to dynamics such as local gauge symmetries and dynamical symmetries in which the algebra of the commutation relation are used to get the eigenvalues and eigenvectors of the Hamiltonian for the system. This latter class of symmetry is considered by Schiff [17] in Sec. 30. In Chapter 13 we will use this dynamic symmetry to generate the spectrum of the one dimensional harmonic oscillator.

Iraj R. Afnan

1. Symmetry under translation in space and time.
2. Symmetry under rotation.
3. Symmetry under space and time inversion.
4. Symmetry under the exchange of protons and neutron – isospin

All but the last of these symmetries are associated with space-time. There are other symmetries in nature (e.g., isospin) which are also associated with conservation laws. These are often referred to as internal symmetries and are as important as the space-time symmetries discussed in this chapter. However, since we have not encountered these symmetries in any of the problems we have studied in previous chapters, we will consider as an illustration of these internal symmetries the isospin symmetry. This symmetry results from the observation that the mass of the proton is approximately the same as the mass of the neutron and the neutron-proton interaction is the same as the neutron-neutron and proton-proton interaction in the absence of the Coulomb force.

12.1 Translation in Space and Time

We will start with the simplest symmetry and the corresponding conservation law. Let us consider a system described by the wave function $\psi_\alpha(x) \equiv \langle x|\alpha\rangle$, in one space dimension. If we now displace the entire physical system by a distance λ, then the state of the system after displacement is given by $\psi_{\alpha'}(x) \equiv \langle x|\alpha'\rangle$. This transformation is illustrated in Figure 12.1

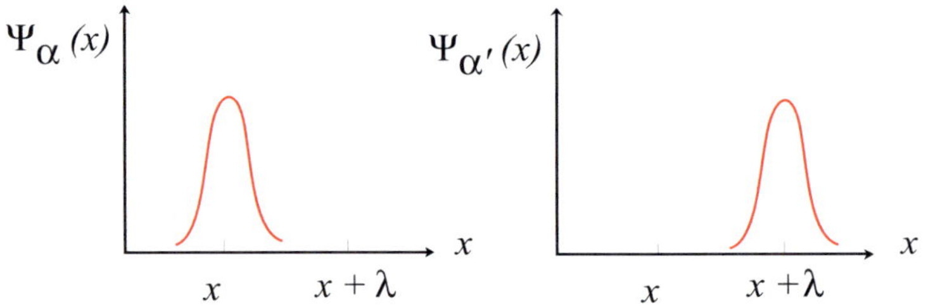

Figure 12.1: An illustration of the translation of the wave function ψ_α along the x-axis by a distance λ.

From the above figure it is clear that

$$\psi_{\alpha'}(x) = \psi_\alpha(x - \lambda) \ . \tag{12.1}$$

This process of displacing the physical system by a distance λ can be represented by an operator $D(\lambda)$ such that

$$|\alpha'\rangle = D(\lambda)\,|\alpha\rangle \ . \tag{12.2}$$

The problem is how to determine $D(\lambda)$. If we expand the right hand side of Eq. (12.1) in a Taylor series about the point x, we get;

$$\begin{aligned}\psi_{\alpha'}(x) = \langle x|\alpha'\rangle &= \psi_\alpha(x) - \lambda\frac{d}{dx}\psi_\alpha(x) + \frac{\lambda^2}{2!}\frac{d^2}{dx^2}\psi_\alpha(x) - \cdots \\ &= \left\{1 - \lambda\frac{d}{dx} + \frac{\lambda^2}{2!}\frac{d^2}{dx^2} + \cdots\right\}\psi_\alpha(x) \qquad (12.3)\end{aligned}$$

and therefore,

$$\begin{aligned}\psi_{\alpha'}(x) &= \exp\left\{-\lambda\frac{d}{dx}\right\}\psi_\alpha(x) \\ &= \langle x|\exp\left\{-i\frac{\lambda}{\hbar}p_x\right\}|\alpha\rangle\ , \qquad (12.4)\end{aligned}$$

where p_x is the momentum operator (i.e., $p_x = -i\hbar\frac{d}{dx}$). We now can write

$$|\alpha'\rangle = \exp\left\{-i\frac{\lambda}{\hbar}p_x\right\}|\alpha\rangle\ , \qquad (12.5)$$

with the operator $D(\lambda)$ given by

$$D(\lambda) = \exp\left\{-i\frac{\lambda}{\hbar}p_x\right\}\ . \qquad (12.6)$$

The generalization of this result to three-dimensions is straightforward. The displacement of the system is now defined by a direction and a magnitude, i.e.,

$$\lambda \to \vec{\lambda}\ ,$$

and the momentum in one dimension is replaced by the momentum vector, i.e.,

$$p \to \vec{p}\ .$$

As a result, the three dimensions displacement operator becomes

$$D(\vec{\lambda}) = \exp\left\{-\frac{i}{\hbar}\vec{\lambda}\cdot\vec{p}\right\}\ . \qquad (12.7)$$

The statement of Eq. (12.1) implies that the system does not change under the operation of translation, i.e., the system is invariant under translation in space.

To study the corresponding conservation law, we take the time derivative of Eq. (12.2), i.e.,

$$i\hbar\frac{d}{dt}|\alpha'\rangle = i\hbar\frac{d}{dt}D(\vec{\lambda})|\alpha\rangle = i\hbar D(\vec{\lambda})\frac{d}{dt}|\alpha\rangle\ . \qquad (12.8)$$

If the state of the system satisfies the Schrödinger equation then

$$i\hbar \frac{d}{dt}\,|\alpha\rangle = H\,|\alpha\rangle \ , \tag{12.9}$$

where H is the Hamiltonian for the system. This allows us to write Eq. (12.8) as

$$\begin{aligned} i\hbar \frac{d}{dt}\,|\alpha'\rangle &= D(\vec{\lambda}\,)\,H\,|\alpha\rangle \\ &= D(\vec{\lambda}\,)\,H\,D^{-1}(\vec{\lambda}\,)\,|\alpha'\rangle \ . \end{aligned} \tag{12.10}$$

Here, $D^{-1}(\vec{\lambda}\,)$ is the inverse operator for translation (i.e., $|\alpha\rangle = D^{-1}(\vec{\lambda}\,)\,|\alpha'\rangle$) and is given by,

$$D^{-1}(\vec{\lambda}\,) = \exp\left\{\frac{i}{\hbar}\,\vec{\lambda}\cdot\vec{p}\right\} \ = \ \left[D(\vec{\lambda}\,)\right]^{\dagger} \ , \tag{12.11}$$

i.e., $D(\vec{\lambda}\,)$ is a unitary operator.

For the system to be invariant under translation, the states $|\alpha\rangle$ and $|\alpha'\rangle = D(\vec{\lambda}\,)\,|\alpha\rangle$ should satisfy the same Schrödinger equation, i.e., if

$$i\hbar \frac{d}{dt}\,|\alpha\rangle = H\,|\alpha\rangle$$

then

$$i\hbar \frac{d}{dt}\,|\alpha'\rangle = H\,|\alpha'\rangle \ .$$

Making use of this result in Eq. (12.10) we get,

$$H\ |\alpha'\rangle = D(\vec{\lambda}\,)\,H\,D^{-1}(\vec{\lambda}\,)\,|\alpha'\rangle$$

or

$$\left[H, D(\vec{\lambda}\,)\right] = 0 \ . \tag{12.12}$$

Therefore, for a system to be invariant under the transformation $D(\vec{\lambda}\,)$, the Hamiltonian for that system should commute with the translation operator $D(\vec{\lambda}\,)$. We note at this point that the only property of $D(\vec{\lambda}\,)$ that we have made use of, is that it has an inverse. If $D(\vec{\lambda}\,)$ is the operator for translation in space then from Eq. (12.7) the commutation relation in Eq. (12.12) is equivalent to

$$[H, \vec{p}\,] = 0 \ , \tag{12.13}$$

i.e., the Hamiltonian commutes with the momentum operator, which in turn implies that we can construct simultaneous eigenstates of the Hamiltonian and the momentum operator. One such Hamiltonian is the free particle Hamiltonian, i.e.,

$$H = \frac{p^2}{2m} \ ,$$

for which it is obvious that $[H, \vec{p}] = 0$, and the eigenstates of H are also eigenstates of $\vec{p}$, i.e.,

$$H\,|\vec{k}\,\rangle = \frac{\hbar^2 k^2}{2m}\,|\vec{k}\,\rangle \qquad \text{and} \qquad \vec{p}\,|\vec{k}\,\rangle = \hbar\vec{k}\;|\vec{k}\,\rangle\ .$$

The second of these two equations can be written in coordinate representation as

$$-i\hbar\vec{\nabla}_r\,\langle\vec{r}\,|\vec{k}\,\rangle = \hbar k\,\langle\vec{r}\,|\vec{k}\,\rangle$$

with the solution of this first order differential equation given by,

$$\langle\vec{r}\,|\vec{k}\,\rangle = \frac{1}{(2\pi)^{3/2}}\;e^{i\vec{k}\cdot\vec{r}}\ . \tag{12.14}$$

Thus in this case the eigenstates of the momentum operator are also eigenstates of the Hamiltonian. Also these eigenstates are such that the operator $D(\vec{\lambda}\,)$ is diagonal in this basis, i.e.,

$$\langle\vec{k}\,|D(\vec{\lambda}\,)|\vec{k}'\,\rangle = \delta(\vec{k} - \vec{k}')\;e^{i\vec{\lambda}\cdot\vec{k}}\ . \tag{12.15}$$

The matrix elements of the operator $D(\vec{\lambda}\,)$ are referred to as the representation of the group of translation. We will come back to this concept later when we are considering rotations.

Before we proceed to a discussion of other symmetries, let us consider some of the general properties of $D(\vec{\lambda}\,)$.

1. If we first translate the system by a displacement $\vec{\lambda}_1$, and then by $\vec{\lambda}_2$, we can write

   $$|\alpha'\rangle = D(\vec{\lambda}_1)\,|\alpha\rangle\ ,$$

   and

   $$|\alpha''\rangle = D(\vec{\lambda}_2)\,|\alpha'\rangle = D(\vec{\lambda}_2)\,D(\vec{\lambda}_1)\,|\alpha\rangle\ .$$

   But

   $$|\alpha''\rangle = D(\vec{\lambda}_1 + \vec{\lambda}_2)\,|\alpha\rangle\ ,$$

   and therefore

   $$D(\vec{\lambda}_2)\,D(\vec{\lambda}_1) = D(\vec{\lambda}_1 + \vec{\lambda}_2)$$

   i.e., the product of two translations is another translation.

2. We can define an identity transformation $D(0)$ such that;

   $$D(0)\,D(\vec{\lambda}\,) = D(\vec{\lambda}\,)\,D(0) = D(\vec{\lambda}\,)\ .$$

3. For every translation $D(\vec{\lambda}\,)$, there is an inverse $D^{-1}(\vec{\lambda}\,)$ such that,

   $$D(\vec{\lambda}\,)\,D^{-1}(\vec{\lambda}\,) = D^{-1}(\vec{\lambda}\,)\,D(\vec{\lambda}\,) = D(0)\ .$$

4. For three successive transformations we have,

$$D(\vec{\lambda}_3)\left\{D(\vec{\lambda}_2)\,D(\vec{\lambda}_1)\right\} = \left\{D(\vec{\lambda}_3)\,D(\vec{\lambda}_2)\right\}\,D(\vec{\lambda}_1)\ .$$

   The above four properties are those of a mathematical structure called a *group*, and $D(\vec{\lambda})$ for a given $\vec{\lambda}$ is an element of this group. If in addition the elements of the group satisfy the condition that:

5. The order in which we perform the transformation is not important, i.e.,

$$D(\vec{\lambda}_1)\,D(\vec{\lambda}_2) = D(\vec{\lambda}_2)\,D(\vec{\lambda}_1)\ ,$$

   then the group is said to be an *Abelian Group.*

6. If the elements of the group $D(\vec{\lambda})$ have a parameter $\vec{\lambda}$ that takes on continuous values, as in the case of translation, then the group is a *continuous group* or *Lie Group.*

Having established that the operators of translation in space $D(\vec{\lambda})$ form a group, let us turn to the symmetry of the system under translation in time. Given the state $\psi_\alpha(t)$, if we displace this system in time by an interval τ, then we get the state $\psi_{\alpha'}(t)$ and

$$\begin{aligned}\psi_{\alpha'}(t) &= \psi_\alpha(t-\tau)\\ &= \left\{1 - \tau\,\frac{d}{dt} + \frac{\tau^2}{2!}\,\frac{d^2}{dt^2} + \cdots\right\}\psi_\alpha(t)\\ &= D(\tau)\,\psi_\alpha(t)\ .\end{aligned}$$

But from the Schrödinger equation we have

$$i\hbar\,\frac{d}{dt}\,\psi_\alpha(t) = H\,\psi_\alpha(t) \qquad \text{or} \qquad \frac{d}{dt} \Rightarrow -\frac{i}{\hbar}\,H$$

so that the operator of translation in time is

$$D(\tau) = \exp\left\{\frac{i}{\hbar}\,H\tau\right\}\ . \tag{12.16}$$

We have established the fact that for a system which is invariant under space translation the momentum operator commutes with the Hamiltonian for that system. Thus at time $t=0$ we can construct a state that is an eigenstate of both the Hamiltonian H and the momentum operator $\vec{p}$. Since the operator that translates the system in time is a function of the Hamiltonian H, the state of the system at time $t>0$ will have the same momentum as the state at time $t=0$. In other words, if

$$H\psi_{n,\vec{k}}(t=0) = E_n\psi_{n,\vec{k}}(t=0) \quad \text{and} \quad \vec{p}\,\psi_{n,\vec{k}}(t=0) = \hbar\vec{k}\,\psi_{n,\vec{k}}(t=0)\ ,$$

then

$$\psi(t) = D(t)\,\psi(0)$$

$$\begin{aligned}\vec{p}\,\psi(t) &= \vec{p}\,D(t)\,\psi(0) \\ &= D(t)\,\vec{p}\,\psi(0) \quad \text{since} \quad [H, \vec{p}] = 0 \\ &= \hbar\vec{k}\,D(t)\,\psi(0) \\ &= \hbar\vec{k}\,\psi(t)\ .\end{aligned}$$

In other words, the momentum is the same at $t > 0$ as it was at $t = 0$, i.e., the momentum is a constant of the motion, or momentum is conserved.

If the Hamiltonian is time dependent, then the operator for translation in time is given by

$$D(\tau) = \exp\left\{-\tau\,\frac{d}{dt}\right\}\ , \tag{12.17}$$

which does not commute with H, and the system is not invariant under translation in time.

12.2 Rotation in Three-Dimensions

So far we have considered the translation of the system in space and time. We found that the operators that generate translation form a group, and that the group is an Abelian group. We now turn our attention to the rotation of the system in three-dimensions. This is the first case of non-trivial symmetry because the group associated with rotation in three-dimensions is not Abelian.

Consider the rotation of the system by an amount ε in the positive sense about a unit vector $\hat{n}$. Then every point labeled $\vec{r}$ in the old system moves to a new point $\vec{r}\,'$ in the new system. For ε small, we can write (see Figure 12.2)

$$\vec{r}\,' = \vec{r} + \varepsilon\,(\hat{n} \times \vec{r})\ , \tag{12.18a}$$

or

$$\vec{r} = \vec{r}\,' - \varepsilon\,(\hat{n} \times \vec{r}\,')\ . \tag{12.18b}$$

If $\hat{n}$ is along the z-axis, then the transformation can be written as

$$\begin{pmatrix} x' \\ y' \\ z' \end{pmatrix} = \begin{pmatrix} 1 & -\varepsilon & 0 \\ \varepsilon & 1 & 0 \\ 0 & 0 & 1 \end{pmatrix} \begin{pmatrix} x \\ y \\ z \end{pmatrix}\ , \tag{12.19}$$

or in general this transformation takes the form

$$\vec{r}\,' = P_\varepsilon \vec{r}\ . \tag{12.20}$$

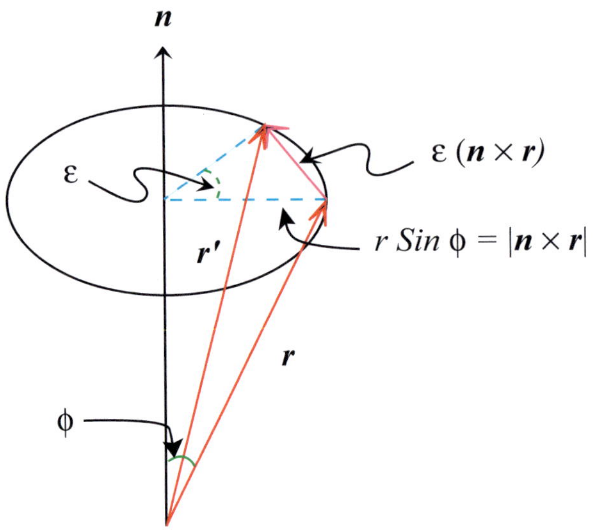

Figure 12.2: The rotation of the vector $\vec{r}$ by the infinitesimal angle ε about the axis $\hat{n}$.

For finite rotation the matrix P_θ is given by

$$P_\theta = \begin{pmatrix} \cos\theta & -\sin\theta & 0 \\ \sin\theta & \cos\theta & 0 \\ 0 & 0 & 1 \end{pmatrix} . \tag{12.21}$$

We note here that the operator rotates the vector $\vec{r}$ by the angle θ leaving the coordinate system fixed.

We now turn our attention to the state of the physical system in order to examine what happens to the state under rotation. We take the system to be initially in a state $\psi_\alpha(\vec{r}) = \langle\vec{r}|\psi_\alpha\rangle$. After rotation, the system is in a state $\psi_{\alpha'}(\vec{r}') = \langle\vec{r}'|\psi_{\alpha'}\rangle$. These two states are related by the operator $\mathcal{R}$, i.e.,

$$|\psi_{\alpha'}\rangle = \mathcal{R}\,|\psi_\alpha\rangle . \tag{12.22}$$

Here the operator $\mathcal{R}$ is analogous to the translation operator $D(\lambda)$ we encountered in the last section.

From Figure 12.3, we observe that if we had a set of axes that rotated with the state, then the new state in the rotated coordinate would be functionally identical to the original state in the old coordinate system. In other words

$$\psi_\alpha(\vec{r}) = \psi_{\alpha'}(\vec{r}\,') . \tag{12.23}$$

Making use of the inverse of the transformation in Eq. (12.18b), i.e., $\vec{r} = \vec{r}\,' - \varepsilon\,(\hat{n}\times\vec{r}\,')$, we can write the left hand side of Eq. (12.23) as

$$\psi_\alpha(\vec{r}) = \psi_\alpha\left(\vec{r}\,' - \varepsilon\hat{n}\times\vec{r}\,'\right) . \tag{12.24}$$

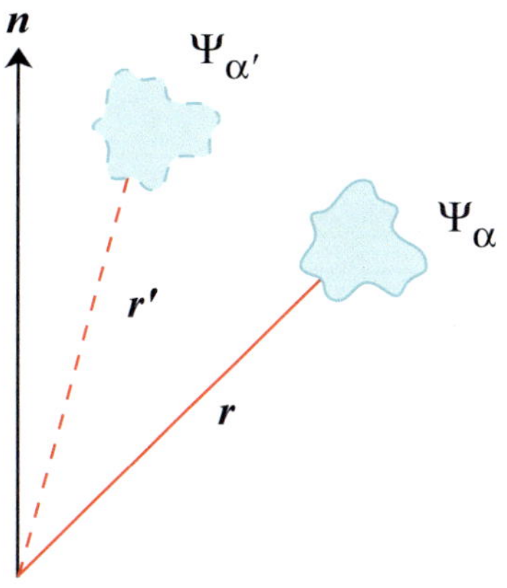

Figure 12.3: The rotation of the state ψ_α about the axes $\hat{n}$ which gives the state $\psi_{\alpha'}$.

For an infinitesimal rotation, $\varepsilon \ll 1$, we have

$$\begin{aligned}\psi_\alpha(\vec{r}\,' - \varepsilon\hat{n}\times\vec{r}\,') &= \psi_\alpha(\vec{r}\,') - \varepsilon(\hat{n}\times\vec{r}\,')\cdot\vec{\nabla}\psi_\alpha(\vec{r}\,') \\ &= \psi_\alpha(\vec{r}\,') - \varepsilon\hat{n}\cdot(\vec{r}\,'\times\vec{\nabla})\psi_\alpha(\vec{r}\,') \\ &= \left[1 - \varepsilon\hat{n}\cdot(\vec{r}\,'\times\vec{\nabla})\right]\psi_\alpha(\vec{r}\,') \ .\end{aligned}$$

Since the momentum, $\vec{p}$, is proportional to the gradient operator ∇, we can replace the operator $\vec{r}\,'\times\nabla$ by the angular momentum, i.e.,

$$\vec{r}\,'\times\vec{\nabla} = \frac{i}{\hbar}\vec{L} \ .$$

With this result in hand we can write the rotated state as

$$\begin{aligned}\psi_{\alpha'}(\vec{r}\,') \equiv \langle\vec{r}\,'|\psi_{\alpha'}\rangle &= \left[1 - \frac{i}{\hbar}\vec{\varepsilon}\cdot\vec{L}\right]\psi_\alpha(\vec{r}\,') \\ &= \langle\vec{r}\,'|\mathcal{R}|\psi_\alpha\rangle \ ,\end{aligned} \tag{12.25}$$

where

$$\mathcal{R}(\vec{\varepsilon}) = 1 - \frac{i}{\hbar}\vec{\varepsilon}\cdot\vec{L} \ , \tag{12.26}$$

with $\vec{\varepsilon}$ defined to be a vector along the axis of rotation $\hat{n}$ with magnitude equal to ε, i.e., $\vec{\varepsilon} = \varepsilon\hat{n}$.

So far we have only considered infinitesimal rotation. For finite rotation we have to consider n infinitesimal rotations of magnitude ε, and then take the limit as $n\to\infty$ $\varepsilon\to 0$ with $n\varepsilon\to\theta$. Thus the rotation operator for finite rotation is given by

$$\mathcal{R}(\vec{\theta}) = \lim_{\varepsilon\to 0}\lim_{n\to\infty}\left[1 - \frac{i}{\hbar}\vec{\varepsilon}\cdot\vec{L}\right]^n = \exp\left[-\frac{i}{\hbar}\vec{\theta}\cdot\vec{L}\right] \ , \tag{12.27}$$

and the rotated state is given by:

$$\psi_{\alpha'}(\vec{r}) = \mathcal{R}(\vec{\theta})\,\psi_\alpha(\vec{r}) \ . \tag{12.28}$$

Here $\mathcal{R}(\vec{\theta})$ is a unitary operator that rotates the system by an angle θ about the axis $\hat{\theta}$.

The rotation operators $\mathcal{R}(\vec{\theta})$ form a group in that

1. The product of two rotations is a third rotation, i.e.,
$$\mathcal{R}(\vec{\theta}_1)\,\mathcal{R}(\vec{\theta}_2) = \mathcal{R}(\vec{\theta}_3) \ .$$
2. There exists a unit operator, $\mathcal{R}(0) = 1$, such that
$$\mathcal{R}(\vec{\theta})\,\mathcal{R}(0) = \mathcal{R}(0)\,\mathcal{R}(\vec{\theta}) = \mathcal{R}(\vec{\theta}) \ .$$
3. There exists an inverse operator $\mathcal{R}^{-1}(\vec{\theta}) = \mathcal{R}(-\vec{\theta})$ that satisfies the condition
$$\mathcal{R}(\vec{\theta})\ \mathcal{R}^{-1}(\vec{\theta}) = \mathcal{R}(0) \ .$$
4. The associative law is satisfied, i.e.,
$$\mathcal{R}(\vec{\theta}_1)\left\{\mathcal{R}(\vec{\theta}_2)\,\mathcal{R}(\vec{\theta}_3)\right\} = \left\{\mathcal{R}(\vec{\theta}_1)\,\mathcal{R}(\vec{\theta}_2)\right\}\mathcal{R}(\vec{\theta}_3) \ .$$

   In addition to the above properties, we have two additional properties which are:

5. In three-dimensions the product of two rotations depends on the order in which the rotations have been carried out, i.e.,
$$\mathcal{R}(\vec{\theta}_1)\,\mathcal{R}(\vec{\theta}_2) \neq \mathcal{R}(\vec{\theta}_2)\,\mathcal{R}(\vec{\theta}_1) \ ,$$
   i.e., the group is a non-Abelian Group.
6. The elements of the group are characterized by a parameter θ that can take on continuous values. Thus, $\mathcal{R}(\vec{\theta})$ forms a *Lie Group*.

In the above definition of $\mathcal{R}(\vec{\theta})$, $\vec{L} = \{L_1, L_2, L_3\}$ are the *generators* of the group and they satisfy the commutation relation

$$[L_i, L_j] = i\hbar\varepsilon_{ijk}\,L_k \ . \tag{12.29}$$

This commutation relation defines the *Lie Algebra* for the rotation group $R(3)$.

To relate the invariance under rotation with the corresponding conservation law, consider the fact that

$$|\psi_{\alpha'}\rangle = \mathcal{R}(\vec{\theta})\,|\psi_\alpha\rangle \tag{12.30}$$

and

$$i\hbar\frac{d}{dt}\left|\psi_\alpha\right\rangle = H\left|\psi_\alpha\right\rangle\ , \tag{12.31}$$

We then have that

$$\begin{aligned}
i\hbar\frac{d}{dt}|\psi_{\alpha'}\rangle &= i\hbar\,\frac{d}{dt}\mathcal{R}(\vec{\theta}\,)|\psi_\alpha\rangle \\
&= i\hbar\,\mathcal{R}(\vec{\theta}\,)\,\frac{d}{dt}|\psi_\alpha\rangle \\
&= \mathcal{R}(\vec{\theta}\,)\,H\,|\psi_\alpha\rangle \\
&= \mathcal{R}(\vec{\theta}\,)\,H\,\mathcal{R}^{-1}(\vec{\theta}\,)|\psi_{\alpha'}\rangle\ .
\end{aligned} \tag{12.32}$$

For the system to be invariant under rotation, both $|\psi_\alpha\rangle$ and $|\psi_{\alpha'}\rangle$ must satisfy the same equation, i.e., if

$$i\hbar\,\frac{d}{dt}|\psi_\alpha\rangle = H\,|\psi_\alpha\rangle\ ,$$

then

$$i\hbar\,\frac{d}{dt}|\psi_{\alpha'}\rangle = H\,|\psi_{\alpha'}\rangle\ .$$

For this to be true, we require that

$$\mathcal{R}(\vec{\theta}\,)\,H\,\mathcal{R}^{-1}(\vec{\theta}\,) = H\ ,$$

or

$$\left[\mathcal{R}(\vec{\theta}\,), H\right] = 0\ . \tag{12.33}$$

Since the rotation operator $\mathcal{R}(\vec{\theta}\,)$ is defined in terms of the angular momentum operator $\vec{L}$, then for the system to be invariant under rotation the angular momentum operator should commute with the Hamiltonian, i.e.,

$$\left[H, \vec{L}\right] = 0\ . \tag{12.34}$$

In other word, the angular momentum of the system must be a constant of the motion.

To use the elements of the group $\mathcal{R}(\vec{\theta}\,)$ to transform the system from one set of coordinates to another, we need to write $\mathcal{R}(\vec{\theta}\,)$ in some representation, e.g.,

$$\langle\alpha|\mathcal{R}(\vec{\theta}\,)|\beta\rangle = \mathcal{D}_{\alpha\beta}(\vec{\theta}\,)\ . \tag{12.35}$$

These matrices are the representation of the rotation group. The group representation is said to be reducible if we can find a unitary transformation on the basis states $|\alpha\rangle$, $|\beta\rangle$, $\cdots$ that will make $\mathcal{D}(\vec{\theta}\,)$ block diagonal. On the other hand, the representation is said to be irreducible if there is no such unitary transformation. For the group $R(3)$, the representation is irreducible if the basis states are taken to be eigenstates of the angular momentum operator square, L^2, along with the 3-component of the angular momentum, L_3, i.e.,

$$\begin{aligned} L^2|\ell, m\rangle &= \hbar^2\ell(\ell+1)|\ell, m\rangle \\ L_3|\ell, m\rangle &= \hbar m|\ell, m\rangle \ . \end{aligned} \tag{12.36}$$

To write the representation of the rotation operator in this basis, we recall that $\vec{L}$ acting on the state $|\ell, m\rangle$ changes m but not ℓ, i.e.,

$$\langle \ell', m'|\vec{L}|\ell, m\rangle = \delta_{\ell'\ell}\ \langle \ell, m'|\vec{L}|\ell, m\rangle \ .$$

This result follows from the fact that $\vec{L}$ is proportional to L_3 and $L_\pm$. The rotation operator, given in Eq.(12.27), has matrix elements given by

$$\begin{aligned} \langle \ell', m'|\mathcal{R}(\vec{\theta})|\ell, m\rangle &= \delta_{\ell'\ell}\ \langle \ell, m'|\mathcal{R}(\vec{\theta})|\ell, m\rangle \\ &= \delta_{\ell'\ell}\ \left[\tilde{D}^\ell(\theta)\right]_{m'm} , \end{aligned} \tag{12.37}$$

where

$$\tilde{\mathcal{D}}^\ell(\theta) = \exp\left\{-\frac{i}{\hbar}\vec{\theta}\cdot\tilde{\vec{\mathrm{L}}}_\ell\right\} \ . \tag{12.38}$$

Here $\tilde{\vec{L}}_\ell$ is the $(2\ell+1)\times(2\ell+1)$ matrix representation of the operator $\vec{L}$, i.e.,

$$\left[\tilde{\vec{L}}_\ell\right]_{m'm} = \langle \ell, m'|\tilde{\vec{L}}|\ell, m\rangle \ .$$

By always specifying the irreducible representation of a group, we have uniquely defined the dimensionality of the representation matrices. However to specify the representation, we need to define the most general rotation possible. This can be achieved in terms of the three Euler angles which define the orientation of a solid body in space. We will define our Euler angles by the follows rotations:

1. A rotation by angle α about the z-axis.

2. Followed by a rotation by an angle β about the new y-axis, y'.

3. Followed by a rotation by an angle γ about the new z-axis, z''.

These rotations should be performed in the order in which they are listed, i.e.,

$$|\psi_{\alpha'}\rangle = e^{-\frac{i}{\hbar}\gamma L_{z''}}\, e^{-\frac{i}{\hbar}\beta L_{y'}}\, e^{-\frac{i}{\hbar}\alpha L_z}|\psi_\alpha\rangle = \mathcal{R}(\alpha, \beta, \gamma)\,|\psi_\alpha\rangle \ . \tag{12.39}$$

This operator is rather difficult to work with as it stands. This is mainly due to the fact that it is written in terms of three different coordinate systems, and to calculate the corresponding representation is impossible with the operator in its present form. But a careful examination of these rotations reveals that a rotation about the new y-axis is equivalent to a rotation about the old z-axis by an angle of $-\alpha$, followed by the rotation about the old y-axis by an angle β and then a rotation by an angle α about the old z axis, i.e.,

$$e^{-\frac{i}{\hbar}\beta L_{y'}} = e^{-\frac{i}{\hbar}\alpha L_z}\, e^{-\frac{i}{\hbar}\beta L_y}\, e^{\frac{i}{\hbar}\alpha L_z} \; . \tag{12.40}$$

In a similar manner we can write the rotation about the z''-axis in terms of rotation about the original set of coordinates. This procedure allows us to rewrite the operator $\mathcal{R}(\alpha, \beta, \gamma)$ in terms or the original coordinates as

$$\mathcal{R}(\alpha, \beta, \gamma) = e^{-\frac{i}{\hbar}\alpha L_z}\, e^{-\frac{i}{\hbar}\beta L_y}\, e^{-\frac{i}{\hbar}\gamma L_z} \; , \tag{12.41}$$

and now the matrix elements of this operator are simple to calculate. Thus the irreducible representation of the rotation group results from taking the matrix elements of $\mathcal{R}(\alpha, \beta, \gamma)$ between the eigenstates of L^2 and L_z, i.e.,

$$\begin{aligned} \langle \ell', m'|\mathcal{R}(\alpha, \beta, \gamma)|\ell, m\rangle &= \delta_{\ell'\ell}\, \langle \ell, m'|\mathcal{R}(\alpha, \beta, \gamma)|\ell, m\rangle \\ &\equiv \delta_{\ell'\ell}\, \mathcal{D}^{\ell}_{m'm}(\alpha, \beta, \gamma) \; , \end{aligned} \tag{12.42}$$

with

$$\begin{aligned} \mathcal{D}^{\ell}_{m'm}(\alpha, \beta, \gamma) &= e^{-i\alpha m'}\, \langle \ell, m'|e^{-\frac{i}{\hbar}\beta L_y}|\ell, m\rangle\, e^{-i\gamma m} \\ &\equiv e^{-i\alpha m'}\, d^{\ell}_{m'm}(\beta)\, e^{-i\gamma m} \; , \end{aligned} \tag{12.43}$$

where

$$\begin{aligned} d^{\ell}_{m'm}(\beta) &= (-1)^{m'-m} \left[\frac{(\ell+m')!\,(\ell-m')!}{(\ell+m)!\,(\ell-m)!}\right]^{1/2} \left(\cos\frac{\beta}{2}\right)^{m'+m} \left(\sin\frac{\beta}{2}\right)^{m'-m} \\ &\quad \times P^{(m'-m, m'+m)}_{\ell-m'}(\cos\beta) \; . \end{aligned} \tag{12.44}$$

Here, $P^{(\alpha,\beta)}_{\ell}$ are the Jacobi polynomials.[31]

Let us now consider the transformation of the state $|\ell, m\rangle$ to the state $|\ell, m\rangle'$, where

$$\begin{aligned} |\ell, m\rangle' &= \mathcal{R}(\alpha, \beta, \gamma)\, |\ell, m\rangle \\ &= \sum_{\ell' m'} |\ell', m'\rangle\, \langle \ell', m'|\mathcal{R}(\alpha, \beta, \gamma)|\ell, m\rangle \\ &= \sum_{m'} |\ell, m'\rangle\, \mathcal{D}^{\ell}_{m'm}(\alpha, \beta, \gamma) \; . \end{aligned} \tag{12.45}$$

In coordinate representation, this reduces to

$$\langle \hat{r}|\ell, m\rangle' = \sum_{m'} \langle \hat{r}|\ell, m'\rangle \, \mathcal{D}^{\ell}_{m'm}(\alpha, \beta, \gamma) \; .$$

But we have that

$$\langle \hat{r}|\ell, m\rangle' = \langle P_R^{-1}\hat{r}|\ell, m\rangle = \langle \hat{r}'|\ell, m\rangle$$

and therefore we can write the transformation of the spherical harmonics associated with the system that is being rotated, as

$$Y_{\ell m}(\hat{r}') = \sum_{m'} Y_{\ell m'}(\hat{r}) \, \mathcal{D}^{\ell}_{m'm}(\alpha, \beta, \gamma) \; . \tag{12.46}$$

In the above discussion we have restricted our analysis to the rotation of a system with no spin, i.e., the angular momentum is the orbital angular momentum. In that case $\ell =$ integer, and therefore a rotation of the system by an angle of 2π about the z-axis is given by

$$\psi' = e^{-\frac{i}{\hbar}2\pi L_z} \, \psi = e^{-i2m\pi} \, \psi = \psi \quad \text{for} \quad m = \text{integer} \; .$$

If we now extend the above analysis of rotation to include spin, and in general to include systems with total angular momentum $\vec{J}$, then the rotation operator can be written as

$$\mathcal{R}(\vec{\theta}) = \exp\left[-\frac{i}{\hbar}\vec{\theta}\cdot\vec{J}\right] \; . \tag{12.47}$$

However, we now have for the case of spin 1/2 that a rotation of the system by 2π does not lead to the original state. In fact, to reach the original state, we have to rotate the system by 4π, since

$$\chi' = e^{-\frac{i}{\hbar}2\pi J_z} \, \chi = e^{-2m\pi} \, \chi = -\chi \quad \text{for} \quad m = \pm\frac{1}{2}$$

while

$$\chi' = e^{-\frac{i}{\hbar}4\pi J_z} \, \chi = e^{-4m\pi} \, \chi = \chi \quad \text{for} \quad m = \pm\frac{1}{2} \; .$$

We therefore have to enlarge our group of rotations to include the half integer angular momentum. This new group is called the *covering group*, and is identical to the group $SU(2)$, the *special unitary* group in two-dimensions.

We now can write the representation of $SU(2)$ as

$$\mathcal{D}^{j}_{m'm}(\alpha, \beta, \gamma) = e^{-i\alpha m'} \, d^{j}_{m'm}(\beta) \, e^{-i\gamma m} \; , \tag{12.48}$$

where

$$\begin{aligned} d^{j}_{m'm}(\beta) &= \langle j, m'|e^{-\frac{i}{\hbar}\beta J_y}|j, m\rangle \\ &= \left\{ e^{-\frac{i}{\hbar}\beta \tilde{J}^j_y} \right\}_{m'm} \; . \end{aligned} \tag{12.49}$$

As an example, let us consider the evaluation of $d^{\frac{1}{2}}_{m'm}(\beta)$. This is given by

$$d^{\frac{1}{2}}_{m'm}(\beta) = \langle \tfrac{1}{2}, m' | e^{-\frac{i}{\hbar}\beta J_y} | \tfrac{1}{2}, m\rangle \ .$$

But for $j = 1/2$ we have $J_y = \frac{\hbar}{2}\sigma_y = \frac{\hbar}{2}\begin{pmatrix} 0 & -i \\ i & 0 \end{pmatrix}$, therefore

$$\begin{aligned} \tilde{d}^{\frac{1}{2}}(\beta) &= e^{-\frac{i}{2}\beta\sigma_y} \\ &= \tilde{1} - \frac{i}{2}\beta\sigma_y + \frac{1}{2!}\left(\frac{i}{2}\beta\sigma_y\right)^2 - \frac{1}{3!}\left(\frac{i}{2}\beta\sigma_y\right)^3 + \cdots \ . \end{aligned}$$

But we have that $(\sigma_y)^2 = \tilde{1}$. This allows us to write $\tilde{d}^{\frac{1}{2}}(\beta)$ as

$$\tilde{d}^{\frac{1}{2}}(\beta) = \tilde{1} - \frac{i}{2}\beta\sigma_y + \frac{1}{2}\left(\frac{i}{2}\beta\right)^2 \tilde{1} - \frac{1}{3!}\sigma_y\left(\frac{i}{2}\beta\right)^3 + \cdots \ .$$

This series can be regrouped to give

$$\begin{aligned} \tilde{d}^{\frac{1}{2}}(\beta) &= \tilde{1}\left\{1 - \frac{1}{2!}\left(\frac{\beta}{2}\right)^2 + \cdots\right\} - i\sigma_y\left\{\frac{\beta}{2} - \frac{1}{3!}\left(\frac{\beta}{2}\right)^3 + \cdots\right\} \\ &= \tilde{1}\cos\frac{\beta}{2} - i\sigma_y \sin\frac{\beta}{2} \\ &= \begin{pmatrix} 1 & 0 \\ 0 & 1 \end{pmatrix}\cos\frac{\beta}{2} - i\begin{pmatrix} 0 & -i \\ i & 0 \end{pmatrix}\sin\frac{\beta}{2} \ . \end{aligned}$$

Therefore

$$\tilde{d}^{\frac{1}{2}}(\beta) = \begin{pmatrix} \cos\frac{\beta}{2} & -\sin\frac{\beta}{2} \\ \sin\frac{\beta}{2} & \cos\frac{\beta}{2} \end{pmatrix} \ . \tag{12.50}$$

In particular $\tilde{d}^{\frac{1}{2}}(\pi)$ when acting on spin up, $\begin{pmatrix} 1 \\ 0 \end{pmatrix}$, gives spin down, $\begin{pmatrix} 0 \\ 1 \end{pmatrix}$, i.e.,

$$\tilde{d}^{\frac{1}{2}}(\pi)\begin{pmatrix} 1 \\ 0 \end{pmatrix} = \begin{pmatrix} 0 & -1 \\ 1 & 0 \end{pmatrix}\begin{pmatrix} 1 \\ 0 \end{pmatrix} = \begin{pmatrix} 0 \\ 1 \end{pmatrix} \ .$$

For a more detailed discussion of the rotation group and its representation, the reader is referred to the book by Edmonds [38].

12.3 Addition of Angular Momentum

In the last section we established that the total angular momentum of the system is conserved in the presence of rotational symmetry, i.e., the angular momentum is a constant of the motion, and the total angular momentum of the system commutes with the Hamiltonian. This can be stated as

$$\left[J^2, H\right] = 0 \qquad \text{and} \qquad [J_z, H] = 0 \ .$$

Because we have three commuting operators, the eigenstates of the system, i.e., the eigenstates of the Hamiltonian H, can also be eigenstates of the total angular momentum square J^2, and its component along the z-axis, J_z. To take full advantage of the symmetry of the problem, we need to write the eigenstates of the system in terms of the eigenstates of J^2 and J_z. In this way we separate the geometry in the problem from the dynamics, as was done for a particle in a central potential in Chapters 8 and 9. There we wrote the total wave function in terms of a radial wave function $R_\ell(r)$ and the angular function $Y_{\ell m}(\theta, \phi)$. In fact the spherical harmonic $Y_{\ell m}(\theta, \phi)$ was an eigenstate of the total angular momentum and its z-component. We also observed that for a system with spherical symmetry, the dependence on the potential, i.e., the dynamics, was only present in the radial equation. The geometry of the problem, which is very important and needs to be taken into consideration exactly, was in the spherical harmonics. Here it should be emphasized that any approximation in the geometry will automatically lead to a violation of the symmetry (conservation law), which in this case is spherical symmetry (angular momentum).

For a system with more than one particle and with spherical symmetry, we need to construct the states of total angular momentum. These will be given in terms of the basis states of the individual particle's angular momentum states. In general, for a system with n particles with angular momentum $\vec{J}_i$, $i = 1, \cdots, n$, the total angular momentum is given by

$$\vec{J} = \vec{J}_1 + \vec{J}_2 + \cdots + \vec{J}_n \ . \tag{12.51}$$

A similar problem arises when we consider a system of one particle with spin $\vec{S}$ and orbital angular momentum $\vec{L}$. In this case the total angular momentum is given by

$$\vec{J} = \vec{L} + \vec{S} \ . \tag{12.52}$$

In this section we will consider the problem of constructing eigenstates of the total angular momentum when $\vec{J}$ is the sum of two angular momenta, e.g., $n = 2$ in Eq. (12.51). This incorporates the case when one of the angular momenta is the spin of the particle while the other is the orbital angular momentum of that particle. The reason why we need not make a distinction between spin and orbital angular momentum is due to the fact that they both satisfy the same algebra, the algebra of $SU(2)$. For two angular momenta, since they are a function of different variables, we have that

$$\left[\vec{J}_1, \vec{J}_2\right] = 0 \ . \tag{12.53}$$

This means we can take as our basis, the eigenstates of the four commuting operators

$$J_1^2\ , \quad J_{1,z}\ , \quad J_2^2\ , \quad J_{2,z}\ .$$

These basis states are

$$|j_1, m_1; j_2, m_2\rangle = |j_1, m_1\rangle\, |j_2, m_2\rangle\ , \tag{12.54}$$

with

$$\begin{aligned}
J_1^2\, |j_1, m_1\rangle &= \hbar^2 j_1(j_1+1)\, |j_1, m_1\rangle & (12.55a)\\
J_{1,z}\, |j_1, m_1\rangle &= \hbar m_1\, |j_1, m_1\rangle & (12.55b)\\
J_2^2\, |j_2, m_2\rangle &= \hbar^2 j_2(j_2+1)\, |j_2, m_2\rangle & (12.55c)\\
J_{2,z}\, |j_2, m_2\rangle &= \hbar m_2\, |j_2, m_2\rangle\ . & (12.55d)
\end{aligned}$$

The disadvantage of this choice of basis states (i.e., $|j_1, m_1; j_2, m_2\rangle$) is that the states are not eigenstates of the total angular momentum J^2 which is obvious from the fact that

$$\begin{aligned}
J^2 &= \left(\vec{J}_1 + \vec{J}_2\right) \cdot \left(\vec{J}_1 + \vec{J}_2\right)\\
&= J_1^2 + J_2^2 + 2\vec{J}_1 \cdot \vec{J}_2\\
&= J_1^2 + J_2^2 + 2\left(J_{1,x}J_{2,x} + J_{1,y}J_{2,y} + J_{1,z}J_{2,z}\right)
\end{aligned}$$

and $|j_1, m_1; j_2, m_2\rangle$ is not an eigenstate of $J_{i,x}$ and $J_{i,y}$. Also, in this basis, the representations of the rotation group are reducible, which implies that there exists a unitary transformation to another basis in which the representations of the rotation group are irreducible, i.e., block diagonal.[2]

An alternative set of bases are the eigenstates of the operators

$$J_1^2\ , \quad J_2^2\ , \quad J^2\ , \quad J_z\ .$$

We can show that these operators commute among themselves, and their eigenstates are given by

$$\begin{aligned}
J_i^2\, |(j_1 j_2) j m\rangle &= \hbar^2 j_i(j_i+1)\, |(j_1 j_2) j m\rangle \quad i = 1, 2 & (12.56a)\\
J^2\, |(j_1 j_2) j m\rangle &= \hbar^2 j(j+1)\, |(j_1 j_2) j m\rangle & (12.56b)\\
J_z\, |(j_1 j_2) j m\rangle &= \hbar m\, |(j_1 j_2) j m\rangle\ . & (12.56c)
\end{aligned}$$

This basis has the advantage that the set of quantum numbers $\{j_1 j_2 j m\}$ are constants of motion because the corresponding operators commute with the spherically symmetric Hamiltonian. Also, in this basis, the representations of the rotation group are irreducible.

---

[2]We will see in Chapter 13 the advantages of working in an irreducible representation to the extent that the matrices we need to work with are smaller in dimension.

The problem now is to find the unitary transformation that will allow us to write the states $|(j_1 j_2)jm\rangle$ in terms of the known states $|j_1, m_1; j_2, m_2\rangle$, i.e.,

$$|(j_1 j_2)jm\rangle = \sum_{\substack{j_1' m_1' \\ j_2' m_2'}} |j_1', m_1'; j_2' m_2'\rangle \langle j_1', m_1'; j_2' m_2'|(j_1 j_2)jm\rangle \ . \tag{12.57}$$

The elements of this unitary transformation are the brackets $\langle j_1', m_1'; j_2' m_2'|(j_1 j_2)jm\rangle$. Before we proceed to determine these brackets, let us determine some of their properties. We have that

$$J_1^2 |(j_1 j_2)jm\rangle = \hbar^2 j_1(j_1+1) |(j_1 j_2)jm\rangle \ , \tag{12.58}$$

and

$$\langle j_1', m_1'; j_2', m_2'| J_1^2 = \hbar^2 j_1'(j_1'+1) \langle j_1', m_1'; j_2', m_2'| \ . \tag{12.59}$$

We now multiply Eq. (12.58) by $\langle j_1', m_1'; j_2', m_2'|$ and Eq. (12.59) by $|(j_1 j_2)jm\rangle$ and subtract to get

$$\hbar^2 \{j_1(j_1+1) - j_1'(j_1'+1)\} \langle j_1', m_1'; j_2', m_2'|(j_1 j_2)jm\rangle = 0 \ .$$

Therefore for $j_1 \neq j_1'$ we have

$$\langle j_1', m_1'; j_2', m_2'|(j_1 j_2)jm\rangle = 0$$

and therefore we have that $j_1 = j_1'$, i.e.,

$$\langle j_1', m_1'; j_2', m_2'|(j_1 j_2)jm\rangle = \delta_{j_1' j_1} \ \langle j_1, m_1'; j_2', m_2'|(j_1 j_2)jm\rangle \ .$$

In a similar way we can show that for $j_2' \neq j_2$ the brackets that form the elements of the unitary transformation are zero. We now can write Eq. (12.57) as

$$|(j_1 j_2)jm\rangle = \sum_{m_1 m_2} |j_1, m_1; j_2, m_2\rangle \langle j_1, m_1; j_2, m_2|(j_1 j_2)jm\rangle \ . \tag{12.60}$$

We now can further reduce this sum by noting that

$$J_z = J_{1,z} + J_{2,z} \ .$$

We then can write

$$J_z |(j_1 j_2)jm\rangle = \hbar m |(j_1 j_2)jm\rangle \ , \tag{12.61}$$

and

$$\langle j_1, m_1; j_2, m_2| \left(J_{1,x} + J_{2,z}\right) = \hbar \left(m_1 + m_2\right) \langle j_1, m_1; j_2, m_2| \ . \tag{12.62}$$

Proceeding as before and multiplying Eq. (12.61) by $\langle j_1, m_1; j_2, m_2|$ and Eq. (12.62) by $|(j_1 j_2)jm\rangle$ and subtracting we get

$$\hbar \{m - (m_1 + m_2)\} \ \langle j_1, m_1; j_2, m_2|(j_1 j_2)jm\rangle = 0$$

and therefore

$$\langle j_1, m_1; j_2, m_2 | (j_1 j_2) j m \rangle = \delta_{m, m_1+m_2} \, (j_1 m_1 j_2 m_2 | j_1 j_2 j m) \ . \tag{12.63}$$

We now can write the unitary transformation from one basis to the other as

$$|(j_1 j_2) j m\rangle = \sum_{\substack{m_1 \; m_2 \\ m_1+m_2=m}} |j_1, m_1; j_2, m_2\rangle \ (j_1 m_1 j_2 m_2 | j_1 j_2 j m) \ . \tag{12.64}$$

The coefficient $(j_1 m_1 j_2 m_2 | j_1 j_2 j m)$ is known as the *Clebsch-Gordan coefficient.* Here we note that in the ket on the left hand side of Eq. (12.64), we have put brackets around the quantum numbers j_1 and j_2. This is to indicate that these two angular momenta are added to get a total angular momentum j with projection m. The inverse transformation is then given by

$$|j_1, m_1; j_2, m_2\rangle = \sum_{\substack{j \; m \\ m=m_1+m_2}} |(j_1 j_2) j m\rangle \ (j_1 m_1 j_2 m_2 | j_1 j_2 j m) \ . \tag{12.65}$$

Since the transformation from one basis to the next is unitary, and the states in each basis are orthonormal, these coefficients satisfy the orthogonality relation

$$\sum_{jm} (j_1 m_1 j_2 m_2 | j_1 j_2 j m) \ (j_1 m_1' j_2 m_2' | j_1 j_2 j m) = \delta_{m_1 m_1'} \ \delta_{m_2 m_2'} \ , \tag{12.66}$$

and

$$\sum_{m_1 m_2} (j_1 m_1 j_2 m_2 | j_1 j_2 j m) \ (j_1 m_1 j_2 m_2 | j_1 j_2 j' m') = \delta_{jj'} \ \delta_{mm'} \ . \tag{12.67}$$

The values of these coefficients can be obtained from standard codes available on most computers or from the package *Mathematica* [39].

Finally, we should note that since $\vec{J}$ is the vector sum of $\vec{J_1}$ and $\vec{J_2}$, the corresponding quantum numbers satisfy the relation

$$|j_1 - j_2| \le j \le |j_1 + j_2| \ .$$

Furthermore, since $\vec{J}$ satisfies the same algebra as the angular momentum operator, we have that

$$-j \le m \le j \ .$$

Example: The simplest example we can consider is that of two particles of spin 1/2, i.e.,

$$\vec{S} = \vec{s_1} + \vec{s_2} \ .$$

In this case

$$|(\tfrac{1}{2}\,\tfrac{1}{2})\,SM\rangle = \sum_{m_1=\pm\frac{1}{2}} (\tfrac{1}{2}\,m_1\,\tfrac{1}{2}\,M-m_1|\tfrac{1}{2}\,\tfrac{1}{2}\,SM)\;|\tfrac{1}{2}\,m_1\rangle\;|\tfrac{1}{2}\,M-m_1\rangle\ .$$

For $S=1$ and $M=+1$, m_1 must be $+\frac{1}{2}$ and we have

$$\begin{aligned}|(\tfrac{1}{2}\,\tfrac{1}{2})\,S=1\,M=+1\rangle &= (\tfrac{1}{2}\,\tfrac{1}{2}\,\tfrac{1}{2}\,\tfrac{1}{2}|\tfrac{1}{2}\,\tfrac{1}{2}\,1\,1)\;|\tfrac{1}{2}\,\tfrac{1}{2}\rangle\;|\tfrac{1}{2}\,\tfrac{1}{2}\rangle\\ &= |\tfrac{1}{2}\,\tfrac{1}{2}\rangle\;|\tfrac{1}{2}\,\tfrac{1}{2}\rangle = \alpha(1)\,\alpha(2)\ .\end{aligned}$$

The normalization of basis required that $(\frac{1}{2}\,\frac{1}{2}\,\frac{1}{2}\,\frac{1}{2}|\frac{1}{2}\,\frac{1}{2}\,1\,1) = 1$. Similarly, for $S=1$ and $M=-1$ we have that $m_1 = -\frac{1}{2}$ and

$$|(\tfrac{1}{2}\,\tfrac{1}{2})\,S=1\,M=-1\rangle = \beta(1)\,\beta(2)\ .$$

Here, α and β correspond to Pauli spinors with spin up and down respectively, while $\alpha(1)$ indicates that particle 1 is in a state with spin up, i.e., along the positive z-axis.

For $M=0$, we have two possible values for the total spin, i.e., $S=1$ or 0. In this case $m_1 = \pm\frac{1}{2}$ and we have two terms in the sum. For $S=1$ we have

$$|(\tfrac{1}{2}\,\tfrac{1}{2})\,S=1\,M=0\rangle = \frac{1}{\sqrt{2}}\left\{\alpha(1)\,\beta(2)+\alpha(2)\,\beta(1)\right\}\ ,$$

while for $S=0$, we have

$$|(\tfrac{1}{2}\,\tfrac{1}{2})\,S=0\,M=0\rangle = \frac{1}{\sqrt{2}}\left\{\alpha(1)\beta(2)-\alpha(2)\,\beta(1)\right\}\ .$$

We observe here that all the $S=1$ *(Spin Triplet)* states have a symmetric wave function under the exchange of the particles coordinates. On the other hand, the $S=0$ *(Spin Singlet)* state is antisymmetric under this exchange. We will come back to this problem at a later stage when we consider the symmetry of the wave function under the permutation of coordinates.

12.4 Space Inversion and Parity

Until 1957, it was always assumed that the observables for any system will not change if we consider the system to have gone under the transformation $\vec{r} \to -\vec{r}$. However, it was realized in 1956 that there has been no experimental test of this symmetry for a certain class of interaction commonly known as the Weak Interaction. The validity of this symmetry corresponds to the question of getting the same result from an experiment and its mirror image. At the suggestion of T. D. Lee and C. N. Yang [40] three experiments were carried out which demonstrated that in Weak Interactions this symmetry is violated at the level of 100%. However for all other interactions, space inversion is a good symmetry, and the corresponding conserved quantity is commonly referred to as *parity*. Thus the states of any system, where the governing interaction is not the Weak Interaction, parity is one of the quantum numbers that labels the states, and in any transition, parity will be conserved.

Let us consider the action of space inversion to be represented by the operator $\mathcal{P}$. We then have that

$$\mathcal{P}\,\psi(\vec{r}) = \psi(-\vec{r})\;. \tag{12.68}$$

This operator has the property that when acting on a state twice, it will give the original state, i.e.,

$$\mathcal{P}^2\,\psi_\alpha(\vec{r}) = \mathcal{P}\,\psi_\alpha(-\vec{r}) = \psi_\alpha(\vec{r})\;. \tag{12.69}$$

This special property of the operator $\mathcal{P}^2$, having eigenvalue one, implies that the operator $\mathcal{P}$ could have eigenvalues ± 1, i.e.,

$$\mathcal{P}\,\psi_\alpha(\vec{r}) = \pm\,\psi_\alpha(\vec{r})\;. \tag{12.70}$$

This in turn means that we can label the state ψ_α by the eigenvalue corresponding to the operator $\mathcal{P}$.

We now examine the condition under which space inversion is a valid symmetry and thus parity is a conserved quantum number. Let us consider the case when the state ψ_α is a solution of the time dependent Schrödinger equation, i.e.,

$$i\hbar\frac{\partial\psi_\alpha}{\partial t} = H\,\psi_\alpha\;. \tag{12.71}$$

Then the action of the space inversion operator on this equation gives

$$i\hbar\mathcal{P}\frac{\partial\psi_\alpha}{\partial t} = \mathcal{P}\,H\,\psi_\alpha\;. \tag{12.72}$$

Taking $\mathcal{P}\,\psi_\alpha = \psi'_\alpha$ and therefore $\psi_\alpha = \mathcal{P}^{-1}\,\psi'_\alpha$, we can write Eq. (12.72) as

$$i\hbar\frac{\partial\psi'_\alpha}{\partial t} = \mathcal{P}H\mathcal{P}^{-1}\psi'_\alpha\;. \tag{12.73}$$

For the theory to be invariant under space inversion, the equation of motion, i.e., the Schrödinger equation, should maintain its form under the transformation. This implies that the Hamiltonian H will commute with the parity operator $\mathcal{P}$, i.e.,

$$[\mathcal{P}, H] = 0\;. \tag{12.74}$$

We thus may conclude that for systems for which the Hamiltonian commutes with the parity operator, the eigenstates of the system can be labeled by the parity quantum number which can take a value of ± 1.

12.5 Time Reversal

We now turn to time reversal which involves the reversal in the direction of time, i.e., it is equivalent to asking the following question. When running a film backwards, are the physical phenomenon observed as valid as those observed when the film runs forward? Although not valid at the macroscopic level, this seems to be true at the microscopic level. To examine this symmetry, first consider the state ψ_α that describes our system. Corresponding to this state then, there is a state ψ'_α that describes the time reversed system. In this time reversed system all momenta and angular momenta point in the opposite direction to the original system, i.e., under time reversal $\vec{p} \to -\vec{p}$ and $\vec{L} \to -\vec{L}$. We now introduce an operator $\mathcal{T}$ such that

$$\mathcal{T}\,\psi_\alpha = \psi'_\alpha \ . \tag{12.75}$$

To the best of our knowledge $\mathcal{T}$ is a symmetry operation that is valid for most systems, i.e., if ψ_α is an eigenstate of the Hamiltonian H, then $\psi'_\alpha = \mathcal{T}\,\psi_\alpha$ is also an eigenstate of H.

We first examine the general property of the operator $\mathcal{T}$ by considering a state ψ_α at time $t = 0$ to be given by

$$\psi_\alpha = \sum_k a_{\alpha k}\,\phi_k \ ,$$

where ϕ_k is an eigenstate of the Hamiltonian H with an eigenvalue E_k, i.e.,

$$H\,\phi_k = E_k\,\phi_k \ .$$

We now would like to follow the time development of this system along two paths:

1. Let the state ψ_α propagate in time from $t = 0$ to time $t = t_1$. Then the state at $t = t_1$ is given by

   $$\psi'_\alpha = e^{-iHt_1/\hbar}\,\psi_\alpha = \sum_k a_{\alpha k}\,e^{-iE_kt_1/\hbar}\,\phi_k \quad \text{at} \quad t = t_1 \ .$$

   We next apply the time reversal operator to this state, i.e., we act on the state with the time reversal operator $\mathcal{T}$ to get

   $$\psi''_\alpha = \mathcal{T}\,\psi'_\alpha = \sum_k a_{\alpha k}\,e^{-iE_kt_1/\hbar}\,\mathcal{T}\phi_k \quad \text{at} \quad t = -t_1 \ .$$

   In writing this result we have assumed $\mathcal{T}$ is a linear operator.

2. Now we can get to the same final state by first applying the time reversal operator to get

   $$\psi'''_\alpha = \sum_k a_{\alpha k}\,\mathcal{T}\phi_k \quad \text{at} \quad t = 0 \ ,$$

and then by letting the state propagate in time to $t = -t_1$, we get

$$\psi''_\alpha = e^{iHt_1/\hbar}\,\psi'''_\alpha = \sum_k a_{\alpha k}\, e^{iE_k t_1/\hbar}\,\mathcal{T}\phi_k \quad \text{at} \quad t = -t_1 \ .$$

To get this result, we have taken $[H, \mathcal{T}] = 0$.

Clearly the final states obtained along the two different paths are different, yet they both should describe the time reversed system at time $t = -t_1$. This suggests that we have made an error in the above analysis. We observe that if the operator $\mathcal{T}$ had the property that for any complex constant c

$$\mathcal{T}\, c\,\Psi = c^*\,\mathcal{T}\,\Psi$$

where Ψ is any state vector, then we would have had the same result for our final state irrespective of the path followed. In fact, with this new definition of the time reversal operator, our two paths are given by:

1. For the first path

$$\begin{aligned}
\psi''_\alpha &= \mathcal{T}\, e^{-iHt_1/\hbar}\,\psi_\alpha \\
&= \mathcal{T}\sum_k a_{\alpha k}\, e^{-iE_k t_1/\hbar}\phi_k \\
&= \sum_k a^*_{\alpha k}\, e^{iE_k t_1/\hbar}\,\mathcal{T}\phi_k \ .
\end{aligned}$$

2. While for the second path we have

$$\begin{aligned}
\psi''_\alpha &= e^{iHt_1/\hbar}\,\mathcal{T}\,\psi_\alpha \\
&= e^{iHt_1/\hbar}\sum_k a^*_{\alpha k}\,\mathcal{T}\phi_k \\
&= \sum_k a^*_{\alpha k}\, e^{iE_k t_1/\hbar}\,\mathcal{T}\phi_k \ .
\end{aligned}$$

We now have the same result for the state at time $t = -t_1$ irrespective of the path followed. From the above property of the time reversal operator we may conclude that this operator is in fact not linear but is referred to as an *anti-linear* operator.

We can now show from the Schrödinger equation that the operator $\mathcal{T}$ has to be anti-linear. The time dependent Schrödinger equation is given by

$$i\hbar\frac{\partial\psi_\alpha}{\partial t} = H\,\psi_\alpha \ .$$

If under time reversal $t \to -t$, then the Schrödinger equation for the time reversed state is given by

$$-i\hbar\frac{\partial\psi'_\alpha}{\partial t} = H\,\psi'_\alpha$$

But this is the Schrödinger equation for the state $\psi'_\alpha = \psi^*_\alpha$. If we operate with $\mathcal{T}$ on the Schrödinger equation, we get

$$-i\hbar\frac{\partial \mathcal{T}\,\psi_\alpha}{\partial t} = \mathcal{T}\,H\,\psi_\alpha\ .$$

If $[\mathcal{T}, H] = 0$, then the above equation becomes

$$-i\hbar\frac{\partial \mathcal{T}\,\psi_\alpha}{\partial t} = H\,\mathcal{T}\,\psi_\alpha$$

and we have the state $\mathcal{T}\,\psi_\alpha$ satisfying the same equation as the state ψ^*_α. Thus if $\mathcal{T}\,\psi_\alpha = \psi^*_\alpha$, then time reversal is a symmetry of the system.

In general, the time reversal operator is of the form

$$\mathcal{T} = K\,U\ , \tag{12.76}$$

where K is the complex conjugation operator, and U is a linear operator. Furthermore, if $|\langle\alpha|\beta\rangle|$ is to remain the same under time reversal, then U must be unitary, i.e., $U^{-1} = U^\dagger$, and in the case of $\mathcal{T} = K\,U$, is an *anti-unitary operator*.

For particles with spin zero, we want to choose $\mathcal{T}$ and thus U such that

$$\mathcal{T}\,\vec{r}\,\mathcal{T}^{-1} = \vec{r} \qquad \text{and} \qquad \mathcal{T}\,\vec{p}\,\mathcal{T}^{-1} = -\vec{p}\ , \tag{12.77}$$

i.e., the time reversal operator changes the direction of the momentum not the coordinate. In coordinate representation, we have that $\vec{r}$ is real while $\vec{p}$ is pure imaginary ($\vec{p} = -i\hbar\nabla$), so that the simplest choice for $\mathcal{T}$ is

$$\mathcal{T} = K\ , \tag{12.78}$$

where K is the complex conjugation operator.

For particles with spin, we also want to change the direction of the spin and angular momentum, i.e., in addition to the conditions in Eq. (12.77) we have

$$\mathcal{T}\,\vec{S}\,\mathcal{T}^{-1} = -\vec{S} \qquad \text{and} \qquad \mathcal{T}\,\vec{J}\,\mathcal{T}^{-1} = -\vec{J}\ , \tag{12.79}$$

where $\vec{J}$ is the angular momentum of the system. As before, if we specify the representation of the spin, we can write the Pauli matrices as

$$\sigma_x = \begin{pmatrix} 0 & 1 \\ 1 & 0 \end{pmatrix}\ , \quad \sigma_y = \begin{pmatrix} 0 & -i \\ i & 0 \end{pmatrix}\ , \quad \sigma_z = \begin{pmatrix} 1 & 0 \\ 0 & -1 \end{pmatrix}\ .$$

In this case $\vec{r}$, σ_x, σ_z are real, while $\vec{p}$ and σ_y are pure imaginary. Thus for $\mathcal{T}\,\vec{S}\,\mathcal{T}^{-1} = -\vec{S}$ and $\mathcal{T} = K\,U$, we require that

$$U\,S_x\,U^{-1} = -S_x \qquad \text{and} \qquad U\,S_z\,U^{-1} = -S_z\ . \tag{12.80}$$

We therefore need to choose U such that it reverses the direction of the x and z axis for the spin matrices, but not the y-axis. This can be achieved by a rotation about the y-axis by an angle π. In this case the rotation operator is

$$U = e^{-i\pi S_y/\hbar}$$

and the time reversal operator is

$$\mathcal{T} = K\,U = e^{-i\pi S_y/\hbar}\,K \; . \tag{12.81}$$

12.6 Isospin and the Pauli Principle

Soon after the neutron was discovered in 1932, Heisenberg [41] suggested that in the absence of electromagnetic interaction, the neutron and the proton are two states of the same system. In fact, the mass of the proton is almost the same as the mass of the neutron since [42]

$$m_p = 938.27203 \pm 0.00008 \text{ MeV.}/c^2 \quad m_n = 939.5636 \pm 0.00008 \text{ MeV.}/c^2 \;, \tag{12.82}$$

where c is the velocity of light in vacuum. Furthermore, the interaction between the neutron and proton is approximately the same as the force between two protons when the Coulomb repulsion is subtracted. This equivalence of the proton and neutron is similar to a degeneracy in the energy spectrum, and therefore, a possible result of a symmetry. In this case the symmetry is not associated with either space or time, but is an internal symmetry. The fact that we have a doublet of states suggested to Heisenberg that he should introduce the same algebra that described spin, because with spin there is also a doublet corresponding to spin up and spin down. Thus *isospin* is the algebra that describes protons and neutrons to which we give the generic name *nucleon* so that

$$\begin{aligned} \text{proton} &= \text{nucleon with isospin up} \\ \text{neutron} &= \text{nucleon with isospin down} \end{aligned}$$

We now can adapt the algebra of spin to isospin by introducing the isospin operators

$$\tau_1 = \begin{pmatrix} 0 & 1 \\ 1 & 0 \end{pmatrix} \quad \tau_2 = \begin{pmatrix} 0 & -i \\ i & 0 \end{pmatrix} \quad \tau_3 = \begin{pmatrix} 1 & 0 \\ 0 & -1 \end{pmatrix} . \tag{12.83}$$

which are the Pauli operators that we used for spin. Since we know the algebra of these operators, we may adapt the result we derived for spin angular momentum to isospin. In particular, we can introduce the operators that convert a neutron to a proton and vice versa in analogy with the operators that flipped the spin angular momentum (see Eqs. (11.126) and (11.127)) as:

$$\frac{1}{2}(\tau_1 + i\tau_2)\; n = p \qquad \frac{1}{2}(\tau_1 - i\tau_2)\; p = n\;, \tag{12.84}$$

where n and p are the states corresponding to a neutron and proton, i.e.,

$$n = \begin{pmatrix} 0 \\ 1 \end{pmatrix} \qquad p = \begin{pmatrix} 1 \\ 0 \end{pmatrix}\;. \tag{12.85}$$

Furthermore, we can use the same Clebsch-Gordan coefficients we used for angular momentum to construct states of total isospin. In fact, since isospin is a good symmetry of the nuclear Hamiltonian, the total isospin operator commutes with the Hamiltonian, and to that extent the eigenstates of total isospin are the proper basis to be working in, as was the case with angular momentum where we needed to construct states of total angular momentum. For the two nucleon system, the states of total isospin consist of a triplet of states with isospin one, which are

$$\begin{aligned} |(\tfrac{1}{2}\tfrac{1}{2})1\; +1\rangle &= p(1)\,p(2) \\ |(\tfrac{1}{2}\tfrac{1}{2})1\; -1\rangle &= n(1)\,n(2) \\ |(\tfrac{1}{2}\tfrac{1}{2})1\; 0\rangle &= \frac{1}{\sqrt{2}}\left[p(1)\,n(2) + n(1)\,p(2)\right]\;, \end{aligned} \tag{ 12.86a}$$

while for the isospin singlet we have

$$|(\tfrac{1}{2}\tfrac{1}{2})0\; 0\rangle = \frac{1}{\sqrt{2}}\left[p(1)\,n(2) - n(1)\,p(2)\right]\;. \tag{ 12.86b}$$

As expected, these states are identical to those we encountered for spin in Sec. 12.3.

The wave function for the two nucleon system is now a product of the space, spin and isospin, i.e.,

$$\Psi(1,2) = \psi(\text{space})\;\phi(\text{spin})\;\chi(\text{isospin})\;. \tag{12.87}$$

Because our two nucleon system now consists of two identical Fermions,[3] their wave function should be antisymmetric, i.e.,

$$\Psi(1,2) = -\,\Psi(2,1)\;. \tag{12.88}$$

This basically states that two Fermions cannot be at the same position in space and at the same time have the same quantum numbers. This statement is known as the *Pauli*

[3]Fermions are particles with half integer spin, and are called by this name because they obey Fermi-Dirac statistics. In contrast, Bosons are particles with integer spin that satisfy Bose-Einstein statistics. As you would see in an introductory course in Statistical Mechanics, the statistic of spin $\frac{1}{2}$ particles is different from the statistic of integer spin particles.

exclusion principle. Since each part of the wave function of the two nucleons can have a definite symmetry under the exchange of the coordinates of the two nucleons, it is clear that only certain combinations are allowed according to this Pauli exclusion principle. From the spin and isospin wave functions we know that the singlet is antisymmetric while the triplet is symmetric. In other words, under the interchange of coordinates we have

$$\phi(2,1) = (-1)^{1+S}\,\phi(1,2) \qquad \chi(2,1) = (-1)^{1+T}\chi(1,2)\ , \tag{12.89}$$

where S and T are the total spin and isospin of the two nucleon system. On the other hand, the space part of the wave function is labeled by the orbital angular momentum ℓ and is a function of the relative coordinates of the two nucleons. Under exchange of coordinates, the space part of the wave function satisfies

$$\psi(2,1) = (-1)^{\ell}\,\psi(1,2)\ . \tag{12.90}$$

Combining Eqs. (12.89) and (12.90), we can see that the total wave function has the symmetry

$$\Psi(2,1) = (-1)^{\ell+S+T}\,\Psi(1,2)\ . \tag{12.91}$$

To satisfy the Pauli exclusion principle, we need to have $\ell+S+T$ to be an odd integer. In

Table 12.1: The partial waves that are allowed by the Pauli exclusion principle and the experiment required to determine the phase shifts in that partial wave.

ℓ	S	T	Spectroscopic Notation	Experiment
0	0	1	${}^1S_0$	$p-p$
	1	0	${}^3S_1$	$p-p$ and $n-p$
1	0	0	${}^1P_1$	$p-p$ and $n-p$
	1	1	${}^3P_0$, ${}^3P_1$, ${}^3P_2$	$p-p$
2	0	1	${}^1D_2$	$p-p$
	1	0	${}^3D_1$, ${}^3D_2$, ${}^3D_3$	$p-p$ and $n-p$

Table 12.1 we present the states that are allowed, and neutron-proton $(n-p)$ or proton-proton $(p-p)$ experiments that have to be done to get information about the scattering amplitude in that partial wave. We have used the spectroscopic notation

$${}^{2S+1}\ell_J$$

where for $\ell = 0, 1, 2, \cdots$ we have S, P, D $\cdots$. Here, J is the total angular momentum.

12.7 Problems

1. The Hamiltonian of a rigid rotator (i.e., quantum mechanical football) in an external magnetic field pointing along the z-axis is given by

$$H = \frac{L^2}{2\mathcal{I}} + B\,L_z \ ,$$

where $\mathcal{I}$ is the moment of inertia of the rotator, and B is a constant. Here $\vec{L}$ is the angular momentum operator. Determine the energy and wave function of the rigid rotator.

2. Consider the infinitesimal rotation operator to be of the form

$$\mathcal{R}(\varepsilon, \hat{n}) = \left\{1 + \varepsilon\hat{n}\cdot\vec{E}\right\} \ .$$

Evaluate:

(a) First, a rotation about the y-axis followed by a rotation about the x-axis, i.e.,

$$\mathcal{R}(\varepsilon, \hat{n}_x)\,\mathcal{R}(\varepsilon, \hat{n}_y)\,\psi_\alpha(\vec{r}).$$

(b) Second a rotation about the x-axis followed by a rotation about the y-axis, i.e.,

$$\mathcal{R}(\varepsilon, \hat{n}_y)\,\mathcal{R}(\varepsilon, \hat{n}_x)\,\psi_\alpha(\vec{r}).$$

Using the definition of rotation about $\hat{n}$ by an angle ε to be

$$\psi_{\alpha'}(\vec{r}\,') = \psi_\alpha(\vec{r}\,' - \varepsilon\hat{n}\times\vec{r}\,') = \mathcal{R}(\varepsilon, \hat{n})\,\psi_\alpha(\vec{r}\,') \ ,$$

determine the commutation relation for E_x, E_y, E_z and relate them to the commutation relation of the angular momentum.

3. Calculate the matrices

$$\langle jm|J_i|jm'\rangle \quad i = 1, 2, 3$$

for $j = 1$. Determine the rotation operator

$$\langle jm|\mathcal{R}(\frac{\pi}{2}, \hat{e}_2)|jm'\rangle = \exp\left\{-\frac{i}{\hbar}\frac{\pi}{2}\,\tilde{J}^j_2\right\} \ ,$$

where $\hat{e}_2$ is a unit vector in the 2 direction, for $j = 1$ as a 3×3 matrix.

4. Prove the following relations for the irreducible representation of the rotation group

$$\sum_\mu \mathcal{D}^\ell_{m\mu}(\alpha,\beta,\gamma)\,\mathcal{D}^{\ell *}_{m'\mu}(\alpha,\beta,\gamma) = \delta_{mm'}$$

and

$$\langle\ell m|\mathcal{R}^{-1}(\alpha,\beta,\gamma)|\ell m'\rangle = \left[\mathcal{D}^\ell_{m'm}(\alpha,\beta,\gamma)\right]^* \ .$$

5. The positronium is a bound state of an electron and a positron (both are spin 1/2 particles). In the ground state, i.e., $n = 1$, $\ell = 0$, the main term in the Hamiltonian, besides the Coulomb attraction is

$$V_1 = \gamma\, \vec{\sigma}_1 \cdot \vec{\sigma}_2 \ .$$

   (a) Calculate the Clebsch-Gordan coefficients for the coupling of spin 1/2 with spin 1/2 by diagonalizing the above interaction in the basis $|s_1 = \frac{1}{2}, m_1; s_2 = \frac{1}{2}, m_2\rangle$.

   (b) Show that

$$|(s_1 s_2)SM\rangle = (-1)^{S+1}\, |(s_2 s_1)SM\rangle \ ,$$

   if $s_1 = s_2 = \frac{1}{2}$.

6. Consider the reaction

$$\pi^- + d \rightarrow n + n \ .$$

   where π^- is the negative pion, d is the deuteron, and n is the neutron. Suppose the π^- has spin 0 and negative intrinsic parity and was captured from an s-orbit.

   (a) What possible states would the two neutrons be in?

   (b) What is the intrinsic parity of the pion?

   Hint: Use conservation of angular momentum, parity and the Pauli exclusion principle.

7. Show that

$$D^{\ell}_{m0}(\alpha, \beta, \gamma) = (-1)^m \sqrt{\frac{4\pi}{2\ell + 1}}\; Y^*_{\ell m}(\beta, \alpha) \ .$$

   Hint Note: The Jacobi polynomial $P^{(m,m)}_{\ell-m}(x)$ is given by

$$P^{(m,m)}_{\ell-m}(x) = (-2)^m \frac{\ell!}{(\ell - m)!}(1 - x^2)^{-m/2} P^{-m}_{\ell}(x) \ .$$

Quantum Mechanics with Applications, 2011, 252-288

CHAPTER 13

Approximation Methods for Bound States

Abstract: With a limit on the small number of problems that admit exact solutions, we turn in this chapter to approximation methods that can be used for bound state problems. The most commonly used method is the perturbation theory for which one use a small parameter to form the basis of a systematic expansion, or a variational method in which one hopes to set the best limit on the energy of a system. We will consider both of these approximation methods with applications to problems in atomic and molecular physics.

In most problems in quantum mechanics we are given a Hamiltonian H for which we need to determine the eigenvalues and the corresponding eigenstates. For most physically interesting systems, the Hamiltonian is complex enough that we cannot solve the Schrödinger equation

$$H\,|\psi\rangle = E\,|\psi\rangle \tag{13.1}$$

in a closed form. In this case, to get any information about the energy spectrum we need to resort to approximation methods. Two such methods that are often used for both the bound state and scattering problem are:

1. Perturbation theory.

2. Variational method.

In the present chapter, we will consider both of these methods within the context of bound state problems.

Iraj R. Afnan

13.1 Perturbation Theory

This method is most useful when we write the Hamiltonian for the system under consideration as the sum of two parts, i.e.,

$$H = H_0 + H_1 \ , \tag{13.2}$$

where H_0 is chosen such that the corresponding Schrödinger equation can be solved exactly, i.e., we can solve the equation

$$H_0|\phi_n\rangle = \varepsilon_n|\phi_n\rangle \ , \tag{13.3}$$

and at the same time H_1 is small enough that the contribution to the energy and wave function from H_1 is relatively small. In particular, the aim of perturbation theory is to write a series solutions for both the eigenvalues and eigenstates of H. The lowest term in the series would be the eigenvalues and eigenstates of H_0. The rest of the terms in the expansion would be a power series in matrix elements of H_1 with respect to the eigenstates of H_0.

Let us first consider the Schrödinger equation for H, i.e.,

$$H|\psi_n\rangle = E_n|\psi_n\rangle \ . \tag{13.4}$$

With the help of Eq. (13.2), this can be written as

$$(\, E_n - H_0 \,)\ |\psi_n\rangle = H_1\ |\psi_n\rangle \ . \tag{13.5}$$

We now multiply, from the left, by the state $\langle\phi_n|$ to get

$$\langle\phi_n|\,(E_n - H_0)\,|\psi_n\rangle = \langle\phi_n|H_1|\psi_n\rangle \ .$$

Making use of the fact that $\langle\phi_n|$ is an eigenstate of H_0, we can solve this equation for the energy, E_n, to get

$$\begin{aligned} E_n &= \varepsilon_n + \frac{\langle\phi_n|H_1|\psi_n\rangle}{\langle\phi_n|\psi_n\rangle} \\ &= \varepsilon_n + \langle\phi_n|H_1|\psi_n\rangle \ , \end{aligned} \tag{13.6}$$

where we have taken the normalization of the wave functions to be such that

$$\langle\phi_n|\psi_n\rangle = 1 \ . \tag{13.7}$$

At the end of this section we will show that this normalization is consistent with our formulation. From Eq. (13.6) it is clear that a series solution for $|\psi_n\rangle$ in powers of H_1

will give us a corresponding series for the eigenvalues E_n, and provided H_1 is small, this series will converge. To get a power series for the eigenstates $|\psi_n\rangle$, we write Eq. (13.5) as

$$(E_n - \Delta - H_0)\;|\psi_n\rangle = (H_1 - \Delta)\;|\psi_n\rangle\;,$$

where the constant Δ, to be determined at a later stage, is chosen such that the operator $(E_n - \Delta - H_0)$ does not have any zero eigenvalues. We then can rewrite the above equation as

$$|\psi_n\rangle = \frac{1}{E_n - \Delta - H_0}\;(H_1 - \Delta)\;|\psi_n\rangle\;. \tag{13.8}$$

Making use of the completeness of the eigenstates of H_0, i.e.,

$$\sum_m |\phi_m\rangle\,\langle\phi_m| = 1\;,$$

we can write Eq. (13.8) as

$$\begin{aligned} |\psi_n\rangle &= \sum_m |\phi_m\rangle\,\langle\phi_m|\frac{1}{E_n - \Delta - H_0}\;(H_1 - \Delta)\;|\psi_n\rangle \\ &= |\phi_n\rangle\,\langle\phi_n|\frac{1}{E_n - \Delta - H_0}\;(H_1 - \Delta)\;|\psi_n\rangle \\ &\quad + \sum_{m\neq n} |\phi_m\rangle\,\langle\phi_m|\frac{1}{E_n - \Delta - H_0}\;(H_1 - \Delta)\;|\psi_n\rangle\;. \end{aligned} \tag{13.9}$$

We now introduce projection operators P_n and Q_n defined in terms of the states $|\phi_n\rangle$ as

$$P_n \equiv |\phi_n\rangle\,\langle\phi_n|\;, \tag{13.10}$$

and

$$\begin{aligned} Q_n &= 1 - P_n \\ &= \sum_{m\neq n} |\phi_m\rangle\,\langle\phi_m|\;. \end{aligned} \tag{13.11}$$

In terms of these projection operators we can write the eigenstate of the full Hamiltonian as

$$\begin{aligned} |\psi_n\rangle &= |\phi_n\rangle\frac{1}{E_n - \Delta - \varepsilon_n}\;(\langle\phi_n|H_1|\psi_n\rangle - \Delta) + Q_n\,\frac{1}{E_n - \Delta - H_0}\;(H_1 - \Delta)\;|\psi_n\rangle \\ &= |\phi_n\rangle + Q_n\,\frac{1}{E_n - \Delta - H_0}\;(H_1 - \Delta)\;|\psi_n\rangle\;, \end{aligned} \tag{13.12}$$

where we have made use of Eq. (13.6) to simplify the first term on the right hand side of the equation. In Eq. (13.12), we have an integral equation that is equivalent to the

Schrödinger equation. We will examine such equations in the Chapter 14 when dealing with scattering theory.

This result for $|\psi_n\rangle$ can be iterated to give a power series in H_1 that is of the form

$$\begin{aligned}|\psi_n\rangle &= |\phi_n\rangle + Q_n \frac{1}{E_n - \Delta - H_0}(H_1 - \Delta)|\phi_n\rangle \\ &+ Q_n \frac{1}{E_n - \Delta - H_0}(H_1 - \Delta)\, Q_n \frac{1}{E_n - \Delta - H_0}(H_1 - \Delta)|\phi_n\rangle \\ &+ \cdots \, . \end{aligned} \tag{13.13}$$

This series can be formally written as

$$|\psi_n\rangle = \sum_{l=0}^{\infty} \left\{ Q_n \frac{1}{E_n - \Delta - H_0}(H_1 - \Delta) \right\}^l |\phi_n\rangle \, . \tag{13.14}$$

At this stage we still have to decide on the value of the constant Δ. Here we observe that the eigenvalues E_n, which are yet to be determined, appear on the right hand side of Eq. (13.14). To overcome this Gordian knot, we take Δ to be

$$\Delta = E_n - \varepsilon_n \, . \tag{13.15}$$

In this case the power series in H_1 for the wave function becomes

$$\begin{aligned}|\psi_n\rangle &= \sum_{l=0}^{\infty} \left\{ Q_n \frac{1}{\varepsilon_n - H_0}(H_1 + \varepsilon_n - E_n) \right\}^l |\phi_n\rangle \\ &= |\phi_n\rangle + \sum_{m \neq n} |\phi_m\rangle \frac{1}{\varepsilon_n - \varepsilon_m} \langle \phi_m | H_1 | \phi_n \rangle + \cdots \, . \end{aligned} \tag{13.16}$$

Although E_n still appears on the right hand side of the above equation, we will see that in fact its first appearance is in the term $O((\langle H_1 \rangle)^3)$ and higher. With this result at hand we can use Eq. (13.6) to write a power series in H_1 for the energy E_n as

$$\begin{aligned}E_n &= \varepsilon_n + \sum_{l=0}^{\infty} \langle \phi_n | H_1 \left\{ Q_n \frac{1}{\varepsilon_n - H_0}(H_1 + \varepsilon_n - E_n) \right\}^l |\phi_n\rangle \\ &= \varepsilon_n + \langle \phi_n | H_1 | \phi_n \rangle + \sum_{m \neq n} \langle \phi_n | H_1 \frac{|\phi_m\rangle\langle\phi_m|}{\varepsilon_n - \varepsilon_m} H_1 |\phi_n\rangle + \cdots \\ &= \varepsilon_n + \langle \phi_n | H_1 | \phi_n \rangle + \sum_{m \neq n} \frac{|\langle \phi_n | H_1 | \phi_m \rangle|^2}{\varepsilon_n - \varepsilon_m} + \cdots \, . \end{aligned} \tag{13.17}$$

As we expected, our choice of Δ gives us a power series in which the first three terms depend only on the eigenvalues and eigenstates of the unperturbed Hamiltonian H_0. The term $O((\langle H_1 \rangle)^3)$ will require a knowledge of E_n, and for that we use the value determined by the first three terms in the series. This procedure can also be followed for the higher order terms in the series. For the ground state, i.e., E_0, the sign of the first order correction is determined by the perturbing potential H_1, while the sign of the second order correction is always negative irrespective of whether H_1 is attractive or repulsive. In most practical problems, only the first few terms in the series are calculated unless there is a compelling reason for including the higher order terms due to the presence of special effects such as collective behavior as might be the case in a many-body system. In these cases we usually sum a certain sub-set of the infinite series to get the effects we are interested in.

Finally, we need to prove that the normalization we chose in Eq. (13.7) is not violated. To prove this we multiply Eq. (13.16) by $\langle\phi_n|$ from the left to get,

$$\begin{aligned}\langle\phi_n|\psi_n\rangle &= \langle\phi_n|\phi_n\rangle + \sum_{l=1}^{\infty}\langle\phi_n|\left\{Q_n\,\frac{1}{\varepsilon_n - H_0}\,(H_1 + \varepsilon_n - E_n)\right\}^l|\phi_n\rangle \\ &= \langle\phi_n|\phi_n\rangle = 1\ ,\end{aligned}\tag{13.18}$$

where to write the second line we made use of the fact that $\langle\phi_n|Q_n = 0$ because of the orthonormality of the eigenstate of H_0.

13.2 The Helium Atom in Perturbation Theory

As a first example of perturbation theory we will consider the ground state energy of the Helium (He) atom. In this case we will consider the He atom to consist of two spinless electrons in motion about an infinitely heavy nucleus.[1] The Hamiltonian for the He atom, in units [2] in which $\hbar = m_e = e = 1$ (see Appendix A), can be written as

$$H = H_0 + H_1\ .\tag{13.19}$$

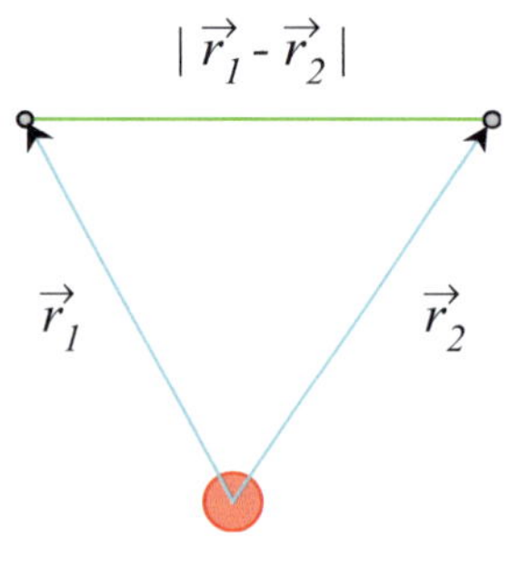

Figure 13.1: Definition of the coordinates for the helium atom.

[1]The ratio of the mass of the electron to the mass of the He nucleus is $\approx 8\times10^4$.

[2]These units in which the Bohr radius $a = 1$ are commonly used by Atomic Physicists (see Appendix A).

The unperturbed Hamiltonian includes the kinetic energy of the two electrons and the attraction due to the nucleus, i.e., H_0 is the sum of two hydrogen-like Hamiltonians, one for each of the electrons, i.e.,

$$\begin{aligned} H_0 &= -\frac{1}{2}\left(\nabla_1^2 + \nabla_2^2\right) - Z\left(\frac{1}{r_1} + \frac{1}{r_2}\right) \\ &= \left(-\frac{1}{2}\nabla_1^2 - \frac{Z}{r_1}\right) + \left(-\frac{1}{2}\nabla_2^2 - \frac{Z}{r_2}\right) . \end{aligned} \tag{13.20}$$

The perturbation in this case is the Coulomb repulsion between the two electrons, i.e.,

$$H_1 = \frac{1}{|\vec{r}_1 - \vec{r}_2|} . \tag{13.21}$$

Since the unperturbed Hamiltonian H_0 is written as the sum of the Hamiltonian for the two electrons, then the eigenstates of H_0 are the product of two eigenstates, one for each of the electrons, i.e.,

$$\begin{aligned} \langle \vec{r}_1; \vec{r}_2 | n_1 \ell_1 m_1; n_2, \ell_2 m_2 \rangle &= \langle \vec{r}_1 | n_1 \ell_1 m_1 \rangle \langle \vec{r}_2 | n_2 \ell_2 m_2 \rangle \\ &= \phi_{n_1 \ell_1 m_1}(\vec{r}_1)\, \phi_{n_2 \ell_2 m_2}(\vec{r}_2) , \end{aligned} \tag{13.22}$$

where $\phi_{n\ell m}(\vec{r})$ is the wave function of hydrogen atom with charge $Z = 2$ on the nucleus. For the ground state we have the two electrons in the lowest energy state of the hydrogenic atom. In this case the unperturbed wave function is given by (see Table 9.2 with $a = 1$)

$$\langle \vec{r}_1; \vec{r}_2 | 100; 100 \rangle = \frac{Z^3}{\pi} e^{-Z(r_1 + r_2)} . \tag{13.23}$$

With this wave function we can calculate the first order correction to the energy, $\Delta E^{(1)}$, which is given by

$$\begin{aligned} \Delta E^{(1)} &= \langle 100; 100 | H_1 | 100; 100 \rangle \\ &= \int d^3 r_1\, d^3 r_2\, \frac{|\phi_{100}(\vec{r}_1)|^2\, |\phi_{100}(\vec{r}_2)|^2}{|\vec{r}_1 - \vec{r}_2|} . \end{aligned} \tag{13.24}$$

To evaluate this integral, we need to expand $|\vec{r}_1 - \vec{r}_2|^{-1}$ in terms of spherical harmonics as [3]

$$\begin{aligned} |\vec{r}_1 - \vec{r}_2|^{-1} &= \sum_\ell \left(\frac{r_<^\ell}{r_>^{\ell+1}}\right) P_\ell(\cos\theta) \\ &= \sum_{\ell m} \frac{4\pi}{2\ell + 1} \left(\frac{r_<^\ell}{r_>^{\ell+1}}\right) Y_{\ell m}(\hat{r}_1)\, Y_{\ell m}^*(\hat{r}_2) , \end{aligned} \tag{13.25}$$

[3]For a proof of this result see Jackson's *Classical Electrodynamics* Eq. (3.70) [43].

where θ is the angle between the vectors $\vec{r}_1$ and $\vec{r}_2$, and

$$\left(\frac{r_<^\ell}{r_>^{\ell+1}}\right) = \begin{cases} \dfrac{r_1^\ell}{r_2^{\ell+1}} & \text{for} \quad r_1 < r_2 \\ \\ \dfrac{r_2^\ell}{r_1^{\ell+1}} & \text{for} \quad r_2 < r_1 \end{cases} . \tag{13.26}$$

With this expansion, and using the orthogonality of the spherical harmonics, we can write this first order correction to the energy as [4]

$$\begin{aligned} \Delta E^{(1)} &= 16Z^6 \int_0^\infty dr_1\, r_1^2\, e^{-2Zr_1} \left[\frac{1}{r_1}\int_0^{r_1} dr_2\, r_2^2\, e^{-2Zr_2} \quad + \int_{r_1}^\infty dr_2\, r_2\, e^{-2Zr_2}\right] \\ &= \frac{5}{8} Z . \end{aligned} \tag{13.27}$$

Therefore, the ground state energy of the helium atom to first order in perturbation theory is given by

$$E_0 = -\frac{Z^2}{2} - \frac{Z^2}{2} + \frac{5}{8}Z = -Z\left(Z - \frac{5}{8}\right) . \tag{13.28}$$

As expected, the repulsion between the electrons reduces the binding energy of the system. The ionization energy, i.e., the energy required to remove one electron from the atom, is now given by

$$\begin{aligned} I &= -E_0(He) + E_0(He^+) = -\left\{-Z\left(Z - \frac{5}{8}\right) + \frac{Z^2}{2}\right\} \\ &= \frac{Z}{2}\left(Z - \frac{5}{4}\right) = \frac{3}{4} \quad \text{for} \quad Z = 2 . \end{aligned} \tag{13.29}$$

To get this ionization energy in eV, we have to multiply the above result by $2R_\infty = 27.2116\,\text{eV}$. This gives us the ionization of He to be $I = 20.4087\,\text{eV}$, which compares well with the experimental value of $24.6401\,\text{eV}$.

13.3 Anharmonic Linear Oscillator

In many problems in physics we are interested in the development of a system around an equilibrium state. In all such systems the lowest order approximation to the behavior of

[4] To evaluate the integrals in Eq. (13.27), we have made use of the fact that

$$\int dx\, x^m\, e^{ax} = e^{ax} \sum_{k=0}^{m} (-1)^k \frac{m!\, x^{m-k}}{(m-k)!a^{k+1}} .$$

the system can be described in terms of a simple harmonic oscillator. This, as we have seen, describes the vibrational states of a diatomic molecule. The advantage of using a harmonic oscillator is the fact that we can write an exact solution to the harmonic oscillator as we demonstrated in Chapter 7. In the present section, we would like to consider problems that require an extension of the analysis beyond the simple harmonic oscillator and include the first anharmonic term. In the process we will re-derive the solution of the harmonic oscillator in occupation number representation.

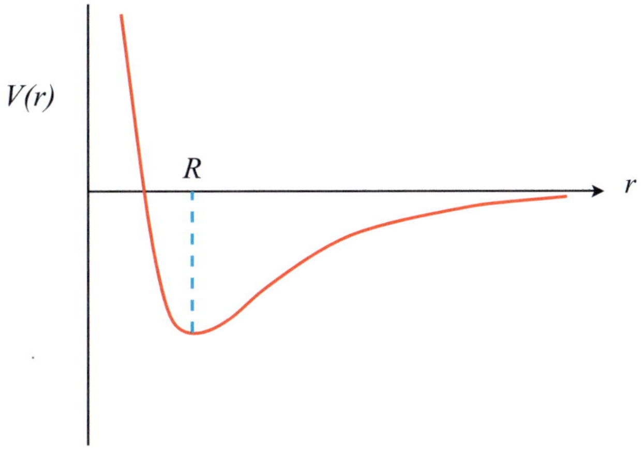

Figure 13.2: The potential between two atoms. Here R is the equilibrium distance.

Let us consider a molecule consisting of two atoms. The potential between the two atoms is presented graphically in Figure 13.2. In the ground state, the two atoms in the molecule have an equilibrium separation $r \approx R$. If we are interested in the low lying states of the molecule, then these correspond to the small oscillation about the equilibrium separation. In this case we can write the potential between the two atoms in terms of a Taylor expansion about the equilibrium separation R, i.e.,

$$\begin{aligned} V(r) &= V(R) + (r-R)\left(\frac{dV}{dr}\right)_{r=R} + \frac{1}{2!}(r-R)^2\left(\frac{d^2V}{dr^2}\right)_{r=R} \\ &+\frac{1}{3!}(r-R)^3\left(\frac{d^3V}{dr^3}\right)_{r=R} + \cdots . \end{aligned} \tag{13.30}$$

Because the potential has a minimum at $r = R$, the second term on the right hand side is zero since $\left(\frac{dV}{dr}\right)_{r=R} = 0$. Furthermore, we can take the zero of our energy to be $V(R)$. In this way, the lowest order approximation to the potential between two atoms is represented by a harmonic oscillator provided the excitation above the ground state is not large. Thus the lowest order Hamiltonian can be written as

$$H_0 = \frac{p^2}{2m} + \frac{1}{2}m\omega^2 x^2 , \tag{13.31}$$

where we have taken $x = (r - R)$, and considered the radial variable as a one-dimensional problem. The frequency of the oscillator can be traced back to the second derivative of the potential, i.e.,

$$\omega^2 = \frac{1}{m}\left(\frac{d^2V}{dr^2}\right)_{r=R}$$

This Hamiltonian gives a spectrum which is of the form (see Eq. (7.19))

$$E_\nu = \hbar\omega\left(\nu + \frac{1}{2}\right) . \tag{13.32}$$

As a first correction to the above Hamiltonian, we have a term that is cubic in the displacement from equilibrium, i.e.,

$$H_1 = \frac{m\omega^2}{2b}\, x^3 , \tag{13.33}$$

where the strength of this perturbation is proportional to the third derivative, i.e.,

$$\frac{m\omega^2}{2b} = \frac{1}{3!}\left(\frac{d^3V}{dr^3}\right)_{r=R} .$$

To determine the effect of this anharmonic term on the vibrational spectrum, we need to determine the eigenstates of H_0. Then for H_1 small, we can use perturbation theory to calculate the shift in energy due to the anharmonicity.

13.3.1 Occupation Number Representation

Although we have solved the Schrödinger equation for the one-dimensional harmonic oscillator in Chapter 7, we propose here to determine the eigenvalues and eigenstates of the harmonic oscillator using the same algebraic method used to find the representation of the angular momenta in Chapter 11 Sec. 11.3.3. Let us define the operator a and $a^\dagger$ as

$$a = \frac{i}{\sqrt{2m\hbar\omega}}\,(p - im\omega x) \qquad a^\dagger = -\frac{i}{\sqrt{2m\hbar\omega}}\,(p + im\omega x) , \tag{13.34}$$

where x and p are the operators for the position and momentum of the particle. We now can write the Hamiltonian for the one-dimensional harmonic oscillator in terms of a and $a^\dagger$ as

$$H_0 = \hbar\omega\left(a^\dagger a + \frac{1}{2}\right) . \tag{13.35}$$

Making use of the commutation relation of the position and momentum operators, i.e., $[x,p] = i\hbar$, we can write the commutation relation between the new operators a and $a^\dagger$ as

$$\left[a, a^\dagger\right] = 1 \qquad [a, a] = 0 = \left[a^\dagger, a^\dagger\right] . \tag{13.36}$$

With these commutation relations we can determine the commutation relations of a and $a^\dagger$ with the Hamiltonian H_0 to be

$$\left[H_0, a^\dagger\right] = \hbar\omega\, a^\dagger \qquad [H_0, a] = -\hbar\omega\, a \ . \tag{13.37}$$

Since the Hamiltonian is a linear Hermitian operator, we can assume that it has eigenstates $|\nu\rangle$ such that

$$H_0 \,|\nu\rangle = \varepsilon_\nu \,|\nu\rangle \ , \tag{13.38}$$

and then using the commutation relation of the Hamiltonian with the operator $a^\dagger$, we can write

$$\left[H_0, a^\dagger\right] \,|\nu\rangle = H_0\, a^\dagger \,|\nu\rangle - a^\dagger\, H_0 \,|\nu\rangle = \hbar\omega\, a^\dagger \,|\nu\rangle \ .$$

This equation can be written as an eigenvalue problem of the form

$$H_0 \left(a^\dagger \,|\nu\rangle\right) = (\varepsilon_\nu + \hbar\omega) \left(a^\dagger \,|\nu\rangle\right) \ . \tag{13.39}$$

In other words, if the state $|\nu\rangle$ has energy ε_ν, then the state $a^\dagger \,|\nu\rangle$ has energy $(\varepsilon_\nu + \hbar\omega)$. This suggests that the operator $a^\dagger$, acting on a state with a given energy, gives us a state of higher energy, and the energy of the new state is always $\hbar\omega$ greater than the original state. Thus the operator $a^\dagger$ seems to have a similar property to the operator J_+ encountered in our discussion of the representation of the angular momentum in Chapter 11 Sec. 11.3.3. In a similar manner we can show, using the commutation of H_0 with a, that

$$H_0 \left(a \,|\nu\rangle\right) = (\varepsilon_\nu - \hbar\omega) \left(a \,|\nu\rangle\right) \ . \tag{13.40}$$

The operators a and $a^\dagger$ lower or raise the energy of a state by an amount $\hbar\omega$. Alternatively, we can refer to the operators $a^\dagger$ (a) as the creation (annihilation) operator since they increase (decrease) the energy of the harmonic oscillator by $\hbar\omega$ or create (annihilate) a quanta of energy equal to $\hbar\omega$. In this way, given one state of the system and its corresponding energy, the creation and annihilation operators allow us to generate all the other states and their corresponding energies. This situation is identical to that encountered in Sec. 11.3.3, where the operators J_+ and J_- raised and lowered the projection of the angular momentum along the z-axis by $\hbar$.

Let us now assume that there is a lowest energy state denoted by $|0\rangle$.[5] Since by definition it is the lowest energy state, we have that

$$a \,|0\rangle = 0 \ . \tag{13.41}$$

This lowest energy state is an eigenstate of the Hamiltonian with eigenvalue $\hbar\omega/2$ since

$$H_0 \,|0\rangle = \frac{1}{2}\hbar\omega \,|0\rangle \ . \tag{13.42}$$

[5] All operators in quantum mechanics should have at least one bound on their spectrum, i.e., either an upper or lower bound. For the Hamiltonian operator we make sure it has a lower bound on its spectrum.

We now can generate all the other states by the repeated application of the operator $a^\dagger$. Thus we have

$$\begin{aligned}
|1\rangle &\equiv N_1 a^\dagger|0\rangle \qquad \text{and} \qquad H_0|1\rangle = \hbar\omega\left(1+\frac{1}{2}\right)|1\rangle \\
|2\rangle &\equiv N_2 \left(a^\dagger\right)^2|0\rangle \qquad \text{and} \qquad H_0|2\rangle = \hbar\omega\left(2+\frac{1}{2}\right)|2\rangle \\
&\vdots \qquad\qquad \vdots \\
|\nu\rangle &\equiv N_\nu \left(a^\dagger\right)^\nu|0\rangle \qquad \text{and} \qquad H_0|\nu\rangle = \hbar\omega\left(\nu+\frac{1}{2}\right)|\nu\rangle \ ,
\end{aligned}$$

where N_ν is the normalization which is to be determined. We now have that the state $|\nu\rangle$ has energy ε_ν with

$$\varepsilon_\nu = \hbar\omega\left(\nu+\frac{1}{2}\right) \ . \tag{13.43}$$

This result is identical to that obtained by solving the Schrödinger equation in coordinate representation. To determine the normalization of our states, let us write the state $|\nu\rangle$ as

$$\begin{aligned}
|\nu\rangle &= N_\nu\left(a^\dagger\right)^\nu|0\rangle \\
&= N_\nu\, a^\dagger\left(a^\dagger\right)^{\nu-1}|0\rangle \\
&= \frac{N_\nu}{N_{\nu-1}}\,a^\dagger|\nu-1\rangle \ .
\end{aligned}$$

Then since the state $|\nu\rangle$ is normalized, we have

$$\begin{aligned}
1 = \langle\nu|\nu\rangle &= \left|\frac{N_\nu}{N_{\nu-1}}\right|^2\langle\nu-1|aa^\dagger|\nu-1\rangle \\
&= \left|\frac{N_\nu}{N_{\nu-1}}\right|^2\langle\nu-1|\left\{1+a^\dagger a\right\}|\nu-1\rangle \ .
\end{aligned}$$

We now can write the operator $a^\dagger a$ in terms of the Hamiltonian H_0, and then make use of the fact that the state $|\nu-1\rangle$ is an eigenstate of the Hamiltonian with an eigenvalue $\hbar\omega(\nu-1/2)$, to write the above expression as

$$1 = \left|\frac{N_\nu}{N_{\nu-1}}\right|^2\nu \ .$$

We thus have that

$$\frac{N_\nu}{N_{\nu-1}} = \frac{1}{\sqrt{\nu}}$$

which allows us to write the state $|\nu\rangle$ as

$$|\nu\rangle = \frac{1}{\sqrt{\nu}}\,a^\dagger|\nu-1\rangle \ . \tag{13.44}$$

In a similar manner we can show that

$$|\nu - 1\rangle = \frac{1}{\sqrt{\nu}}\, a|\nu\rangle \ . \tag{13.45}$$

Finally we can write the state $|\nu\rangle$ in terms of the lowest energy state as

$$|\nu\rangle = \frac{1}{\sqrt{\nu!}} \left(a^\dagger\right)^\nu |0\rangle \ . \tag{13.46}$$

In this way we have generated all the eigenstates and eigenvalues of the Hamiltonian by writing H_0 in terms of the operators a and $a^\dagger$ and making use of the algebra of the commutation relation of these operators. Here we note that since the position and momentum operators can be written in terms of a and $a^\dagger$, then any operator that is a function of x and p can also be written in terms of a and $a^\dagger$. This in turn implies that the space we have defined is closed under the action of any operator that is a function of x and p, and to that extent the eigenstates $|\nu\rangle$ form a complete set. One final observation is the fact that the Hamiltonian in this case plays the same role as J_z played in the angular momentum case. This algebraic method can be used to solve a number of other problems, the most famous being the solution of the Schrödinger equation for the Coulomb potential. In fact it was first solved using this algebraic method by Pauli. [33]

13.3.2 Lowest Order Contribution

Having determined the eigenenergies and corresponding eigenstates of H_0, we are in a position to examine the contribution of the anharmonic term to the energy and wave function of the vibrational motion of a molecule. To calculate the matrix elements of the perturbing Hamiltonian H_1, we need to write this perturbation in terms of the creation and annihilation operators $a^\dagger$ and a. From the definition of $a^\dagger$ and a we have that

$$a + a^\dagger = \frac{2m\omega}{\sqrt{2m\hbar\omega}}\, x \ ,$$

and therefore

$$x = \sqrt{\frac{\hbar}{2m\omega}} \left(a + a^\dagger\right) \ . \tag{13.47}$$

Using this expression for x, we can calculate the matrix element of x between the eigenstates of H_0, i.e.,

$$\begin{aligned}
\langle\nu'|x|\nu\rangle &= \sqrt{\frac{\hbar}{2m\omega}} \left\{ \langle\nu'|a|\nu\rangle + \langle\nu'|a^\dagger|\nu\rangle \right\} \\
&= \sqrt{\frac{\hbar}{2m\omega}} \left\{ \sqrt{\nu}\langle\nu'|\nu - 1\rangle + \sqrt{\nu+1}\langle\nu'|\nu+1\rangle \right\} \\
&= \sqrt{\frac{\hbar}{2m\omega}} \left\{ \sqrt{\nu}\,\delta_{\nu',\nu-1} + \sqrt{\nu+1}\,\delta_{\nu',\nu+1} \right\} \ .
\end{aligned} \tag{13.48}$$

We observe at this stage that if the perturbing Hamiltonian was proportional to x, there would be no first order contribution to the energy from this perturbation, and we would need to go to the next order in perturbation theory to observe a shift in the energy. Furthermore, this shift would lower the energy of the ground state as pointed out previously. With this result as an illustration of how we may determine the contribution of the perturbation to the energy and wave function of the system, we can proceed to write H_1 in terms of our creation and annihilation operators as

$$\begin{aligned} H_1 = \frac{m\omega^2}{2b}\,x^3 &= \hbar\omega\gamma(a+a^\dagger)\,(a+a^\dagger)\,(a+a^\dagger) \\ &= \hbar\omega\gamma\left\{3aa^\dagger a + 3a^\dagger a a^\dagger + a^3 + (a^\dagger)^3\right\}\ , \end{aligned} \tag{13.49}$$

where we have used the commutation relations of the operators a and $a^\dagger$ to write the second line of Eq. (13.49). Here the constant γ is given by

$$\gamma = \frac{1}{4b}\sqrt{\frac{\hbar}{2m\omega}}\ , \tag{13.50}$$

From the structure of H_1, as given in Eq. (13.49) in term of the creation and annihilation operators $a^\dagger$ and a, it is clear that the contribution to the energy of the anharmonic oscillator, in first order perturbation theory, is zero, i.e., $\langle\nu|H_1|\nu\rangle = 0$. Thus the lowest order correction to the energy comes in second order perturbation theory, and is of the form

$$\sum_{\substack{\nu' \\ \nu'\neq\nu}} \frac{|\langle\nu'|H_1|\nu\rangle|^2}{\varepsilon_\nu - \varepsilon_{\nu'}}\ .$$

In this sum the only non-zero matrix elements are;

$$\begin{aligned} \langle\nu+1|H_1|\nu\rangle &= 3\hbar\omega\gamma\langle\nu+1|a^\dagger a a^\dagger|\nu\rangle = 3\hbar\omega\gamma(\nu+1)^{3/2} && (\text{13.51a}) \\ \langle\nu-1|H_1|\nu\rangle &= 3\hbar\omega\gamma\langle\nu-1|aa^\dagger a|\nu\rangle = 3\hbar\omega\gamma\nu^{3/2} && (\text{13.51b}) \\ \langle\nu+3|H_1|\nu\rangle &= \hbar\omega\gamma\langle\nu+3|(a^\dagger)^3|\nu\rangle = \hbar\omega\gamma\left[(\nu+1)(\nu+2)(\nu+3)\right]^{1/2} && (\text{13.51c}) \\ \langle\nu-3|H_1|\nu\rangle &= \hbar\omega\gamma\langle\nu-3|a^3|\nu\rangle = \hbar\omega\gamma\left[\nu(\nu-1)(\nu-2)\right]^{1/2}\ . && (\text{13.51d}) \end{aligned}$$

This means that our energy is given by

$$E_\nu = \hbar\omega\left(\nu+\frac{1}{2}\right) + \sum_{\substack{\nu' \\ \nu'\neq\nu}} \frac{|\langle\nu|H_1|\nu'\rangle|^2}{\hbar\omega(\nu-\nu')}\ . \tag{13.52}$$

Using Eq. (13.51), we can write Eq. (13.52) as

$$\begin{aligned} E_\nu &= \hbar\omega\left(\nu+\frac{1}{2}\right)+\hbar\omega\gamma^2\Big\{\frac{1}{3}|\langle\nu-3|a^3|\nu\rangle|^2+9|\langle\nu-1|aa^\dagger a|\nu\rangle|^2 \\ &\quad -9|\langle\nu+1|a^\dagger aa^\dagger|\nu\rangle|^2-\frac{1}{3}|\langle\nu+3|(a^\dagger)^3|\nu\rangle|^2\Big\} \\ &= \hbar\omega\left(\nu+\frac{1}{2}\right)+\hbar\omega\gamma^2\Big\{\frac{1}{3}\nu(\nu-1)(\nu-2)+9\nu^3-9(\nu+1)^3 \\ &\quad -\frac{1}{3}(\nu+1)(\nu+2)(\nu+3)\Big\} \\ &= \hbar\omega\left(\nu+\frac{1}{2}\right)-\hbar\omega\gamma^2\left(30\nu^2+30\nu+11\right)\,. \end{aligned} \tag{13.53}$$

This is the energy of the anharmonic oscillator to second order in perturbation theory. We note here that this result is valid for $\nu \geq 3$. For $\nu < 3$, we have to take into consideration the fact that the state $|\nu\rangle$ is zero for $\nu < 0$ in our calculation of the matrix elements of H_1, e.g., to calculate the shift to the ground state energy ($\nu = 0$) in second order perturbation theory, we only need to consider the terms with $a^\dagger aa^\dagger$ and $(a^\dagger)^3$ in Eq. (13.53). In this case the ground state energy to second order in perturbation theory is given by

$$E_0 = \hbar\omega\left[\frac{1}{2}-\gamma^2\left(9+\sqrt{\frac{2}{3}}\right)\right],$$

which is less than the ground state energy of the harmonic oscillator $\frac{\hbar\omega}{2}$.

The wave function to the same order in perturbation theory can be written as

$$\begin{aligned} |\psi_\nu\rangle &= |\phi_\nu\rangle+\sum_{\substack{\nu' \\ \nu'\neq\nu}}|\phi_{\nu'}\rangle\frac{\langle\nu'|H_1|\nu\rangle}{\hbar\omega(\nu-\nu')} \\ &= |\phi_\nu\rangle+\gamma\Big\{|\phi_{\nu-3}\rangle\frac{1}{3}\sqrt{\nu(\nu-1)(\nu-2)}+|\phi_{\nu-1}\rangle 3\nu^{3/2} \\ &\quad -|\phi_{\nu+1}\rangle 3(\nu+1)^{3/2}-|\phi_{\nu+3}\rangle\frac{1}{3}\sqrt{(\nu+1)(\nu+2)(\nu+3)}\Big\}\,. \end{aligned} \tag{13.54}$$

Let us see if this result agrees with what we intuitively expect. If we take the Hamiltonian for the two atoms to be H_0, then we have taken a harmonic oscillator about $r = R$, where R is the equilibrium distance between the two atoms. However the potential $V(r)$ is wider than the harmonic oscillator, and thus the addition of the contribution of H_1 to the energy should lower the eigenstates, which is exactly what we derived, and is illustrated in Figure 13.3.

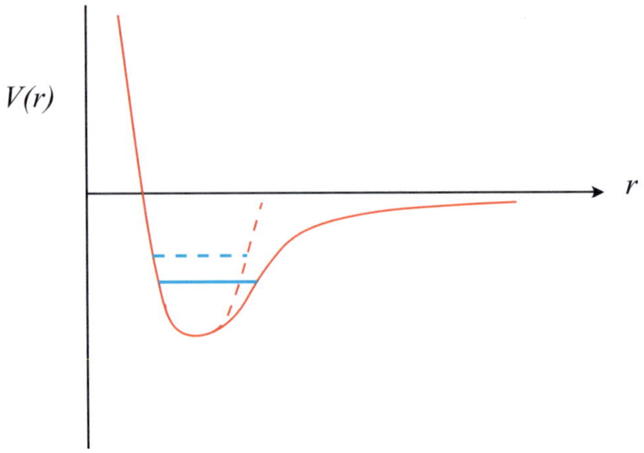

Figure 13.3: A comparison of the harmonic oscillator potential with the potential between two atoms.

13.4 Van der Waals' Potential

Let us consider the force between two hydrogen like atoms, i.e., two atoms each having one valence electron which plays an active role in giving rise to the potential between the two atoms. If the distance between the centers of the two atoms is much larger than the distance of each electron from its nucleus, i.e., $R \gg r_1$ and r_2, then the energy of the system as a function of R will be the interatomic potential. The Hamiltonian for the system can now be written as

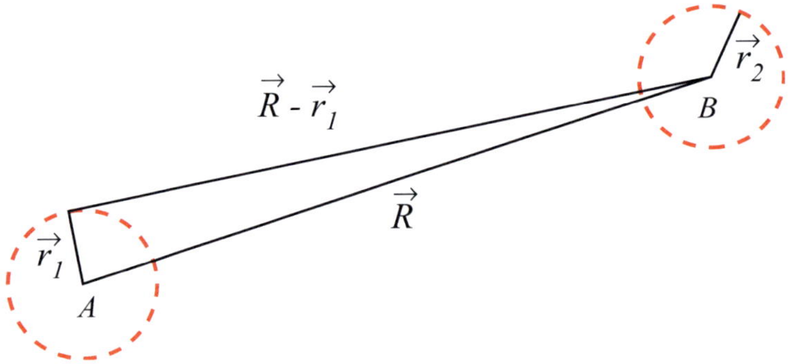

Figure 13.4: Definition of the coordinates for the hydrogen molecule.

$$H = H_0 + H_1 \ , \tag{13.55}$$

where the unperturbed Hamiltonian, in units of $\hbar = m = e = 1$, is given by

$$H_0 = -\frac{1}{2}\left(\nabla_1^2 + \nabla_2^2\right) - \left(\frac{1}{r_1} + \frac{1}{r_2}\right) \ , \tag{13.56}$$

i.e., it is the sum of the Hamiltonians for the two individual atoms, while the perturbing Hamiltonian is given by

$$H_1 = \frac{1}{R} + \frac{1}{r_{12}} - \frac{1}{r_{1B}} - \frac{1}{r_{2A}} . \tag{13.57}$$

This includes the interaction of the electron and nucleus of one atom with the electron and nucleus of the other atom (see Figure 13.4). Since we are interested in the long range behavior of the potential between the two atoms, i.e., when $R \gg r_1$ and r_2, we will write the perturbing Hamiltonian in terms of R, r_1, and r_2 as

$$\begin{aligned} H_1 &= \frac{1}{R} + \frac{1}{|\vec{R} - \vec{r}_1 + \vec{r}_2|} - \frac{1}{|\vec{R} - \vec{r}_1|} - \frac{1}{|\vec{R} + \vec{r}_2|} \\ &= \frac{1}{R}\left\{1 + \left[1 + \frac{2\hat{R}\cdot(\vec{r}_2 - \vec{r}_1)}{R} + \frac{(\vec{r}_2 - \vec{r}_1)^2}{R^2}\right]^{-1/2} - \left[1 - 2\frac{\hat{R}\cdot\vec{r}_1}{R} + \frac{r_1^2}{R^2}\right]^{-1/2}\right. \\ &\quad \left. - \left[1 + 2\frac{\hat{R}\cdot\vec{r}_2}{R} + \frac{r_2^2}{R^2}\right]^{-1/2}\right\} , \end{aligned} \tag{13.58}$$

and then take the limit as the distance between the two atoms becomes larger than the size of the individual atoms, i.e., $R \gg r_1$ and r_2. To get the first non-zero term in the series in $\frac{r_i}{R}$, $i = 1, 2$ we expand the terms in square bracket to get [6]

$$H_1 \approx -\frac{1}{R^3}\left[3\left(\hat{R}\cdot\vec{r}_1\right)\left(\hat{R}\cdot\vec{r}_2\right) - (\vec{r}_1\cdot\vec{r}_2)\right] . \tag{13.59}$$

The first term on the right hand side corresponds to the interaction of two dipoles. In fact what is happening is that the lowest order interaction corresponds to the situation when each atom gets polarized by the field of the other atom. This results in a configuration where we have two dipoles a distance R apart. We can now proceed to calculate the perturbation due to H_1 as given in Eq. (13.59), since we know the eigenstates of H_0, which are hydrogen-like wave functions. However to get the R dependence of the potential between the two atoms, let us replace the two atoms by two one-dimensional harmonic oscillators a distance R apart. This will give us the correct R dependence, and at the same time simplify the computation. Taking the vector $\vec{R}$ to be along the x-axis, our perturbing Hamiltonian becomes

$$H_1 = -\frac{2}{R^3}\, x_1\, x_2 , \tag{13.60}$$

---

[6] Here we have made use of the fact that

$$(1 + x)^{-1/2} = 1 - \frac{x}{2} + \frac{3x^2}{8} - \frac{5x^3}{16} + \cdots ,$$

to derive Eq. (13.59).

while the unperturbed Hamiltonian H_0 is modeled to be

$$H_0 = \left(\frac{p_1^2}{2m} + \frac{1}{2} m\omega_1^2 x_1^2\right) + \left(\frac{p_2^2}{2m} + \frac{1}{2} m\omega_2^2 x_2^2\right) . \tag{13.61}$$

The eigenstates of H_0 can be written as (see Sec. 13.3.2),

$$|\nu_1;\nu_2\rangle = |\nu_1\rangle\,|\nu_2\rangle . \tag{13.62}$$

This factorization is possible only because the Hamiltonian H_0 is the sum of two harmonic oscillator Hamiltonians with no interaction between them. If we take $\omega_1 = \omega_2 = \omega$ then the eigenvalues of H_0 are

$$H_0|\nu_1;\nu_2\rangle = \hbar\omega\,(\nu_1+\nu_2+1)\,|\nu_1;\nu_2\rangle \equiv \varepsilon_{\nu_1\nu_2}|\nu_1;\nu_2\rangle . \tag{13.63}$$

Taking the two atoms to be in their ground state, i.e., $\nu_1 = \nu_2 = 0$, the contribution to the energy from first order perturbation theory is given by

$$\langle 0;0|H_1|0;0\rangle = -\frac{2}{R^3}\,\langle 0|x_1|0\rangle\,\langle 0|x_2|0\rangle = 0 . \tag{13.64}$$

Thus the lowest order correction to the ground state energy comes from second order perturbation theory, and this is given by

$$\begin{aligned}
\Delta E^{(2)} &= \sum_{\nu_1\nu_2} \frac{\langle 0;0|H_1|\nu_1;\nu_2\rangle\,\langle \nu_1;\nu_2|H_1|0;0\rangle}{\varepsilon_{00}-\varepsilon_{\nu_1\nu_2}} \\
&= \left(\frac{2}{R^3}\right)^2 \sum_{\nu_1\nu_2} \frac{|\langle 0|x_1|\nu_1\rangle|^2\,|\langle 0|x_2|\nu_2\rangle|^2}{\varepsilon_{00}-\varepsilon_{\nu_1\nu_2}} \\
&= -\left(\frac{2}{R^3}\right)^2 \sum_{\nu_1\nu_2} \frac{|\langle 0|x_1|\nu_1\rangle|^2\,|\langle 0|x_2|\nu_2\rangle|^2}{\hbar\omega(\nu_1+\nu_2)} .
\end{aligned} \tag{13.65}$$

Making use of the matrix element of x as given in Eq. (13.48), we can reduce the sum to just one term corresponding to $\nu_1 = \nu_2 = 1$. This gives us

$$\Delta E^{(2)} = -\frac{1}{2}\hbar\omega\left(\frac{1}{m\omega^2}\right)^2 \frac{1}{R^6} = -\frac{A}{R^6} . \tag{13.66}$$

The above result gives the correct R dependence of the potential between two atoms. To derive a more accurate determination of A, we need to use an atomic wave function for the electrons rather than the harmonic oscillator wave function. However in the atomic case, the sum in the calculation of $\Delta E^{(2)}$ is infinite and needs to be truncated.

13.5 The Variational Method

An alternative approximation to perturbation theory is the variational method, and like perturbation theory it has its advantages and disadvantages. Its main advantages are:

1. The method does not require that we write the Hamiltonian as the sum of two parts, e.g., $H = H_0 + H_1$, with H_1 small and the eigenvalues and eigenstates of H_0 known.
2. In many problems the variational results are better than the result we get with perturbation theory with little additional effort.

On the other hand, the disadvantages of the variational method are:

1. The computation can be very difficult if one requires a good result.
2. The starting point and success of the calculation depends to a large extent on how clever we can be. This last point will become clearer when we discuss the method.

Theorem: The ground state energy E_0 of a system with a Hamiltonian H satisfies the relation

$$E_0 \leq \langle\psi_0|H|\psi_0\rangle \ , \tag{13.67}$$

where ψ_0 is any function that is normalized, i.e.,

$$\langle\psi_0|\psi_0\rangle = 1 \ . \tag{13.68}$$

Proof: Let us introduce the eigenstates of H as $|\phi_n\rangle$, i.e.,

$$H|\phi_n\rangle = E_n|\phi_n\rangle \ . \tag{13.69}$$

Since the states $|\phi_n\rangle$ form a complete set of states, we can expand $|\psi_0\rangle$ in terms of the eigenstates of H, i.e.,

$$|\psi_0\rangle = \sum_n a_n|\phi_n\rangle \ . \tag{13.70}$$

Then the requirement that $|\psi_0\rangle$ be normalized means

$$\begin{aligned}\langle\psi_0|\psi_0\rangle \; = \; 1 &= \sum_{nm} a_n^*\, a_m\langle\phi_n|\phi_m\rangle \\ &= \sum_n |a_n|^2 \ .\end{aligned} \tag{13.71}$$

We now can write the matrix element of H as

$$\begin{aligned}\langle\psi_0|H|\psi_0\rangle &= \sum_{nm} a_n^*\,\langle\phi_n|H|\phi_m\rangle\, a_m \\ &= \sum_n |a_n|^2\, E_n \ .\end{aligned} \tag{13.72}$$

If we take $E_0 < E_1 < E_2 \cdots$, then we have that

$$\begin{aligned}\langle\psi_0|H|\psi_0\rangle &= \sum_n E_n\,|a_n|^2 \\ &\geq \sum_n E_0\,|a_n|^2 = E_0\ . \end{aligned} \tag{13.73}$$

Therefore we have that

$$E_0 \leq \langle\psi_0|H|\psi_0\rangle\ . \tag{13.74}$$

The fact that $E_0 \leq \langle\psi_0|H|\psi_0\rangle$ means that if we choose a $|\psi_0\rangle$ such that $\langle\psi_0|H|\psi_0\rangle$ is a minimum, then we have an upper bound on E_0, and as we improve our wave function $|\psi_0\rangle$, we get better bounds on E_0.

To achieve this upper bound, we take ψ_0 to be a function of a set of parameters, e.g.,

$$|\psi_0\rangle = |\psi_0(\alpha,\beta,\gamma,\cdots)\rangle\ . \tag{13.75}$$

We then define the function $J(\alpha,\beta,\gamma,\cdots)$ as

$$J(\alpha,\beta,\gamma,\cdots) \equiv \langle\psi_0|H|\psi_0\rangle\ . \tag{13.76}$$

We now can find the minimum of the function $J(\alpha,\beta,\gamma,\cdots)$ with respect to the parameters $\alpha,\beta,\gamma,\cdots$ by taking

$$\frac{\partial J}{\partial\alpha} = 0\ , \qquad \frac{\partial J}{\partial\beta} = 0\ , \qquad \cdots\ . \tag{13.77}$$

This method of getting the upper bound on the ground state energy E_0 is known as *Ritz variational method.*

So far we have considered the ground state energy of the system described by the Hamiltonian H. We can in principle extend the method to determine bounds on the excited states of the system. Thus to calculate the first excited state, we construct a function ψ_1 that is orthogonal to the ground state ψ_0, i.e.,

$$\langle\psi_1|\psi_0\rangle = 0\ , \tag{13.78}$$

and this wave function is normalized so that

$$\langle\psi_1|\psi_1\rangle = 1\ . \tag{13.79}$$

Then we can calculate an upper bound on the first excited state as

$$E_1 \leq \langle\psi_1|H|\psi_1\rangle\ . \tag{13.80}$$

Here we note that in calculating an upper bound on a given state, we have to make sure that the symmetry of our trial function ψ_n is the same as the eigenstate corresponding to the eigenvalue we are calculating the upper bound on. For example, if the ground state is a state of zero angular momentum and positive parity, then our trial function should correspond to a positive parity state with zero angular momentum. If we ignore this and take a wave function with negative parity, we would be calculating an upper bound on the lowest negative parity state of the system which could be an excited state and not a ground state.

13.6 The Helium Atom - Variational Method

To illustrate the use of the variational method, and at the same time compare the method of perturbation theory with the variational approach, we consider the helium atom for a second time. Here again, we take the helium atom as a system of two electrons with no spin in order to avoid the problem of angular momentum addition. In this case we expect the wave function for the two electrons to be the product of two hydrogen-like wave functions with the added feature that the electron does not see the full charge of the nucleus because of the screening by the other electron. This means we will take the charge in the trial wave function to be a variational parameter. In other words, the trial function for the ground state is taken as

$$\psi_0(\vec{r}_1, \vec{r}_2) = \frac{\beta^3}{\pi} e^{-\beta(r_1+r_2)} . \tag{13.81}$$

This wave function, which is properly normalized, assumes that the two electrons are in a state with angular momentum zero, i.e., s-state. Furthermore, each electron sees a charge β on the nucleus. To optimize our wave function we have to minimize

$$J(\beta) = \langle\psi_0|H|\psi_0\rangle , \tag{13.82}$$

with respect to β. Here, H is the full Hamiltonian for the helium atom as given by Eqs. (13.19) - (13.21). To minimize $J(\beta)$ we write it as the sum of three terms, i.e.,

$$J(\beta) = J_1(\beta) + J_2(\beta) + J_3(\beta) , \tag{13.83}$$

where $J_1(\beta)$ is the kinetic energy of the system and is given by

$$\begin{aligned} J_1(\beta) &= -\frac{1}{2}\int d^3r_1\, d^3r_2\, \psi_0^*(\vec{r}_1, \vec{r}_2) \left(\nabla_1^2 + \nabla_2^2\right) \psi_0(\vec{r}_1, \vec{r}_2) \\ &= \beta^2 . \end{aligned} \tag{13.84}$$

To perform this integral we have made use of the fact that the trial wave function has no angle dependence, and therefore if we take ∇_i^2 in spherical coordinates, the only contribution comes from the radial part of ∇_i^2 which is given by

$$\frac{1}{r_i^2}\frac{\partial}{\partial r_i}\left(r_i^2 \frac{\partial}{\partial r_i}\right) .$$

This result allows us to do the angle integration in a trivial manner since the integrand is angle independent. The radial integral then is a standard integral to be found in any table of integrals.[7] Here, we observe that the kinetic energy is as usual a positive quantity.

[7] We have that

$$\int_0^\infty x^n\, e^{-ax}\, dx = \frac{\Gamma(n+1)}{a^{n+1}} \quad (n > -1,\ a > 0) .$$

The second term on the right hand side of Eq. (13.83) corresponds to the attraction of the electrons by the nucleus. This integral is given by

$$\begin{aligned} J_2(\beta) &= -Z \int d^3r_1\, d^3r_2\, \psi_0^*(\vec{r}_1, \vec{r}_2) \left(\frac{1}{r_1} + \frac{1}{r_2}\right) \psi_0(\vec{r}_1, \vec{r}_2) \\ &= -2Z\beta\ . \end{aligned} \tag{13.85}$$

Here again the integrand has no angle dependence allowing us to do the angle integration, while the radial integrals are standard integrals. This term, as we would expect, is negative because of the attraction between the electrons and the nucleus.

Finally, the third term on the right hand side of Eq. (13.83) gives the contribution due to the repulsion between the two electrons, and therefore should be positive. It is given by

$$\begin{aligned} J_3(\beta) &= \int d^3r_1\, d^3r_2\, \psi_0^*(\vec{r}_1, \vec{r}_2) \frac{1}{|\vec{r}_1 - \vec{r}_2|} \psi_0(\vec{r}_1, \vec{r}_2) \\ &= \frac{5}{8}\beta\ . \end{aligned} \tag{13.86}$$

This integral was evaluated previously and given in Eq. (13.27). We now combine the results of Eqs. (13.84) to (13.86) to write

$$J(\beta) = \beta^2 - 2Z\beta + \frac{5}{8}\beta\ . \tag{13.87}$$

To determine the minimum of this function with respect to β, we differentiate it and set the derivative to zero, i.e.,

$$\frac{\partial J}{\partial \beta} = 2(\beta - Z) + \frac{5}{8} = 0$$

or

$$\beta = Z - \frac{5}{16}\ . \tag{13.88}$$

This is the effective charge on the nucleus that the electron sees, and as expected it is always less than the full charge on the nucleus. With this value of β we now can calculate the upper bound on the energy to be

$$\begin{aligned} E_0 \le E_U &= J(\beta)|_{\beta = Z - \frac{5}{16}} \\ &= -\left(Z^2 - \frac{5}{8}Z + \frac{25}{256}\right)\ . \end{aligned} \tag{13.89}$$

If we compare this result with the result of first order perturbation theory $E_0^{(1)}$ as given in Eq. (13.28), we find that

$$E_U = E_0^{(1)} - \frac{25}{256}$$

indicating that the variational result gives a better upper bound on the ground state energy. Since the wave function used in the two cases is of the same form, the advantage of the variational method is the use of the effective charge $\beta = Z - \frac{5}{16}$ in the variational calculation.

Table 13.1: Ionization energy in atomic units for two electron systems.

Atom or Ion	Experimental results	Perturbation theory	Variational method
He	0.9055	0.75	0.85
Li^+	2.7798	2.62	2.72
Be^{++}	5.655	5.5	5.60
C^{++++}	14.4070	14.25	14.35

The ionization energy of He is now given by

$$\begin{aligned} I &= -E_U(He) + E_U(He^+) \\ &= \frac{1}{2}\left(Z^2 - \frac{5}{4} + \frac{25}{128}\right) . \end{aligned} \tag{13.90}$$

In Table 13.1 we present the energy needed to remove an electron from a two electron atom. Here we compare the result of experiment, perturbation theory, and the variational method. We find that in general, the variational method gives a result that is in better agreement with experiment than the result of first order perturbation theory. The result of perturbation theory can be improved if we include the higher order terms in the expansion, but then we have a more difficult computational problem. We also can improve the variational result by taking a better trial wave function. In particular, in the above analysis we made use of a product wave function that was very simple and had only one parameter. We could take as a trial wave function the sum of such product, e.g.,

$$\psi_0(\vec{r}_1, \vec{r}_2) = \sum_{n\,n'} A_{n\,n'}\, \phi_{n00}(\vec{r}_1, \beta)\, \phi_{n'00}(\vec{r}_2, \beta) , \tag{13.91}$$

where $A_{n\,n'}$ are variational parameters adjusted to minimize the upper bound on the energy, and $\phi_{n00}(\vec{r}, \beta)$ are hydrogenic wave functions with an effective charge β on the nucleus, with β as a parameter. Another direction we can take to improve our trial wave function is to deviate from the product ansates by multiplying the wave function we used, by a function of the relative coordinates of the two electrons. All these changes can in principle improve the upper bound on the ground state energy at the cost of more parameters and in general, more complicated calculations.

13.7 Degenerate Perturbation Theory

In our discussion of perturbation theory we have implicitly assumed that the eigenstates of the unperturbed Hamiltonian H_0 all have different energies, i.e., there are no degeneracies. In particular, we have assumed that there are no degeneracies that are removed by the perturbation H_1. On the other hand, all problems with spherical symmetry give us a wave function that is proportional to the spherical harmonics, i.e., $\psi(\vec{r}) = R_\ell(r)\, Y_{\ell m}(\hat{r})$, and the corresponding eigenenergies do not depend on the quantum number m. This implies that for systems with spherical symmetry we have at least a degeneracy of $(2\ell + 1)$. To appreciate the problem with perturbation theory in the presence of this degeneracy, let us consider the case when we have two eigenstates of H_0, $|a\rangle$ and $|b\rangle$, with the same energy $\varepsilon_a = \varepsilon_b = \varepsilon$. When we turn on the perturbation one of two things can take place:

1. If $\langle a|H_1|b\rangle = 0$, then the perturbation may remove the degeneracy to the extent that to first order in perturbation theory we have

$$\begin{aligned} E_a &= \varepsilon_a + \langle a|H_1|a\rangle = \varepsilon + \langle a|H_1|a\rangle \\ E_b &= \varepsilon_b + \langle b|H_1|b\rangle = \varepsilon + \langle b|H_1|b\rangle \end{aligned}$$

with $E_a \neq E_b$. Here if we could turn on the interaction, H_1, by slowly increasing its strength (e.g., $H_1 \to \lambda\, H_1$, with $\lambda \leq 1$, and change λ from zero to 1), we could see that the degenerate states $|a\rangle$ and $|b\rangle$ slowly develop to have different energies. This is illustrated in Figure 13.5. In this case we do not have a problem in using the results of perturbation theory presented in Sec. 13.1.

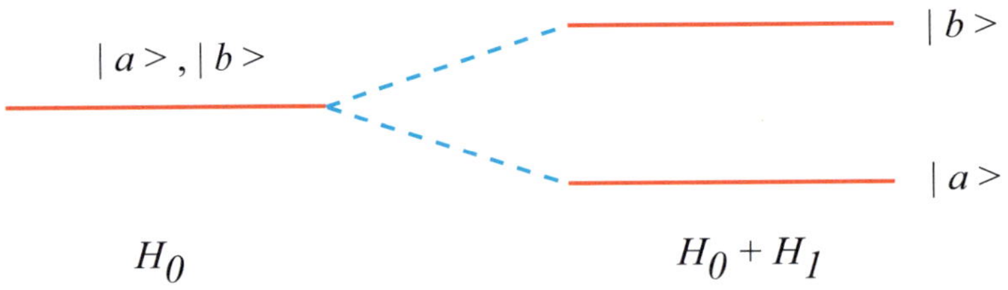

Figure 13.5: The splitting of energy levels for the case when the perturbation H_1 is diagonal in the basis considered.

2. In this case H_1 is not diagonal in our basis, i.e.,

$$\langle a|H_1|b\rangle \neq 0 \ ,$$

and we get the situation where the states $|a\rangle$, and $|b\rangle$ are split by the perturbation into two different states $|A\rangle$ and $|B\rangle$, such that

$$\begin{aligned} |A\rangle &= \alpha\,|a\rangle + \beta\,|b\rangle \\ |B\rangle &= \gamma\,|a\rangle + \delta\,|b\rangle \end{aligned}$$

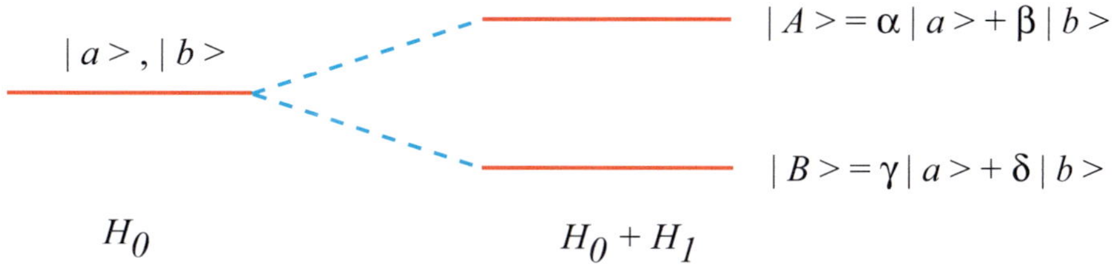

Figure 13.6: The splitting of the energy levels for the case when the perturbation H_1 is not diagonal in the basis considered.

In this case the instant the perturbation is turned on we have the states $|A\rangle$ and $|B\rangle$ irrespective of the strength of the perturbation due to H_1. To avoid that, we have to choose our basis, which are eigenstates of H_0, such that H_1 is diagonal in this basis, i.e., we have to diagonalize the matrix

$$\begin{pmatrix} \langle a|H_1|a\rangle & \langle a|H_1|b\rangle \\ \langle b|H_1|a\rangle & \langle b|H_1|b\rangle \end{pmatrix} .$$

Furthermore, since $|a\rangle$ and $|b\rangle$ are eigenstates of H_0 with the same eigenvalue ε, then $|A\rangle$ and $|B\rangle$ are also eigenstates of H_0 with eigenvalue ε.

From the above simple example we may conclude that the correct procedure for implementing perturbation theory is: (i) We divide the Hamiltonian into the unperturbed part H_0, and the perturbation H_1. (ii) We choose the basis for the eigenstates of H_0 such that the perturbing Hamiltonian H_1 is diagonal in this basis. (iii) If that is not possible, then we need to actually diagonalize our full Hamiltonian $H = H_0 + H_1$ in the basis we have chosen. This diagonalization gives us not only the eigenvalues including the perturbation to first order, but also gives us a new set of eigenstates to use in calculating the higher order terms in the perturbation series.

13.8 The Spin-Orbit Interaction

As an example of a physical problem where we have a degenerate basis for the eigenstates of H_0, and where the perturbation removes part of this degeneracy, let us consider the contribution of the spin-orbit interaction to the energy spectrum of the hydrogen atom.

If the electron in the hydrogen atom was stationary relative to the proton, then the only interaction would be the Coulomb force. However since the electron is moving relative to the proton, then by sitting on the electron we observe a magnetic field generated by the motion of the proton. This magnetic field is given by [8]

$$\vec{B} = e\left(\frac{\mu_0}{4\pi}\right)\frac{\vec{r}\times\vec{v}}{r^3} , \tag{13.92}$$

where μ_0 is the permeability of free space, v is the velocity of the proton relative to the electron, and e is the charge on the proton. This expression can be rearranged so that

$$\begin{aligned} \vec{B} &= \frac{e}{m}\left(\frac{\mu_0}{4\pi}\right)\frac{\vec{r}\times\vec{p}}{r^3} \\ &= \frac{e}{m}\frac{\mu_0}{4\pi}\frac{\vec{L}}{r^3} . \end{aligned} \tag{13.93}$$

However, we have that

$$\frac{e^2}{r^3} = \frac{1}{r}\frac{d}{dr}\left(-\frac{e^2}{r}\right) = \frac{4\pi\epsilon_0}{r}\frac{dV_c}{dr} ,$$

where ϵ_0 is the permittivity of free space, and V_c is the Coulomb potential between the proton and electron. We now can write the magnetic field as

$$\vec{B} = \frac{\mu_0\epsilon_0}{me}\frac{1}{r}\frac{dV_c}{dr}\vec{L} = \frac{1}{mec^2}\frac{1}{r}\frac{dV_c}{dr}\vec{L} . \tag{13.94}$$

where $c = \frac{1}{\sqrt{\mu_0\epsilon_0}}$ is the velocity of light in vacuum. The interaction between the magnetic moment of the electron, $\vec{M}$, and the magnetic field generated by the proton will be an interaction present in the hydrogen atom in addition to the Coulomb interaction. This additional interaction is given by

$$\begin{aligned} V_{LS} &= -\vec{M}\cdot\vec{B} \\ &= \frac{e}{m}\vec{S}\cdot\vec{B} \\ &= \frac{1}{m^2c^2}\frac{1}{r}\frac{dV_c}{dr}\left(\vec{S}\cdot\vec{L}\right) , \end{aligned} \tag{13.95}$$

where $\vec{S}$ is the spin of the electron. This derivation, which is non-relativistic, is off by a factor of two when compared to the relativistic result, which is the correct one to use, i.e., the spin orbit interaction is given by

$$V_{LS} = \frac{1}{2m^2c^2}\frac{1}{r}\frac{dV_c}{dr}\vec{L}\cdot\vec{S} . \tag{13.96}$$

With this result in hand we can write the Hamiltonian for the hydrogen atom as

$$H = H_0 + H_1 , \tag{13.97}$$

[8]We have used SI system of units in the derivation of the spin-orbit interaction so that the student can make use of results derived in a standard course in electromagnetic theory (see Jackson [43]).

where the unperturbed Hamiltonian is the one discussed in Chapter 9 Sec. 9.3, and referred to as the Coulomb Hamiltonian, i.e.,

$$H_0 = -\frac{\hbar^2}{2m}\nabla^2 - \frac{e^2}{r} \; , \qquad (13.98)$$

and the perturbation $H_1 = V_{LS}$, where V_c is given by

$$V_c = -\frac{e^2}{r} \; . \qquad (13.99)$$

Since this Hamiltonian for the hydrogen atom now includes the spin of the electron, the corresponding eigenstates of H_0 will need to include the spin wave function of the electron, if we are to use them in perturbation theory to calculate the correction to the spectrum due to the spin-orbit interaction. In other words, the wave function that are eigenstates of H_0 are of the form

$$\langle \vec{r}\,;\xi | n\ell m_\ell ; s m_s\rangle = R_{n\ell}(r)\, Y_{\ell m_\ell}(\hat{r})\, \chi_{s m_s}(\xi) \; , \qquad (13.100)$$

where $\chi_{sm_s}(\xi)$ is the spin wave function with ξ playing the role of the coordinate.[9] Since this wave function is an eigenstate of H_0, with the corresponding eigenvalue depending on n but not ℓ and m_ℓ or the spin quantum numbers, we have in general more than one eigenstate for a given eigenvalue, i.e., we have degenerate eigenstates of H_0. This requires that we examine our basis states to make sure that the perturbing Hamiltonian H_1 is diagonal in this basis. Since H_1 involves the operator $\vec{L}\cdot\vec{S}$, and we know that

$$\vec{L}\cdot\vec{S} = \frac{1}{2}\left(L_+ S_- + L_- S_+\right) + L_z S_z$$

and the operators $L_\perp$ change the quantum number m_ℓ, we can conclude that in this basis, H_1 is not diagonal. We now have a choice of either diagonalizing this Hamiltonian in the basis given in Eq. (13.100), or finding an alternative basis set of eigenstates of H_0. Here we can resort to our discussion of symmetry and conservation where we found that rotational invariance implies that the total angular momentum is a constant of the motion. This implies that the total angular momentum commutes with the Hamiltonian. In other words, the Hamiltonian is diagonal in a basis in which the total angular momentum is a good quantum number. The basis for which the total angular momentum $\vec{J}$ ($\vec{J} = \vec{L} + \vec{S}$) is a good quantum number is related to the basis defined in Eq. (13.100) by a unitary transformation whose elements are the Clebsch-Gordan coefficients of the group $SU(2)$

[9] The spin wave function $\chi_{sm_s}(\xi)$ is nothing more than the Pauli spinors encountered in Chapters 11 and 12 (see Eq. (11.137)), and can be written as

$$\chi_{sm_s}(\xi) = \delta_{m_s \xi} \; ,$$

where ξ can take on two values $+\frac{1}{2}$ and $-\frac{1}{2}$.

(see Chapter 12 Sec. 12.3). In other words, the basis for treating the spin-orbit interaction in perturbation theory are the eigenstates of the operators S^2, L^2, J^2, J_z as well as the Hamiltonian H_0. These states can be written as

$$\langle \vec{r}, \xi | n(\ell s) j m_j \rangle = R_{n\ell}(r)\, \mathcal{Y}_{(\ell s) j m_j}(\hat{r}, \xi) \ , \tag{13.101}$$

with

$$\mathcal{Y}_{(\ell s) j m_j}(\hat{r}, \xi) = \sum_{m_\ell m_s} (\ell m_\ell\, s m_s | \ell s\, j m_j)\, Y_{\ell m_\ell}(\hat{r})\, \chi_{s m_s}(\xi) \ . \tag{13.102}$$

To show that in this basis the perturbation H_1 is diagonal, we need to write the operator $\vec{L}\cdot\vec{S}$ in terms of the set of operators (J^2, L^2, S^2). To achieve this we consider J^2 as

$$J^2 = (\vec{L}+\vec{S})\cdot(\vec{L}+\vec{S}) = L^2 + S^2 + 2\vec{L}\cdot\vec{S} \ ,$$

and then solving for $\vec{L}\cdot\vec{S}$, we get

$$\vec{L}\cdot\vec{S} = \frac{1}{2}\left(J^2 - L^2 - S^2\right) \ . \tag{13.103}$$

Thus, H_1 is diagonal in this basis, and the matrix elements of $\vec{L}\cdot\vec{S}$ in this new basis are given by

$$\begin{aligned} \langle (\ell s) j m_j | 2\vec{L}\cdot\vec{S} | (\ell s) j m_j \rangle &= \hbar^2 \left[j(j+1) - \ell(\ell+1) - \frac{3}{4} \right] \\ &= \hbar^2 \begin{cases} \ell & \text{for} \quad j = \ell + \frac{1}{2} \\ -(\ell+1) & \text{for} \quad j = \ell - \frac{1}{2} \end{cases} \ . \end{aligned} \tag{13.104}$$

With this result we can write the first order correction to the energy due to the spin-orbit interaction to be

$$\begin{aligned} \Delta E^{(1)} &= \langle n(\ell s) j m_j | H_1 | n(\ell s) j m_j \rangle \\ &= \left\langle n\ell \left| \frac{1}{2m^2c^2}\, \frac{1}{r}\frac{dV_c}{dr} \right| n\ell \right\rangle \langle (\ell s) j m_j | \vec{L}\cdot\vec{S} | (\ell s) j m_j \rangle \ , \end{aligned} \tag{13.105}$$

where

$$\begin{aligned} \left\langle n\ell \left| \frac{1}{2m^2c^2}\, \frac{1}{r}\frac{dV_c}{dr} \right| n\ell \right\rangle &= \frac{1}{2m^2c^2} \int_0^\infty dr\, r^2 R_{n\ell}(r) \left(\frac{1}{r}\frac{dV_c}{dr} \right) R_{n\ell}(r) \\ &\equiv \frac{2}{\hbar^2}\, \alpha_{n\ell} > 0 \ . \end{aligned} \tag{13.106}$$

Since the radial integral is positive, and making use of the matrix elements of the $\vec{L}\cdot\vec{S}$ as given by Eq. (13.104), we may conclude that the effect of the perturbation is to lower the energy of the state with $j = \ell - \frac{1}{2}$ while at the same time increase the energy of the state with $j = \ell + \frac{1}{2}$. This is illustrated in Figure 13.7 for the case of $n = 3$, where we demonstrate how the the spin-orbit interaction removes the degeneracy between states with different j. However there is still the degeneracy that is the result of spherical symmetry which requires that each state with total angular momentum j has a degeneracy of $(2j + 1)$. To remove this degeneracy we need to break the isotropy of space. In the case of atoms, this can be achieved by placing the atom in a magnetic or electric field (see problems at the end of the chapter.)

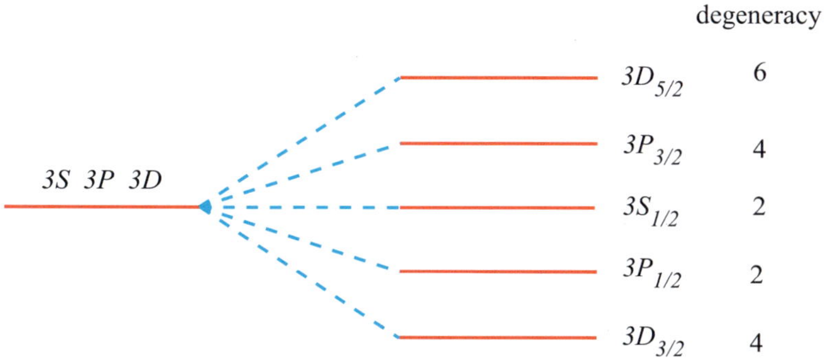

Figure 13.7: The splitting of the $n = 3$ energy levels as a result of the spin-orbit interaction.

In the above discussion we have only considered the correction to the energy levels of the hydrogen atom in lowest order perturbation theory. The second and higher order correction require non-diagonal matrix elements of H_1. Since H_1 is diagonal in spin s, orbital angular momentum ℓ, and total angular momentum j, the only non-zero matrix elements are those with a different principle quantum number n, and in this case the radial matrix elements are small and differ in energy substantially due to the shell structure of atomic energy levels.

Had we attempted to diagonalize our Hamiltonian in the basis given in Eq. (13.100), the result of the diagonalization would have been the states which are eigenstates of (S^2, L^2, J^2, J_z), and the unitary matrix of transformation would have had for its elements the Clebsch-Gordan coefficients of the group $SU(2)$ (see problem 2 at the end of this chapter).

13.9 The Zeeman Effect

In the last section we found, by examining the symmetry of the system, that we could choose a basis in which the Hamiltonian is diagonal. In this way we avoided the problem of diagonalizing the Hamiltonian in the basis given in Eq. (13.100). In the present section we will consider the problem where the choice of basis states for perturbation is not as simple.

Let us consider the case when the isotropy of space is broken by introducing an external magnetic field. In particular, let us place a hydrogenic atom in a uniform magnetic field. In this case the total angular momentum of the atom is no longer a constant of the motion. However, if we take our constant magnetic field to be pointing along the z-axis, i.e., we have a uniform field in the $x - y$ plane, then we have rotational invariance for those rotations about the z-axis. This implies that

$$\mathcal{R}(\hat{z}, \theta) = \exp\left[-\frac{i}{\hbar}\theta L_z\right]$$

commutes with the Hamiltonian, or that

$$[L_z, H] = 0 \ .$$

In other words, the z-component of the angular momentum is the constant of motion. In this case, the basis we used to calculate the contribution of the spin-orbit interaction, i.e., the eigenstates of L^2, S^2, J^2, J_z are not the proper basis, and we will need to diagonalize our Hamiltonian in order to determine the correct basis states for perturbation theory.

Let us consider the case when both the spin-orbit interaction and the interaction of the atom with a constant magnetic field along the z-direction are present. In this case the perturbing Hamiltonian is given by [10]

[10] The Zeeman Hamiltonian is given by

$$H_Z = -\vec{M} \cdot \vec{B} \ ,$$

where $\vec{M}$ is the magnetic moment of the electron. There are two contributions to this magnetic moment, one is from the spin, i.e.,

$$\vec{M}_S = -\frac{e}{mc}\vec{S} \ ,$$

while the other is due to the velocity of the electron, which gives rise to a current that produces a magnetic moment

$$\begin{aligned}\vec{M}_L &= \frac{1}{2c}\int d^3r\, \vec{r} \times \vec{J}(\vec{r}) \\ &= -\frac{e}{2mc}\vec{L} \ .\end{aligned}$$

To write this final result, we have taken the current, corresponding to an electron with velocity v, to be $\vec{J}(\vec{r}) = -ev\delta(\vec{r} - \vec{r}_0)$. We now can write the Zeeman Hamiltonian as

$$H_Z = -\vec{M} \cdot \vec{B} = (\vec{M}_L + \vec{M}_S) \cdot \vec{B} = \frac{e}{2mc}\left(\vec{L} + 2\vec{S}\right) \cdot \vec{B} \ ,$$

where $\vec{B}$ is the external magnetic field. The term involving $\vec{L} \cdot \vec{B}$ can be derived using the idea of minimal electromagnetic coupling which involves the substitution $\vec{p} \to \vec{p} - \frac{e}{c}\vec{A}$, where $\vec{A}$ is the vector potential.

$$H_1 = \frac{1}{2m^2c^2}\frac{1}{r}\frac{dV_c}{dr}\vec{L}\cdot\vec{S} + \frac{e}{2mc}\left(\vec{L}+2\vec{S}\right)\cdot\vec{B}\ , \tag{13.107}$$

where $\vec{B}$ is the magnetic field. This second term is known as the Zeeman term named after the Dutch physicist who first observed the line splitting in the spectrum when the source of the radiation was placed in a magnetic field [44]. Here we observe that the first term is diagonal in the basis of eigenstates of $S^2,\ L^2,\ J^2,\ J_z$. On the other hand, for a magnetic field pointing along the z-axis, the second term is diagonal in the basis which are eigenstates of $L^2,\ L_z,\ S^2,\ S_z$. In other words, neither bases we are familiar with are satisfactory for perturbation theory, and we need to diagonalize H_1 in one of the two bases. Let us take as our basis the eigenstates of the total angular momentum, i.e., $|(\ell s)jm\rangle$. We then have

$$\begin{aligned}\langle n,(\ell s)jm_j|H_1|n,(\ell' s)j'm_j'\rangle &= \left(\frac{\hbar}{2mc}\right)^2\left\langle n\ell\Big|\frac{1}{r}\frac{dV_c}{dr}\Big|n\ell\right\rangle \\ &\quad\times\left[j(j+1)-\ell(\ell+1)-\frac{3}{4}\right]\delta_{\ell\ell'}\,\delta_{jj'}\,\delta_{m_jm_j'} \\ &\quad+\frac{eB_0}{2mc}\,\langle(\ell s)jm_j|J_z+S_z|(\ell' s)j'm_j'\rangle\ ,\end{aligned} \tag{13.108}$$

where we have taken the magnetic field to be along the z-axis, i.e., $\vec{B}=B_0\hat{z}$, and made use of the fact that $J_z = L_z + S_z$. In writing the above result for the matrix element of H_1 we take into consideration the fact that the spin of the electron s is $\frac{1}{2}$ and can not be changed by H_1. As a result, the matrix element of the Zeeman Hamiltonian is diagonal in s. To determine the second term on the right hand side of the above equation, we need to evaluate

$$\langle(\ell s)jm_j|S_z|(\ell' s)j'm_j'\rangle\ .$$

To calculate this matrix element, we write the states of good total angular momentum in terms of eigenstates of $L^2,\ L_z,\ S^2,\ S_z$ as

$$\begin{aligned}\frac{1}{\hbar}\langle(\ell s)jm_j|S_z|(\ell' s)j'm_j'\rangle &= \frac{1}{\hbar}\sum_{m_\ell m_s}\sum_{m_\ell' m_s'}(\ell m_\ell\, s m_s|\ell s j m_j)\,(\ell' m_\ell'\, s m_s'|\ell' s j' m_j') \\ &\qquad\times\langle\ell m_\ell|\ell' m_\ell'\rangle\,\langle s m_s|S_z|s m_s'\rangle \\ &= \delta_{\ell\ell'}\,\delta_{m_jm_j'}\sum_{m_\ell m_s} m_s\,(\ell m_\ell\, s m_s|\ell s j m_j)\,(\ell m_\ell\, s m_s|\ell s j' m_j) \\ &= \delta_{\ell\ell'}\,\delta_{m_jm_j'}\sum_{m_s=\pm\frac{1}{2}} m_s\,(\ell m - m_s\, s m_s|\ell s j m_j) \\ &\qquad\times(\ell m - m_s\, s m_s|\ell s j' m_j)\end{aligned} \tag{13.109}$$

Table 13.2: Table of Clebsch-Gordan coefficients $(\ell\, m_\ell\, s\, m_s|\ell\, s\, j\, m_j)$ for the addition of orbital angular momentum ℓ to spin $s = \frac{1}{2}$.

	$m_s = \frac{1}{2}$	$m_s = -\frac{1}{2}$
$j = \ell + \frac{1}{2}$	$\sqrt{\frac{\ell+m_j+\frac{1}{2}}{2\ell+1}}$	$\sqrt{\frac{\ell-m_j+\frac{1}{2}}{2\ell+1}}$
$j = \ell - \frac{1}{2}$	$-\sqrt{\frac{\ell-m_j+\frac{1}{2}}{2\ell+1}}$	$\sqrt{\frac{\ell+m_j+\frac{1}{2}}{2\ell+1}}$

From this result we can conclude that the Zeeman Hamiltonian is diagonal in the quantum number corresponding to the projection of the total angular momentum along the z axis, i.e., J_z. This result is expected, considering the fact that the system is invariant under rotation about the z-axis.

For the maximum value of m_j ($m_j = \ell + \frac{1}{2}$) we have that $j = j' = \ell + \frac{1}{2}$. In this case there is only one term in the sum in Eq. (13.109) with the value of the Clebsch-Gordan involved being one. As a result, the perturbing Hamiltonian given in Eq. (13.107) is diagonal in our basis, and therefore no diagonalization is required to calculate the correction to the energy, which is given by

$$\begin{aligned}\Delta E^{(1)} &= \langle n, (\ell s)j = \ell + \tfrac{1}{2};\, m_j = \pm(\ell + \tfrac{1}{2})|H_1|n, (\ell s)j = \ell + \tfrac{1}{2};\, m_j = \pm(\ell + \tfrac{1}{2})\rangle \\ &= \alpha_{n\ell}\left[\ell \pm (\ell + 1)x\right] \ , \end{aligned} \tag{13.110}$$

where $\alpha_{n\ell}$ is defined in Eq. (13.106), and x is given by

$$x = \frac{\beta}{2\alpha_{n\ell}} \ , \tag{13.111}$$

with β given in terms of the magnetic field by the relation

$$\beta = \frac{eB_0\hbar}{mc} \ . \tag{13.112}$$

To illustrate the result we have so far, let us consider the $1S_{1/2}$ level of Hydrogen. In this case $\ell = 0$ and

$$\Delta E^{(1)} = \pm\frac{\beta}{2} = \pm\frac{eB_0\hbar}{2mc} \ .$$

Thus the $1S_{1/2}$ level is split into two levels, one corresponding to $m_j = +\frac{1}{2}$, the other is for $m_j = -\frac{1}{2}$.

For all other values of m_j, the sum in Eq. (13.109) has only two terms and can be written as

$$\frac{1}{\hbar}\langle(\ell s)jm_j|S_z|(\ell s)j'm_j\rangle = \frac{1}{2}\Big[(\ell m-\tfrac{1}{2}s\tfrac{1}{2}|\ell sjm_j)\,(\ell m-\tfrac{1}{2}s\tfrac{1}{2}|\ell sj'm_j) \\ -(\ell m+\tfrac{1}{2}s-\tfrac{1}{2}|\ell sjm_j)\,(\ell m+\tfrac{1}{2}s-\tfrac{1}{2}|\ell sj'm_j)\Big] \quad (13.113)$$

We now make use of Table 13.2 for the Clebsch-Gordan coefficient to calculate the matrix elements of S_z. This gives the 2×2 matrix in Table 13.3 which has the matrix elements of S_z.

Table 13.3: The matrix $\frac{1}{\hbar}\langle(\ell s)jm_j|S_z|(\ell s)j'm_j\rangle$.

	$j=\ell-\frac{1}{2}$	$j=\ell+\frac{1}{2}$
$j=\ell-\frac{1}{2}$	$-\frac{m_j}{2\ell+1}$	$-\frac{\sqrt{\ell(\ell+1)+\frac{1}{4}-m_j^2}}{2\ell+1}$
$j=\ell+\frac{1}{2}$	$-\frac{\sqrt{\ell(\ell+1)+\frac{1}{4}-m_j^2}}{2\ell+1}$	$\frac{m_j}{2\ell+1}$

Using the results in Table 13.3 we can write the matrix element of the perturbing Hamiltonian H_1 as

$$\langle n,(\ell s)jm_j|H_1|n,(\ell s)j'm_j\rangle\ .$$

This 2×2 matrix is tabulated in Table 13.4

Table 13.4: The matrix $\langle n,(\ell s)jm_j|H_1|n,(\ell s)j'm_j\rangle$

	$j=\ell-\frac{1}{2}$	$j=\ell+\frac{1}{2}$
$j=\ell-\frac{1}{2}$	$-\alpha_{n\ell}(\ell+1)+\beta\frac{\ell m_j}{2\ell+1}$	$-\frac{\beta}{2}\frac{\sqrt{\ell(\ell+1)+\frac{1}{4}-m_j^2}}{2\ell+1}$
$j=\ell+\frac{1}{2}$	$-\frac{\beta}{2}\frac{\sqrt{\ell(\ell+1)+\frac{1}{4}-m_j^2}}{2\ell+1}$	$\alpha_{nl}\,\ell+\beta\frac{m_j(\ell+1)}{2\ell+1}$

To diagonalize the perturbation H_1 in this basis we need to solve the equation

$$\det\left|\langle n,(\ell s)jm_j|H_1|n,(\ell s)j'm_j\rangle - \lambda\,\delta_{jj'}\right| = 0\,. \tag{13.114}$$

This gives a quadratic equation in λ with a solution given by

$$\begin{aligned}\lambda &= \alpha_{n\ell}\left\{m_j x - \tfrac{1}{2} \pm \left[(\ell+\tfrac{1}{2})^2 + m_j x + \frac{x^2}{4}\right]^{\frac{1}{2}}\right\}\\ &\equiv \Delta E^{(1)}\,,\end{aligned} \tag{13.115}$$

For a weak magnetic field, i.e., $x \ll 1$ the above expression for $\Delta E^{(1)}$ reduces to

$$\begin{aligned}\Delta E^{(1)} &\approx \alpha_{n\ell}\left\{m_j x - \tfrac{1}{2} \pm (\ell+\tfrac{1}{2})\left[1 + \frac{m_j x}{(\ell+\frac{1}{2})^2}\right]^{\frac{1}{2}}\right\}\\ &\approx \alpha_{n\ell}\left\{m_j x - \tfrac{1}{2} \pm (\ell+\tfrac{1}{2})\left[1 + \frac{m_j x}{2(\ell+\frac{1}{2})^2}\right]\right\}\\ &= \alpha_{n\ell}\left\{\begin{array}{l}\ell + m_j x\left(1 + \frac{1}{2\ell+1}\right)\\[2ex] -(\ell+1) + m_j x\left(1 - \frac{1}{2\ell+1}\right)\end{array}\right. \qquad \text{Zeeman Effect}\,.\end{aligned} \tag{13.116}$$

Here we observe that the splitting in the hydrogen spectrum due to the spin-orbit interaction is further split by the presence of the magnetic field. This is in fact what we expected, since the magnetic field has broken the isotropic (roational) symmetry of space. For small magnetic field, i.e., $x \ll 1$, the splitting is such that the spin orbit effect is still recognized (see Figure 13.8). On the other hand, for strong magnetic fields, i.e., $x \gg 1$, we have that

$$\begin{aligned}\left[(\ell+\tfrac{1}{2})^2 + m_j x + \frac{x^2}{4}\right]^{\frac{1}{2}} &= \left[(\ell+\tfrac{1}{2})^2 - m_j^2 + (m_j + \tfrac{1}{2}x)^2\right]^{\frac{1}{2}}\\ &= (m_j + \tfrac{1}{2}x)\left[1 + \frac{(\ell+\frac{1}{2})^2 - m_j^2}{(m_j+\frac{1}{2}x)^2}\right]^{\frac{1}{2}}\\ &\approx (m_j + \tfrac{1}{2}x)\,.\end{aligned}$$

With this result in hand, we can write the correction to the energy in first order perturbation theory in a strong magnetic field as

$$\Delta E^{(1)} \approx \alpha_{n\ell}\left\{\begin{array}{l}(m_j - \frac{1}{2}) + (m_j + \frac{1}{2})x\\[2ex] -(m_j + \frac{1}{2}) + (m_j - \frac{1}{2})x\end{array}\right. \qquad \text{Paschen-Back Effect}\,. \tag{13.117}$$

Here the splitting due to the magnetic field is so large that the level spacing has no resemblance to the original spin-orbit splitting. In Figure 13.8 we sketch the levels for the P-state, i.e., $\ell = 1$ and indicate how the spacing changes as we increase the magnetic field or x.

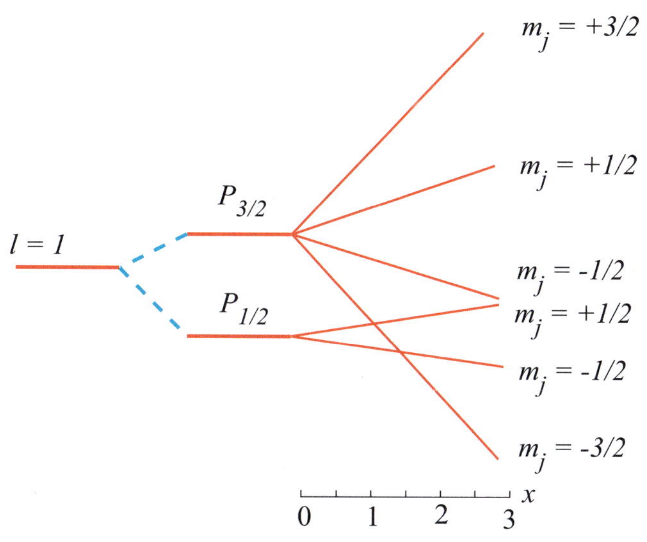

Figure 13.8: The splitting of the $\ell = 1$ levels as a result of the spin-orbit interaction and a magnetic field. In particular we demonstrate how the spectrum changes with x which is proportional to the magnetic field.

13.10 Problems

1. To first order in perturbation theory, calculate the correction to the ground state of the hydrogen-like atom due to the finite size of the nucleus. For simplicity, assume that the nucleus is spherical, of radius R, and that its charge Ze is uniformly distributed throughout its volume.

   Hint: Use Gauss's law to determine the potential due to a uniformly charged sphere as a function of the distance from the center of the sphere.

2. The positronium is a bound state of an electron and a positron (both spin $\frac{1}{2}$ particles). In the ground state, i.e., $n = 1, \ell = 0$, the main term in the Hamiltonian, besides the Coulomb attraction is

$$H_1 = \gamma\, \vec{\sigma}_1 \cdot \vec{\sigma}_2 \; .$$

   (a) Calculate the Clebsch-Gordan coefficients for the coupling of spin $\frac{1}{2}$ with spin $\frac{1}{2}$ by diagonalizing the above interaction, H_1, in the basis

$$|s_1 = \tfrac{1}{2}\, m_1; s_2 = \tfrac{1}{2}\, m_2\rangle \; .$$

   (b) Show that

$$|(s_1 s_2) SM\rangle = (-1)^{S+1}\, |(s_2 s_1) SM\rangle \; ,$$

   if $s_1 = s_2 = \frac{1}{2}$.

(c) Calculate the matrix elements of H_1 in this basis, and thus predict, to first order in perturbation theory, the energy splitting of the spin singlet and spin triplet states.

(d) If we put the positronium atom in the $1S$ state, in an external magnetic field B along the z-axis, we get an additional term in the Hamiltonian of the form

$$V_z = \frac{eB}{\mu c}\left(S_{1z} - S_{2z}\right)$$

where S_{1z} is the projection of the spin of the electron along the z-axis, while S_{2z} is the projection of the spin of the positron along the z-axis. Calculate the eigenvalues and eigenstates of the total Hamiltonian, and classify the states according to the quantum numbers associated with the constants of motion.

3. Consider a charged one-dimensional harmonic oscillator in a constant external electric field E. The potential energy of such a charged particle in an electric field is given by $-eEx$, where x and e are the position and the charge of the particle.

(a) Find the ground state energy of this system to second order in perturbation theory.

(b) Show that for this problem the Schrödinger equation can be solved exactly.

(c) Compare the result of the exact solution with that of perturbation theory.

4. Consider the nucleus ^{3}He to consist of a deuteron of spin one, and a proton of spin $\frac{1}{2}$ in relative angular momentum zero.

(a) What are the possible spin states of ^{3}He? Write these states in terms of the spin states of the deuteron and proton using the table of Clebsch-Gordan coefficients given in Table 13.2.

(b) Take the interaction between the deuteron and proton as

$$H_1 = V_0\, \vec{S}_d \cdot \vec{S}_p\ ,$$

where $\vec{S}_d$ is the spin of the deuteron and $\vec{S}_p$ the spin of the proton. What should the sign of V_0 be, for ^{3}He to have a ground state spin of $\frac{1}{2}$?

5. The Hamiltonian for a system is given as

$$H = -\frac{\hbar^2}{2m}\nabla^2 - A\, e^{-ar}\ .$$

Using the trial function

$$\psi(r) = B\, e^{-\beta r}\ ,$$

calculate the upper bound on the ground state energy.

6. Given a one-dimensional harmonic oscillator with Hamiltonian

$$H = -\frac{\hbar^2}{2m}\frac{d^2}{dx^2} + \frac{1}{2}m\omega^2 x^2 .$$

Assume a trial wave function

$$\psi_0(x, a) = A\, e^{-ax^2/2} \quad \text{with} \quad A = \left(\frac{a}{\pi}\right)^{1/4} ,$$

for the ground state.

(a) Calculate the ground state energy using the above trial function. Compare your result with the exact eigenvalue.

(b) Find a trial function for the first excited state, and calculate the energy of that state.

7. Calculate the upper bound on the ground state energy of the Coulomb Hamiltonian using the trial functions

$$\psi_0(r, a) = N\, e^{-ar^2} \quad , \quad \psi_0(r, a) = \frac{N}{r^2 + a^2} \quad \text{and} \quad \psi_0(r, a) = N\, r e^{-ar} .$$

Compare your results with the exact answer.

8. Using the variational method, find an approximate energy and wave function for the $2S$ state of the hydrogen atom.

9. A trial function Ψ differs from an eigenfunction ϕ_E,

$$H\phi_E = E\phi_E,$$

by a small amount, so that $\Psi = \phi_E + \varepsilon\psi$, where ϕ_E and ψ are both normalized and $\varepsilon \ll 1$. Show that the upper bound on the energy $\langle\, H\,\rangle$ differs from E only by a term of order ε^2, and find the term.

Note the function Ψ is not normalized.

10. Calculate the splitting in the $n = 3$ levels of hydrogen due to the spin-orbit and Zeeman interaction with a constant magnetic field in the z-direction of magnitude B_0.

11. If the general form of a spin-orbit interaction for a particle of mass m and spin $\vec{S}$ moving in a potential $V(r)$ is

$$V_{LS} = \frac{1}{2m^2c^2}\,\frac{1}{r}\frac{dV}{dr}\,\vec{L}\cdot\vec{S} ,$$

what is the effect of this interaction on the spectrum of the three-dimensional harmonic oscillator?

12. Calculate the perturbation of the first two energy levels of a hydrogen atom placed in a constant electric field $\vec{E}$ along the z-axis. (This effect is known as the Stark effect).

13. The *normal* Zeeman effect involves the interaction of the magnetic moment of the electron due to its angular momentum with an external magnetic field, i.e., the spin of the electron is neglected. In this case the Hamiltonian for the hydrogen atom in a magnetic field $\vec{B}$ is given by

$$H = \frac{1}{2}\nabla^2 - \frac{1}{r} + \frac{1}{2}\vec{L} \cdot \vec{B} \ ,$$

in atomic units.

(a) Calculate the splitting of the $n = 2$ levels in hydrogen, to first order in perturbation theory, for the case when the magnetic field is in the z-direction, i.e., $\vec{B} = B_0\hat{z}$.

(b) How many spectral lines does the $n = 2$ to $n = 1$ transition split into?

CHAPTER 14

Scattering Theory; Revisited

Abstract: In this chapter we return to scattering theory and set the foundation for approximation methods in calculating the scattering amplitude. We also address the problem of solving the Schrödinger equation for scattering in other than the coordinate representation. We convert the Schrödinger equation into an integral equation for the scattering amplitude and discuss ways of solving such integral equations. In the process we apply the formalism to electron-atom scattering and nucleon-nucleon scattering.

In Chapter 10 we showed how we can calculate the cross section for two particle scattering in terms of the scattering amplitude $f(k,\theta)$, and how this scattering amplitude is related to the phase shifts δ_ℓ. In particular, we found that we needed to solve the Schrödinger equation for the wave function $\psi_\ell(r)$ for all r and then extract the phase shifts, knowing that for $r \to \infty$

$$\psi_\ell(r) \to \frac{1}{kr}\sin(kr - \tfrac{1}{2}\pi\ell + \delta_\ell) \; .$$

In other words, the asymptotic wave function determines the scattering amplitude and, therefore the cross section, yet we need to calculate the wave function for all r. Since the wave function is not the observable we measure, we would like to set up an equation for the scattering amplitude $f(k,\theta)$ which is the observable.

In this chapter we will first derive an equation, the Lippmann-Schwinger [45] equation, which can be solved for the scattering amplitude. We will find that this equation allows us not only to derive the Optical Theorem, but also to consider the perturbation expansion for the scattering amplitude in the form of the Born series [46]. Finally we will consider the special class of potentials referred to separable potentials, where we can study the properties of the scattering amplitude analytically.

Iraj R. Afnan

14.1 Formal Theory of Scattering

To derive an equation for the scattering amplitude $f(k, \theta)$, we need to convert the Schrödinger equation plus boundary condition into an integral equation which incorporates these boundary conditions. Consider the Hamiltonian

$$H = H_0 + V \ , \tag{14.1}$$

where H_0 is taken to be the kinetic energy in the center of mass, i.e., $H_0 = \frac{p^2}{2\mu}$ with μ the reduced mass of the system given by

$$\mu = \frac{m_1 m_2}{m_1 + m_2} \ .$$

Here m_1 and m_2 are the masses of the two particles in the collision. In Eq. (14.1), V is the interaction between the two particles. We also introduce the eigenstates of H_0 to be

$$H_0|\phi_{\vec{k}}\rangle = E|\phi_{\vec{k}}\rangle \ , \tag{14.2}$$

where $E = \frac{\hbar^2 k^2}{2\mu}$. These states are labeled by the initial momentum $\vec{p} = \hbar\vec{k}$. In coordinate representation, the state $|\phi_{\vec{k}}\rangle$ is a plane wave, i.e.,

$$\langle\vec{r}\,|\phi_{\vec{k}}\rangle = \frac{1}{(2\pi)^{3/2}}\, e^{i\vec{k}\cdot\vec{r}} \ . \tag{14.3}$$

We note here that this state is identical to the eigenstate of the momentum operator as given in Eq. (11.93), i.e., $\langle\vec{r}\,|\vec{k}\rangle = \langle\vec{r}\,|\phi_{\vec{k}}\rangle$. The Schrödinger equation for the full Hamiltonian H can now be written as

$$(E - H_0)\ |\psi\rangle = V\,|\psi\rangle \ . \tag{14.4}$$

This can be thought of as an inhomogeneous differential equation if we take it in coordinate representation, and in this case, the solution can be written as the sum of a particular solution plus a solution to the homogeneous equation. The particular solution is given by

$$|\psi\rangle = \frac{1}{E - H_0}\, V\,|\psi\rangle \ . \tag{14.5}$$

In writing the operator $(E - H_0)^{-1}$, we have assumed that the operator $(E - H_0)$ has no zero eigenvalues. In fact, as we will find, the operator $(E - H_0)$ does have zero eigenvalues and therefore its inverse which appears in Eq. (14.5) is singular. We will next show how the boundary conditions in the Schrödinger equation are used to overcome these singularities. The general solution of Eq. (14.4) is then given by

$$|\psi_{\vec{k}}\rangle = |\phi_{\vec{k}}\rangle + \frac{1}{E - H_0}\, V\,|\psi_{\vec{k}}\rangle \ . \tag{14.6}$$

We have labeled our solution $|\psi_{\vec{k}}\rangle$ by the vector $\vec{k}$ to indicate that the momentum of the incident beam is $\hbar\vec{k}$. This in turn means that the energy of the incident beam in the center of mass is given by $E = \frac{\hbar^2 k^2}{2\mu}$.

To examine the singularities of $(E - H_0)^{-1}$, we write Eq. (14.6) in coordinate representation as

$$\langle \vec{r}\,|\psi_{\vec{k}}\rangle = \langle \vec{r}\,|\phi_{\vec{k}}\rangle + \int d^3r\, \langle \vec{r}\,|(E - H_0)^{-1}|\vec{r}\,'\rangle\, \langle \vec{r}\,'|V|\psi_{\vec{k}}\rangle\ . \tag{14.7}$$

We are now in a position to examine the singularities of the *Green's function.*[1] Making use of the fact that the complete set of eigenstates of the momentum operator are also eigenstates of H_0, we can write the Green's function in coordinate space as

$$\begin{aligned} \langle \vec{r}\,|(E - H_0)^{-1}|\vec{r}\,'\rangle &= \int d^3k\, \langle \vec{r}\,|(E - H_0)^{-1}|\vec{k}\,\rangle\, \langle \vec{k}\,|\vec{r}\,'\rangle \\ &= \int d^3k\, \frac{\langle \vec{r}\,|\vec{k}\,\rangle\, \langle \vec{k}\,|\vec{r}\,'\rangle}{E - \frac{\hbar^2 k^2}{2\mu}}\ , \end{aligned} \tag{14.8}$$

since the eigenvalue of H_0 is $\frac{\hbar^2 k^2}{2\mu}$. Taking the energy E, which is the energy of the initial incident particle in the two-body center mass, to be related to the incident momentum, $\hbar k_0$, then the integral in Eq. (14.8) can be written as

$$\begin{aligned} \langle \vec{r}\,|(E - H_0)^{-1}|\vec{r}\,'\rangle &= \frac{1}{(2\pi)^3}\frac{2\mu}{\hbar^2}\int d^3k\, \frac{e^{i\vec{k}\cdot(\vec{r}-\vec{r}\,')}}{k_0^2 - k^2} \\ &= \frac{1}{(2\pi)^3}\frac{2\mu}{\hbar^2}\int_0^\infty dk\, \frac{k^2}{k_0^2 - k^2}\int_0^\pi d\theta\, \sin\theta\, e^{ik|\vec{r}-\vec{r}\,'|\cos\theta}\int_0^{2\pi} d\phi \\ &= \frac{1}{(2\pi)^2}\frac{2\mu}{\hbar^2}\int_0^\infty dk\, \frac{k^2}{k_0^2 - k^2}\, \frac{e^{ik|\vec{r}-\vec{r}\,'|} - e^{-ik|\vec{r}-\vec{r}\,'|}}{ik|\vec{r} - \vec{r}\,'|} \\ &= \frac{1}{i(2\pi)^2}\frac{2\mu}{\hbar^2}\frac{1}{|\vec{r} - \vec{r}\,'|}\int_{-\infty}^{+\infty} dk\, k\, \frac{e^{ik|\vec{r}-\vec{r}\,'|}}{k_0^2 - k^2}\ . \end{aligned} \tag{14.9}$$

However this integral is not defined for $k_0^2 > 0$, (i.e., $E > 0$) because the integrand has a pole at $k = \pm k_0$, which is on the integration path. To overcome this problem, we need to go around these poles. This can be achieved by taking [2]

$$k_0^2 \longrightarrow k_0^2 \pm i\epsilon\ , \tag{14.10}$$

[1]This Green's function is identical to the Green's function encountered in the theory of differential equations. After all, we are solving the Schrödinger equation which is a second order differential equation. The Green's function is named after George Green (1793-1841) who was surprisingly a miller in Nottingham by day and a mathematician/physicist by night.

[2]This is one choice of integration path. Other choices will entail taking the integration path below or above both poles. These correspond to other boundary conditions.

where ϵ is an infinitesimal quantity. We then perform the integral and take the limit of $\epsilon \to 0$. As we will see, the choice of sign in $k_0^2 \pm i\epsilon$ will determine the boundary condition. Let us take the positive sign, i.e., $k_0^2 + i\epsilon$, for the present. Then our Green's function has the integral representation given by

$$\langle \vec{r}|(E - H_0)^{-1}|\vec{r}\,'\rangle = \frac{1}{i(2\pi)^2}\frac{2\mu}{\hbar^2}\frac{1}{|\vec{r}-\vec{r}\,'|}\int_{-\infty}^{+\infty} dk\, k\, \frac{e^{ik|\vec{r}-\vec{r}\,'|}}{k_0^2 + i\epsilon - k^2} \ . \tag{14.11}$$

The integrand now has two poles at

$$k = \pm(k_0 + i\epsilon) \ .$$

We now can perform the integral in Eq. (14.11) by making use of Cauchy's Theorem (see Appendix B). For k in the upper half of the complex k-plane, i.e., $Im(k) > 0$, the integrand has the property that it goes to zero as $|k| \to \infty$. This condition allows us to convert the integral over the interval $-\infty$ to $+\infty$ to an integral along a contour C in Figure 14.1 where the semi-circle corresponds to $|k| \to \infty$. Because the integrand is zero on the infinite semi-circle it does not contribute to the value of the Green's function. This allows us to write the Green's function as

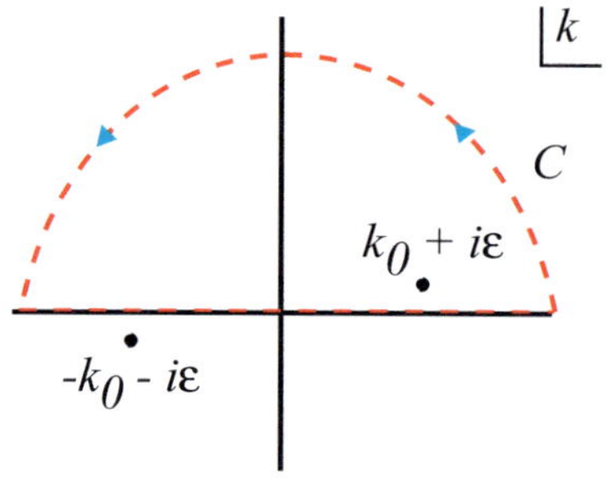

Figure 14.1: The contour of integration used to calculate the Green's function with an outgoing spherical wave.

$$\begin{aligned}\langle \vec{r}|(E - H_0)^{-1}|\vec{r}\,'\rangle &= \frac{1}{i(2\pi)^2}\frac{2\mu}{\hbar^2}\frac{1}{|\vec{r}-\vec{r}\,'|}\oint_C \frac{dk\, k\, e^{ik|\vec{r}-\vec{r}\,'|}}{(k_0 + i\epsilon - k)(k_0 + i\epsilon + k)} \\ &= -\frac{1}{4\pi}\frac{2\mu}{\hbar^2}\frac{e^{ik_0|\vec{r}-\vec{r}\,'|}}{|\vec{r}-\vec{r}\,'|} \ ,\end{aligned} \tag{14.12}$$

where we have made use of Cauchy's Residue Theorem (see Appendix B Sec. B.3) and have taken the limit $\epsilon \to 0$.

Had we chosen $k_0^2 \rightarrow k_0^2 - i\epsilon$, the poles in the integrand would have been at

$$k = \pm(k_0 - i\epsilon) \ .$$

In this case the contour C would enclose the pole at $k = -k_0 + i\epsilon$, and the Green's function would have the factor of

$$e^{-ik_0|\vec{r}-\vec{r}\,'|}$$

in Eq. (14.12) instead of

$$e^{ik_0|\vec{r}-\vec{r}\,'|} \ .$$

This we will see corresponds to spherical waves which are moving towards the scattering center, while the Green's function in Eq. (14.12) corresponds to spherical outgoing waves. These two Green's functions can now be written as

$$\begin{aligned} \langle \vec{r}\,|\,(E^{\pm} - H_0)^{-1}|\vec{r}\,'\rangle &= \lim_{\epsilon\rightarrow 0} \langle \vec{r}\,|\,(E \pm i\epsilon - H_0)^{-1}|\vec{r}\,'\rangle \\ &= \mp \frac{1}{4\pi}\frac{2\mu}{\hbar^2}\frac{e^{\pm ik_0|\vec{r}-\vec{r}\,'|}}{|\vec{r}-\vec{r}\,'|} \ . \end{aligned} \tag{14.13}$$

With this result in hand we can write the scattering wave function, in coordinate representation, by substituting the Green's function $\langle \vec{r}\,|\,(E^{+} - H_0)^{-1}|\vec{r}\,'\rangle$ in Eq. (14.7) to get

$$\langle \vec{r}\,|\psi^{(+)}_{\vec{k}_0}\rangle = \langle \vec{r}\,|\phi_{\vec{k}_0}\rangle - \frac{1}{4\pi}\frac{2\mu}{\hbar^2}\int d^3r'\,\frac{e^{ik_0|\vec{r}-\vec{r}\,'|}}{|\vec{r}-\vec{r}\,'|}\,\langle \vec{r}\,'|V|\psi^{(+)}_{\vec{k}_0}\rangle \ , \tag{14.14}$$

where the (+) superscript on the wave function indicates that the boundary condition corresponds to an outgoing spherical wave. To extract the scattering amplitude from this wave function we need to: (i) Take the limit as $r \rightarrow \infty$ to determine the coefficient of the outgoing spherical wave. (ii) Compare the resultant asymptotic wave function with Eq. (10.3) to determine the scattering amplitude. To take the limit as $r \rightarrow \infty$ we make use of the fact that

$$\begin{aligned} |\vec{r}-\vec{r}\,'| &= \sqrt{r^2 + r'^2 - 2\vec{r}\cdot\vec{r}\,'} \\ &= r\left\{1 + \left(\frac{r'}{r}\right)^2 - 2\frac{\vec{r}\cdot\vec{r}\,'}{r^2}\right\}^{\frac{1}{2}} \\ &\rightarrow r - \hat{r}\cdot\vec{r}\,' \quad \text{for } r \rightarrow \infty \ , \end{aligned}$$

where $\hat{r}$ is a unit vector in the $\vec{r}$ direction. We now can write the asymptotic wave function (i.e., $r \rightarrow \infty$) as

$$\langle \vec{r}\,|\psi^{(+)}_{\vec{k}_0}\rangle \rightarrow \frac{1}{(2\pi)^{3/2}}\,e^{i\vec{k}_0\cdot\vec{r}} - \frac{1}{4\pi}\frac{2\mu}{\hbar^2}\frac{e^{ik_0 r}}{r}\int d^3r'\,e^{-ik_0\hat{r}\cdot\vec{r}\,'}\,\langle \vec{r}\,|V|\psi^{(+)}_{\vec{k}_0}\rangle \ . \tag{14.15}$$

Since the unit vector $\hat{r}$ is in the same direction as the final momentum of the scattered particle, and the magnitude of the initial and final momentum are the same due to energy conservation, we can take the final momentum to be $\vec{k}_f = k_0\hat{r}$. Making use of this result and the fact that

$$\langle\phi_{\vec{k}_f}|\vec{r}\,'\rangle = \frac{1}{(2\pi)^{3/2}}\,e^{-i\vec{k}_f\cdot\vec{r}\,'} ,$$

we can write the asymptotic wave function as

$$\langle\vec{r}|\psi^{(+)}_{\vec{k}_i}\rangle \rightarrow \frac{1}{(2\pi)^{3/2}}\left\{e^{i\vec{k}_i\cdot\vec{r}} - \frac{4\pi^2\mu}{\hbar^2}\,\langle\phi_{\vec{k}_f}|V|\psi^{(+)}_{\vec{k}_i}\rangle\,\frac{e^{ikr}}{r}\right\} . \tag{14.16}$$

In writing the above scattering wave function, we have labeled the initial momentum, i.e., the momentum of the incident beam, by $\vec{k}_i$, while the final momentum is labeled as $\vec{k}_f$. Conservation of energy then requires that the magnitude of the initial and final momentum be equal if the scattering is elastic, i.e., $|\vec{k}_i| = |\vec{k}_f| \equiv k$. Comparing Eqs. (14.16) and (10.3) allows us to write the scattering amplitude in terms of the matrix elements of the potential as

$$f(k,\theta) = -\frac{4\pi^2\mu}{\hbar^2}\,\langle\phi_{\vec{k}_f}|V|\psi^{(+)}_{\vec{k}_i}\rangle , \tag{14.17}$$

where $\cos\theta = \hat{k}_f\cdot\hat{k}_i$. Since the scattering amplitude is basically an observable, we will introduce a corresponding operator which we will call the T-matrix, and which is defined by the relation

$$V|\psi^{(+)}_{\vec{k}_i}\rangle \equiv T(E^+)|\phi_{\vec{k}_i}\rangle , \tag{14.18}$$

where the boundary condition is now specified by labeling the T-matrix with $E^+ = E + i\epsilon$. We now can write the scattering amplitude in terms of the T-matrix as

$$\begin{aligned} f(k,\theta) &= -\frac{4\pi^2\mu}{\hbar^2}\,\langle\phi_{\vec{k}_f}|T(E^+)|\phi_{\vec{k}_i}\rangle & (14.19\text{a}) \\ &= -\frac{4\pi^2\mu}{\hbar^2}\,\langle\vec{k}_f|T(E^+)|\vec{k}_i\rangle . & (14.19\text{b}) \end{aligned}$$

Therefore, the problem of finding an equation for the scattering amplitude has been reduced to the problem of finding an equation for the T-matrix. Making use of the definition of the T-matrix as given in Eq. (14.18), we can write the scattering wave function given in Eq. (14.6) as

$$|\psi^{(+)}_{\vec{k}_i}\rangle = |\phi_{\vec{k}_i}\rangle + G_0(E^+)\,T(E^+)|\phi_{\vec{k}_i}\rangle , \tag{14.20}$$

where $G_0(E^+) = (E^+ - H_0)^{-1}$ is the Green's function in operator form. This Green's function is often referred to as the free-particle Green's function since it is the Green's function for the Schrödinger equation in the absence of any interaction. We now multiply this equation from the left by the potential V, to get

$$V|\psi^{(+)}_{\vec{k}_i}\rangle = V|\phi_{\vec{k}_i}\rangle + V\,G_0(E^+)\,T(E^+)|\phi_{\vec{k}_i}\rangle \ . \tag{14.21}$$

Applying the definition of the T-matrix, Eq. (14.18), to the left hand side of Eq. (14.21) gives us the equation

$$T(E^+)|\phi_{\vec{k}_i}\rangle = V|\phi_{\vec{k}_i}\rangle + V\,G_0(E^+)\,T(E^+)|\phi_{\vec{k}_i}\rangle \ . \tag{14.22}$$

Since this is valid for any state $|\phi_{\vec{k}_i}\rangle$, we can write an operator equation for the T-matrix which is of the form

$$T(E^+) = V + V\,G_0(E^+)\,T(E^+) \ . \tag{14.23}$$

We have taken the operator T to be a function of the energy E^+ since the Green's function $G_0(E^+)$ is a function of the energy, and we need to know the Green's function to determine the T-matrix. Also, by specifying the energy in the form of E^+, we have specified the boundary condition that we have spherically outgoing waves. Equation (14.23) is known as the *Lippmann-Schwinger* equation [45], and is equivalent to the Schrödinger equation including the boundary conditions. The solution of this equation will give us the scattering amplitude directly and thus the cross section.

To solve Eq. (14.23), we need to write the equation in a given representation. In this case the natural representation is the momentum representation [3] which gives us $\langle\vec{k}\,|T(E^+)|\vec{k}\,'\rangle$ and then the scattering amplitude, as measured in elastic scattering, is given by

$$\langle\vec{k}\,|T(E^+)|\vec{k}\,'\rangle \quad \text{with} \quad |\vec{k}| = |\vec{k}\,'| \quad \text{and} \quad E = \frac{\hbar^2 k^2}{2\mu} \ . \tag{14.24}$$

The Lippmann-Schwinger equation, Eq. (14.23), can be written in momentum space as

$$\begin{aligned} \langle\vec{k}\,|T(E^+)|\vec{k}\,'\rangle &= \langle\vec{k}\,|V|\vec{k}\,'\rangle + \langle\vec{k}\,|V\,G_0(E^+)\,T(E^+)|\vec{k}\,'\rangle \\ &= \langle\vec{k}\,|V|\vec{k}\,'\rangle + \int d^3k''\,\frac{\langle\vec{k}\,|V|\vec{k}\,''\rangle\,\langle\vec{k}\,''|T(E^+)|\vec{k}\,'\rangle}{E + i\epsilon - \frac{\hbar^2 k''^2}{2\mu}} \ . \end{aligned} \tag{14.25}$$

In writing the second line of the above equation we have made use of the fact that the eigenstates of the momentum operator are complete, and that these states are also eigenstates of H_0 with eigenvalue $\frac{\hbar^2 k''^2}{2\mu}$. In Eq. (14.25), we have an integral equation in three-dimensions which is very difficult to solve as it stands. To reduce the dimensionality

---

[3] Note that the eigenstates of the momentum operator and the free Hamiltonian H_0 are identical with our choice of normalization, i.e.,

$$|\phi_{\vec{k}}\rangle = |\vec{k}\,\rangle \ .$$

of the equation, we need to partial wave expand the matrix elements of the potential V and the T-matrix $T(E^+)$. This expansion involves the introduction of eigenstates of the total angular momentum and its projection along the z-axis. In the case of the scattering of spin-less particles, the total angular momentum is the orbital angular momentum. For the potential V, the partial wave expansion involves writing the matrix elements of V as [4]

$$\begin{aligned}\langle\vec{k}|V|\vec{k}'\rangle &= \langle\hat{k};k|V|k';\hat{k}'\rangle \\ &= \sum_{\ell m}\langle\hat{k}|\ell m\rangle\,\langle\ell m,k|V|k',\ell m\rangle\,\langle\ell m|\hat{k}'\rangle \\ &\equiv \sum_{\ell m}\langle\hat{k}|\ell m\rangle\,\langle k|V_\ell|k'\rangle\,\langle\ell m|\hat{k}'\rangle\ . \end{aligned} \tag{14.26}$$

In writing the first line of the above equation we have separated the radial from the angular part of the momentum eigenstates. We have also taken into consideration that the potential is spherically symmetric by taking the matrix elements of V to be diagonal in ℓ and independent of m, i.e., since $[H,L^2]=[H,L_z]=0$, we have

$$\langle\ell m,k|V|k',\ell' m'\rangle = \delta_{\ell\ell'}\,\delta mm'\,\langle k|V_\ell|k'\rangle\ . \tag{14.27}$$

The independence of the matrix elements from m is a result of the fact that the radial wave functions for a spherically symmetric potential are independent of m. In Eq. (14.26) the bracket $\langle\hat{k}|\ell m\rangle$ is nothing other than the spherical harmonic in momentum representation, i.e.,

$$\langle\hat{k}|\ell m\rangle = Y_{\ell m}(\hat{k})\ .$$

In other words the expansion in Eq. (14.26) is similar to that in Eq. (10.45) if we take into consideration that the Legendre polynomial can be written in terms of spherical harmonics using the addition theorem, i.e.,

$$\sum_m Y_{\ell m}(\hat{k})\,Y^*_{\ell m}(\hat{k}') = \frac{2\ell+1}{4\pi}\,P_\ell(\cos\theta)\ . \tag{14.28}$$

In this case we can write the potential in momentum representation as

$$\langle\vec{k}\,|V|\vec{k}'\rangle = \frac{1}{4\pi}\sum_\ell(2\ell+1)\,V_\ell(k,k')\,P_\ell(\cos\theta)\ , \tag{14.29}$$

with

$$V_\ell(k,k') = \langle k|V_\ell|k'\rangle\ , \tag{14.30}$$

[4]In writing Eq. (14.26), we are making use of the notation

$$|\vec{k}\,\rangle = |k;\hat{k}\rangle = |k\rangle\,|\hat{k}\rangle$$

to separate the radial from the angular variables.

and $\cos\theta = \hat{k}\cdot\hat{k}'$. In a similar manner we can write a partial wave expansion for the T-matrix which is of the form

$$\begin{aligned}\langle\vec{k}\,|T(E^+)|\vec{k}\,'\rangle &= \sum_{\ell m}\langle\hat{k}|\ell m\rangle\,\langle k|T_\ell(E^+)|k'\rangle\,\langle\ell m|\hat{k}'\rangle \\ &= \frac{1}{4\pi}\sum_\ell (2\ell+1)\,T_\ell(k,k';E^+)\,P_\ell(\cos\theta)\,. \end{aligned} \tag{14.31}$$

With these partial wave expansions for the potential V and the T-matrix $T(E^+)$, and using the orthogonality of the spherical harmonics, i.e.,

$$\int d\hat{k}\,\langle\ell m|\hat{k}\rangle\,\langle\hat{k}|\ell'm'\rangle = \delta_{\ell\ell'}\,\delta_{mm'}\,, \tag{14.32}$$

we can write the Lippmann-Schwinger equation as a one dimensional integral equation of the form

$$\langle k|T_\ell(E^+)|k'\rangle = \langle k|V_\ell|k'\rangle + \int_0^\infty dk''k''^2\,\frac{\langle k|V_\ell|k''\rangle\,\langle k''|T_\ell(E^+)|k'\rangle}{E+i\epsilon-\frac{\hbar^2k''^2}{2\mu}}\,, \tag{14.33a}$$

or

$$T_\ell(k,k';E^+) = V_\ell(k,k') + \int_0^\infty dk''\,\frac{k''^2\,V_\ell(k,k'')}{E+i\epsilon-\frac{\hbar^2k''^2}{2\mu}}\,T_\ell(k'',k';E^+)\,. \tag{14.33b}$$

This equation can be solved numerically on any present day laptop computer. This is achieved by replacing the integral by a sum and in this way we turn the integral equation into a set of linear algebraic equations. In converting the above integral equation into a set of algebraic equations we should note that the energy E and the initial momentum k' are parameters and play no role in the solution of the equation. In fact, to get the physical amplitude we should take

$$E = \frac{\hbar^2k_0^2}{2\mu} \quad \text{and} \quad k' = k_0$$

where k_0 is referred to as the *on-shell* momentum. Here we observe that the solution of Eq. (14.33b) gives us the T-matrix $T_\ell(k,k_0;E^+)$ as function of k, yet to calculate the scattering amplitude we only need the on-shell T-matrix, i.e., $T_\ell(k,k_0;E^+)$. This is similar to the fact that we only need the asymptotic wave function to calculate the phase shift, but to calculate the asymptotic wave function in coordinate representation, we need to integrate the differential equation from the origin to the asymptotic region.

By comparing the partial wave expansion of the scattering amplitude, $f(k,\theta)$, as given in Eq. (10.45) with the expansion of the T-matrix as given in Eq. (14.31), and making use of Eq. (14.19), we can write the partial wave scattering amplitude $f_\ell(k)$ in terms of the partial wave T-matrix $T_\ell(k_0, k_0; E^+)$ as

$$\begin{aligned} f_\ell(k_0) = \frac{1}{2i}\left(e^{2i\delta_\ell} - 1\right) = e^{i\delta_\ell}\,\sin\delta_\ell &= -\frac{\pi\mu k_0}{\hbar^2}\,T_\ell(k_0, k_0; E^+) \\ &\equiv -\frac{\pi\mu k_0}{\hbar^2}\,T_\ell(k_0)\ . \end{aligned} \tag{14.34}$$

To solve Eq. (14.33) for the scattering amplitude, we need to determine the input partial wave potential $V_\ell(k, k')$, and the Green's function $G_0(E^+)$ in momentum representation. To determine the partial wave potential we first need to write the potential in momentum representation in terms of the potential in coordinate space as we have used it in previous chapters. All the potentials we have encountered to date are diagonal in coordinate representation, i.e.,

$$\langle\vec{r}\,|V|\vec{r}\,'\rangle = \delta(\vec{r} - \vec{r}\,')\,V(\vec{r})\ . \tag{14.35}$$

These potentials are known as local potentials in contrast to non-local potentials that are not diagonal in coordinate representation. For local potentials, we have in momentum space

$$\begin{aligned} \langle\vec{k}\,|V|\vec{k}\,'\rangle &= \int d^3r\,d^3r'\,\langle\vec{k}\,|\vec{r}\,\rangle\,\langle\vec{r}\,|V|\vec{r}\,'\rangle\,\langle\vec{r}\,'|\vec{k}\,'\rangle \\ &= \int d^3r\,\langle\vec{k}\,|\vec{r}\,\rangle\,V(\vec{r})\,\langle\vec{r}\,|\vec{k}\,'\rangle\ . \end{aligned} \tag{14.36}$$

Making use of the partial wave expansion of the eigenstates of the momentum operator, i.e.,

$$\begin{aligned} \langle\vec{r}\,|\vec{k}\,\rangle &= \frac{1}{(2\pi)^{3/2}}\,e^{i\vec{k}\cdot\vec{r}} \\ &= \sqrt{\frac{2}{\pi}}\sum_{\ell m} i^\ell\,j_\ell(kr)\,Y_{\ell m}(\hat{r})\,Y^*_{\ell m}(\hat{k})\ , \end{aligned} \tag{14.37}$$

and the orthogonality of the spherical harmonics in Eq. (14.32) allows us to write the potential in momentum space as

$$\langle\vec{k}\,|V|\vec{k}\,'\rangle = \sum_{\ell m} Y_{\ell m}(\hat{k})\,V_\ell(k, k')\,Y^*_{\ell m}(\hat{k}')\ , \tag{14.38}$$

where

$$V_\ell(k, k') = \frac{2}{\pi}\int_0^\infty dr\,r^2\int_0^\infty dr'\,r'^2\,j_\ell(kr)\,V_\ell(r, r')\,j_\ell(k'r') \tag{14.39a}$$

$$= \frac{2}{\pi}\int_0^\infty dr\,r^2\,j_\ell(kr)\,V_\ell(r)\,j_\ell(k'r)\ . \tag{14.39b}$$

Here, Eq. (14.39b) is to be used for the special case when the potential is local and central, in which case $V_\ell(r) = V(r)$. Thus for any local central potential, Eq. (14.39b) defines the momentum space partial wave potential for use in the integral equation given in Eq. (14.33). The solution of this Lippmann-Schwinger equation then gives us the scattering amplitude for a given angular momentum ℓ. To get the differential cross section, we need to solve Eq. (14.33) for all partial waves that are important. For finite range potentials, the maximum value of ℓ can be estimated by taking the classical argument that the maximum angular momentum is the cross product of the incident momentum, times the maximum impact parameter, which in this case is the range of the potential, i.e., $\ell_{\max} = r_0 k$, where r_0 is the range of the potential.[5]

14.2 The Born Approximation

The solution of the Lippmann-Schwinger equation, Eq. (14.33), is often not simple to get, and one may need to resort to approximation methods. One approximation often used at high energies or weak potentials in both atomic and nuclear physics is the Born approximation. This is the first term in a series similar to the perturbation series we developed in the last chapter. To get the series, we iterate Eq. (14.23) to get, in operator form, the *Born series* [46] given by

$$T(E^\pm) = V + V\,G_0(E^\pm)\,V + V\,G_0(E^\pm)\,V\,G_0(E^\pm)\,V + \cdots\ . \tag{14.40}$$

This is a power series in the potential, and therefore for a potential that is weak, the series converges. Here, as in the case of perturbation theory for bound states, we can define $\tilde{H}_0$ to include some of the interaction and in this way, the remaining interaction $(H - \tilde{H}_0)$ is weak enough for the Born series to converge. However, in this case, to get the amplitude or T-matrix, we need to take the matrix element of $(H - \tilde{H}_0)$ with respect to the solution of the Schrödinger equation with the Hamiltonian $\tilde{H}_0$. Also the Green's function $G_0(E^\pm)$ has to be replaced by the Green's function for the Hamiltonian $\tilde{H}_0$ which includes the interaction. This procedure has been implemented for the Coulomb plus short range potential, where we know the solution for the Coulomb Hamiltonian analytically.

Another condition under which the above Born series converges is when the energy E is high.[6] In that case the series is again a power series in E^{-1}, and the first few terms of the series give a good approximation.

The Born approximation is used when the first term in the Born series is taken to be

[5]Here, we should note that for the case of the Coulomb potential, where the range of the interaction is infinite, the partial wave sum should be examined carefully.

[6]By high energy we mean high, relative to the depth of the potential V, or the energy of bound states in the potential V. For example, for electron scattering in atomic physics, high is more than 10^3 eV, while for proton scattering in nuclear physics, high is more than 10^3 MeV.

the amplitude for scattering. For this case, we have

$$\langle \vec{k}\,|T(E^+)|\vec{k}\,'\rangle \approx \langle \vec{k}\,|V|\vec{k}\,'\rangle \equiv \langle \vec{k}\,|T_B|\vec{k}\,'\rangle \ . \tag{14.41}$$

For the case of a local central potential, the Born approximation reduces to

$$\begin{aligned} \langle \vec{k}_f|T_B|\vec{k}_i\rangle &= \langle \phi_{\vec{k}_f}|T_B|\phi_{\vec{k}_i}\rangle \\ &= \int d^3r\,\phi^*_{\vec{k}_f}\,V(r)\,\phi_{\vec{k}_i} \\ &= \frac{1}{(2\pi)^3}\int d^3r\,e^{i\vec{q}\cdot\vec{r}}\,V(r)\ , \end{aligned} \tag{14.42}$$

where $\vec{q} = \vec{k}_i - \vec{k}_f$ is the *momentum transfer* in the scattering.

As an example of the application of the Born approximation, let us consider the problem of scattering by a Coulomb potential, and in particular the potential due to the nucleus with charge Ze, i.e.,

$$V_C(r) = -\frac{Ze^2}{r}\ . \tag{14.43}$$

Here the Born approximation is given by

$$\begin{aligned} \langle \phi_{\vec{k}_f}|T_B|\phi_{\vec{k}_i}\rangle &= -\frac{Ze^2}{(2\pi)^3}\int d^3r\,e^{i\vec{q}\cdot\vec{r}}\,\frac{1}{r} \\ &= -\frac{Ze^2}{(2\pi)^2}\int_0^\infty dr\,r\int_{-1}^{+1} dx\,e^{iqrx} \\ &= -\frac{Ze^2}{(2\pi)^2}\,\frac{2}{q}\int_0^\infty dr\,\sin qr\ . \end{aligned} \tag{14.44}$$

This integral, as it stands is not well defined. However, if we replace the Coulomb potential by the screened Coulomb potential, i.e.,

$$V(r) = -Ze^2\,\frac{e^{-ar}}{r}\ ,$$

the integral becomes well defined, and then we can take the limit as $a \to 0$ to get the Born amplitude for the Coulomb potential, i.e.,[7]

$$\langle \phi_{\vec{k}_f}|T_B|\phi_{\vec{k}_i}\rangle = -\frac{Ze^2}{2\pi^2}\,\frac{1}{q}\,\lim_{a\to 0}\int_0^\infty dr\,e^{-ar}\,\sin qr$$

---

[7]We have made use of the integral

$$\int_0^\infty dx\,e^{-ax}\,\sin qx = \frac{q}{a^2+q^2}$$

to get the Born amplitude for Coulomb scattering.

$$= -\frac{Ze^2}{2\pi^2}\frac{1}{q}\lim_{a\to 0}\frac{q}{q^2+a^2}$$
$$= -\frac{Ze^2}{2\pi^2}\frac{1}{q^2}\,, \tag{14.45}$$

where the momentum transfer square is given by

$$q^2 = 2k^2(1-\cos\theta)\,, \tag{14.46}$$

and $\cos\theta = \hat{k}_f \cdot \hat{k}_i$, and $|\vec{k}_f| = |\vec{k}_i| = k$. The scattering amplitude can now be written as

$$f(k,\theta) = \frac{Ze^2\mu}{\hbar^2}\frac{1}{k^2(1-\cos\theta)}\,. \tag{14.47}$$

This scattering amplitude is real, and to that extent does not satisfy unitarity. This is true for the Born approximation for any real potential. With this result, we can calculate the differential cross section, given by

$$\frac{d\sigma}{d\Omega} = |f(k,\theta)|^2$$
$$= \frac{Z^2e^4\mu^2}{\hbar^4}\frac{1}{k^4(1-\cos\theta)^2}$$
$$= \frac{Z^2e^4}{16E^2\sin^4(\theta/2)}\,, \tag{14.48}$$

which is the Rutherford cross section we got in Eq. (10.85) This cross section goes to infinity when $\cos\theta = 1$, i.e., for $\theta = 0$ and $\theta = \pi$. In fact, the total cross section is infinite due to the fact that the Coulomb potential is infinite in range. In nature, we always have screened potentials in the sense that proton-proton scattering is a problem of two charge particles when the two hydrogen atoms overlap. For large distances we have neutral hydrogen atoms. Although the Born amplitude for the Coulomb potential is real and does not satisfy unitarity, the cross section we get is the exact classical Rutherford cross section.

14.3 Electron Atom scattering

As a second application of the Born approximation, let us consider electron atom scattering, and in particular, electron scattering from hydrogen. This approximation is very good at high energies where the second term in the Born series, which goes to zero as $\frac{1}{E}$, becomes negligible. The potential that describes the interaction between the incident electron and the target atom is the sum of two terms. The first term describes the interaction of the electron with the point nucleus, is attractive, and of the form

$$V_1(r) = -\frac{Z}{r}\,, \tag{14.49}$$

where Z is the charge on the nucleus. Here we are using atomic units. The second term corresponds to the interaction of the incident electron with the charge distribution resulting from the bound electrons in the atom. This is of the from

$$V_2(\vec{r}) = \int d^3r' \, \frac{\rho(\vec{r}\,')}{|\vec{r}-\vec{r}\,'|} \,, \tag{14.50}$$

where $\rho(\vec{r}\,')$ is the charge distribution of the bound electrons and is given by

$$\rho(\vec{r}) = \sum_{n=1}^{Z} |\phi_n(\vec{r})|^2 \,. \tag{14.51}$$

Here we note that the sum runs over the Z electrons, each in a state $\phi_n(\vec{r}), n = 1, \cdots, Z$.

For the case of electron hydrogen scattering, $Z = 1$ and the bound state electron wave function can be taken as the ground state of the hydrogen atom, i.e., $\phi_{100}(\vec{r})$ (see Table 9.2). In this case V_2 is given by

$$V_2(\vec{r}) = \int d^3r' \, \frac{|\phi_{100}(\vec{r}\,')|^2}{|\vec{r}-\vec{r}\,'|} \,, \tag{14.52}$$

with

$$\phi_{100}(\vec{r}) = R_{10}(r)\, Y_{00}(\hat{r}) = \frac{1}{\sqrt{4\pi}}\, R_{10}(r) \,, \tag{14.53}$$

and the radial wave function is given in Sec. 9.3, which in atomic units is given by

$$R_{10}(r) = 2\, e^{-r} \,. \tag{14.54}$$

We now can write the total potential for electron hydrogen scattering as

$$V(\vec{r}) = -\frac{1}{r} + \int d^3r' \, \frac{|\phi_{100}(\vec{r}\,')|^2}{|\vec{r}-\vec{r}\,'|} \,. \tag{14.55}$$

To calculate the second term on the right hand side of the above expression for the potential, we need to expand the factor of $|\vec{r}-\vec{r}\,'|^{-1}$ in terms of spherical harmonics (see Eq. (13.25), i.e.,

$$\frac{1}{|\vec{r}-\vec{r}\,'|} = 4\pi \sum_{\ell m} \frac{1}{2\ell+1} \frac{r_<^\ell}{r_>^{\ell+1}}\, Y_{\ell m}(\hat{r})\, Y_{\ell m}(\hat{r}') \,. \tag{14.56}$$

This allows us to write the potential as

$$V(\vec{r}) = -\frac{1}{r} + \sum_{\ell m} \frac{4\pi}{2\ell+1} \int_0^\infty dr' \, r'^2 \, \frac{|R_{10}(r)|^2}{4\pi} \, \frac{r_<^\ell}{r_>^{\ell+1}}\, Y_{\ell m}(\hat{r}) \int d\hat{r}\, Y_{\ell m}(\hat{r}') \,. \tag{14.57}$$

Making use of the orthogonality of the spherical harmonics, we have only one term in the sum corresponding to $\ell = m = 0$. This reduces the potential to just a radial integral of the form

$$V(\vec{r}) = -\frac{1}{r} + \int_0^\infty dr'\, r'^2 \frac{|R_{10}(r')|^2}{r_>} \ . \tag{14.58}$$

At this stage we note that the potential we have is central, i.e., it is a function of $r = |\vec{r}|$. To evaluate the integral we need to first divide the domain of integration into two parts corresponding to $r' < r$ and $r' > r$ which determines the value of $r_>$. We then make use of the fact that the radial wave function is normalized to write our potential as

$$V(r) = \int_r^\infty dr'\, r'^2 \left(\frac{1}{r'} - \frac{1}{r}\right) |R_{10}(r')|^2 \ . \tag{14.59}$$

To simplify this result we need to calculate the integral using the radial wave function for the ground state of hydrogen. This gives [8]

$$\begin{aligned} V(r) &= 4 \int_r^\infty dr'\, r'^2 \left(\frac{1}{r'} - \frac{1}{r}\right) e^{-2r'} \\ &= -(1+r)\frac{e^{-2r}}{r} \ . \end{aligned} \tag{14.60}$$

Here we observe that the range of the potential is determined by the charge distribution due to the bound electron. Thus we expect that the incident electron sees a neutral atom, and therefore no potential energy until the electron gets within the charge distribution of the atom.

We now turn to the calculation of the Born amplitude for the above potential. In the Born approximation the T-matrix is given by

$$\begin{aligned} \langle \phi_{\vec{k}_f} | T_B | \phi_{\vec{k}_i} \rangle &= \frac{1}{(2\pi)^3} \int d^3r\, e^{-i\vec{q}\cdot\vec{r}}\, V(r) \\ &= \frac{2}{(2\pi)^2} \int_0^\infty dr\, r^2\, V(r) \frac{\sin qr}{qr} \ . \end{aligned} \tag{14.61}$$

Making use of the fact that

$$\int_0^\infty dx x^n\, e^{-\beta x} \sin qx = (-1)^n \frac{\partial^n}{\partial \beta^n} \left(\frac{q}{q^2+\beta^2}\right) \ ,$$

[8]To calculate the radial integral we have made use of the fact that

$$\int dx\, x^m\, e^{-ax} = -e^{-ax} \sum_{l=0}^{m} \frac{m!\, x^{m-l}}{(m-l)!\, a^{l+1}} \ .$$

we can evaluate the above integral and write the Born T-matrix as

$$\langle\phi_{\vec{k}_f}|T_B|\phi_{\vec{k}_i}\rangle = -\frac{1}{2\pi^2}\frac{q^2+8}{(q^2+4)^2} \ . \tag{14.62}$$

Here the momentum transfer q is given in terms of the center of mass scattering angle and the momentum of the incident electron by the relation

$$q^2 = 2k^2(1-\cos\theta) \ . \tag{14.63}$$

The scattering amplitude and cross section can then be written as

$$\begin{aligned} f(k,\theta) &= 2\,\frac{q^2+8}{(q^2+4)^2} \\ &= 2\,\frac{2k^2(1-\cos\theta)+8}{[2k^2(1-\cos\theta)+4]^2} \end{aligned} \tag{14.64}$$

and

$$\begin{aligned} \frac{d\sigma}{d\Omega} &= |f(k,\theta)|^2 \\ &= 4\left\{\frac{2k^2(1-\cos\theta)+8}{[2k^2(1-\cos\theta)+4]^2}\right\}^2 \ . \end{aligned} \tag{14.65}$$

This cross section is in units of Bohr radii square, and illustrated in Fig 14.2

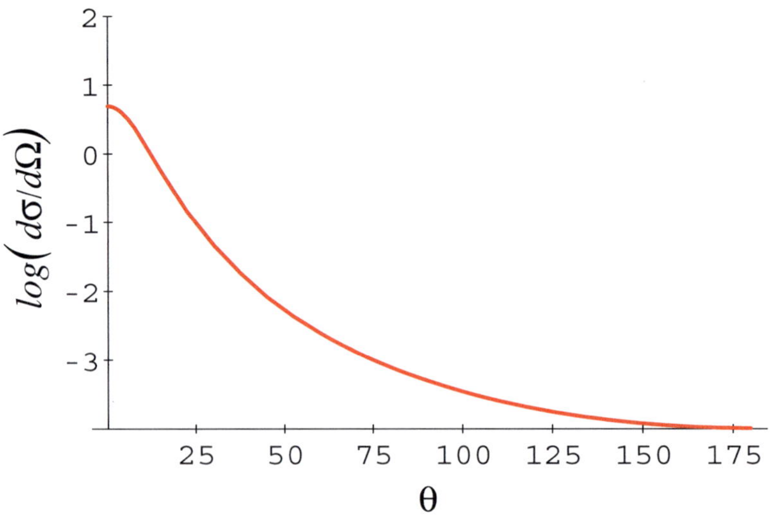

Figure 14.2: The Born cross section for electron hydrogen scattering at 100 eV.

14.4 Unitarity of the T-matrix

In Chapter 10, we demonstrated that the scattering amplitude $f(k, \theta)$ satisfied the Optical Theorem, and as a result, the S-matrix in a given partial wave is unitary, i.e., it is represented by a phase. In this section we will derive this unitarity relation from the Lippmann-Schwinger equation.

In operator from, the Lippmann-Schwinger equation is given by

$$T(E^{\pm}) = V + V\, G_0(E^{\pm})\, T(E^{\pm}) \ . \tag{14.66}$$

We now multiply this equation from the left by V^{-1} and from the right by $T^{-1}(E^{\pm})$. This gives us the equation

$$T^{-1}(E^{\pm}) = V^{-1} - G_0(E^{\pm}). \tag{14.67}$$

If we consider $T(z)$ as an analytic function in the complex z-plane, then

$$T^{-1}(E^{+}) - T^{-1}(E^{-}) = -\left[G_0(E^{+}) - G_0(E^{-})\right] \equiv -\Delta G_0(E) \ , \tag{14.68}$$

assuming the potential V is an analytic function of E for $E > 0$. For the free Green's function we have that

$$G_0(E^{+}) = \frac{1}{E + i\epsilon - H_0} = \frac{(E - H_0)}{(E - H_0)^2 + \epsilon^2} - \frac{i\epsilon}{(E - H_0)^2 + \epsilon^2} \ .$$

But [9]

$$\lim_{\epsilon \to 0(+)} \frac{\epsilon}{(E - H_0)^2 + \epsilon^2} = \pi\, \delta(E - H_0) \ . \tag{14.69}$$

Using this result we can write

$$G_0(E^{\pm}) = \frac{\mathcal{P}}{E - H_0} \mp i\pi\delta(E - H_0) \ , \tag{14.70}$$

where the $\mathcal{P}$ indicates that we should calculate the *principle value integral* when the operator is written in terms of the eigenstates of H_0. Making use of Eq. (14.70), we can write ΔG_0 as

$$\Delta G_0(E) = -2i\pi\, \delta(E - H_0) \ . \tag{14.71}$$

We now can write Eq. (14.68) as

$$T^{-1}(E^{-})\, \left[T(E^{+}) - T(E^{-})\right]\, T^{-1}(E^{+}) = -2i\pi\, \delta(E - H_0) \ ,$$

and therefore

$$\Delta T(E) \equiv T(E^{+}) - T(E^{-}) = -2i\pi\, T(E^{-})\, \delta(E - H_0)\, T(E^{+}) \ , \tag{ 14.72a}$$

[9] For a proof of this result see Gel'fand and Shilov [50].

or

$$\mathcal{I}m\, T(E) = -\pi\, T(E^-)\, \delta(E - H_0)\, T(E^+)\ . \tag{14.72b}$$

This is the operator form of the unitarity equation, and when taken in momentum representation it gives the *off-shell* unitarity equation if the initial and final momenta are different and not related to the energy E. The diagonal elements of this operator will give us the Optical Theorem, i.e.,

$$\langle \vec{k}_f | \mathcal{I}m\{T(E)\} | \vec{k}_i \rangle = -\pi\, \langle \vec{k}_f | T(E^-)\, \delta(E - H_0)\, T(E^+) | \vec{k}_i \rangle\ . \tag{14.73}$$

We now introduce a complete set of eigenstates of the momentum operator, which are also eigenstates of H_0, next to the δ-function on the right hand side of Eq. (14.73). This allows us to evaluate the radial integral over the δ-function which sets the magnitude of the intermediate momentum to k_0 such that $E = \frac{\hbar^2 k_0^2}{2\mu}$. We thus can write

$$\begin{aligned} \langle \vec{k}_f | \mathcal{I}m\{T(E)\} | \vec{k}_i \rangle &= -\pi \int d^3k\, \langle \vec{k}_f | T(E^-) | \vec{k} \rangle\, \delta\left(E - \frac{\hbar^2 k^2}{2\mu}\right) \langle \vec{k} | T(E^+) | \vec{k}_i \rangle \\ &= -\pi \int d\hat{k}_0\, \langle \vec{k}_f | T(E^-) | \vec{k}_0 \rangle\, \frac{\mu k_0}{\hbar^2}\, \langle \vec{k}_0 | T(E^+) | \vec{k}_i \rangle\ . \end{aligned} \tag{14.74}$$

If we now take $\vec{k}_f = \vec{k}_i$, i.e., the forward direction, then the above unitarity equation reduces to

$$\langle \vec{k}_i | \mathcal{I}m\{T(E)\} | \vec{k}_i \rangle = -\frac{\pi \mu k_0}{\hbar^2} \int d\hat{k}_0\, |\langle \vec{k}_i | T(E) | \vec{k}_0 \rangle|^2\ . \tag{14.75}$$

Taking $|\vec{k}_i| = k_0$, the T-matrices in the above expression are then on-shell and we can write this equation in terms of the scattering amplitude $f(k, \theta)$. This is achieved by multiplying the above unitarity equation by the factor $-\frac{4\pi^2\mu}{\hbar^2}$, and then making use of Eq. (14.19) to relate the T-matrix to the scattering amplitude. This gives

$$-\frac{4\pi^2\mu}{\hbar^2}\, \langle \vec{k}_i | \mathcal{I}m\{T(E)\} | \vec{k}_i \rangle = \frac{4\pi^2\mu}{\hbar^2}\, \frac{\pi \mu k_0}{\hbar^2} \int d\hat{k}_0\, |\langle \vec{k}_i | T(E) | \vec{k}_0 \rangle|^2\ ,$$

or

$$\mathcal{I}m\, f(k_0, 0) = \frac{k_0}{4\pi}\, \sigma_T\ . \tag{14.76}$$

This result is identical to the Optical Theorem derived in Sec. 10.5.

14.5 The T-matrix for Separable Potentials

In Sec. 14.1 we showed that the Schrödinger equation plus the boundary condition can be replaced by an integral equation commonly known as the Lippmann-Schwinger equation.

After partial wave expansion, this equation can be written as

$$\langle k|T_\ell(E^+)|k'\rangle = \langle k|V_\ell|k'\rangle + \int_0^\infty dk''\, k''^2 \frac{\langle k|V_\ell|k''\rangle\, \langle k''|T_\ell(E^+)|k'\rangle}{E^+ - \frac{\hbar^2 k''^2}{2\mu}} \; . \tag{14.77}$$

In this section we will solve this equation for a special class of potentials known as separable potentials. These have the special feature that the Lippmann-Schwinger equation can be solved analytically because the kernel of the integral equation is separable. Because we can get an analytic solution to our equations, we will be able to present a general discussion of the properties of the scattering amplitude that are valid in general for a large class of potentials.

Before we proceed with our discussion, we will introduce some definitions that will simplify the algebra without obscuring the physics. These are:

- We will take $\frac{\hbar^2}{2\mu} = 1$.
- We will restrict our discussion to S-wave scattering, i.e., $\ell = 0$, and therefore drop any reference to the angular momentum.
- Matrix elements of operators are understood in the sense that

$$\langle \phi|\mathcal{O}|\psi\rangle \equiv \int_0^\infty dk\, k^2\, \phi^*(k)\, \mathcal{O}(k)\, \psi(k) \; . \tag{14.78}$$

  The observation that we only have a radial integration in Eq. (14.78) is due of the fact that we have restricted our analysis to just one partial wave.

With these definitions we can write the operator form of the partial wave Lippmann-Schwinger equation as

$$T_\ell(E^+) = V_\ell + V_\ell\, G_0(E^+)\, T_\ell(E^+) \; , \tag{14.79a}$$

or with no reference to ℓ as

$$T(E^+) = V + V\, G_0(E^+)\, T(E^+) \; . \tag{14.79b}$$

The separable potential we are considering is given in momentum representation for the ℓth partial wave as

$$\langle k|V_\ell|k'\rangle = \langle k|g_\ell\rangle\, \lambda_\ell\, \langle g_\ell|k'\rangle \tag{14.80a}$$

$$= g_\ell(k)\, \lambda_\ell\, g_\ell(k') \; , \tag{14.80b}$$

or in operator form with no reference to ℓ as

$$V = |g\rangle\, \lambda\, \langle g| \; . \tag{14.80c}$$

Here the function $g_\ell(k)$ is referred to as the form factor. We now can write the Lippmann-Schwinger equation, Eq. (14.79b), for this potential as

$$\begin{aligned} T(E^+) &= |g\rangle\,\lambda\,\langle g| + |g\rangle\,\lambda\,\langle g|G_0(E^+)\,T(E^+) \\ &= |g\rangle\,\lambda\,\left\{\langle g| + \langle g|G_0(E^+)\,T(E^+)\right\} \\ &\equiv |g\rangle\,\lambda\,\langle \mathcal{F}|\ , \end{aligned} \tag{14.81}$$

where the bra $\langle\mathcal{F}|$ is defined in terms of the T-matrix as

$$\langle\mathcal{F}| = \langle g| + \langle g|G_0(E^+)\,T(E^+)\ . \tag{14.82}$$

Making use of Eq. (14.81) we can write

$$\langle\mathcal{F}| = \langle g| + \langle g|G_0(E^+)|g\rangle\,\lambda\,\langle\mathcal{F}|\ . \tag{14.83}$$

If we now solve this equation for $\langle\mathcal{F}|$ and substitute in Eq. (14.81), we get the result that the T-matrix has the form

$$T(E^+) = |g\rangle\,\tau(E^+)\,\langle g|\ , \tag{14.84}$$

where

$$\begin{aligned} \tau(E^+) &= \left\{\lambda^{-1} - \langle g|G_0(E^+)|g\rangle\right\}^{-1} \\ &= \left\{\lambda^{-1} - \int_0^\infty dk\,k^2\,\frac{[g(k)]^2}{E^+ - k^2}\right\}^{-1}\ . \end{aligned} \tag{14.85}$$

In momentum representation this T-matrix takes the form

$$\langle k|T(E^+)|k'\rangle = \langle k|g\rangle\,\tau(E^+)\,\langle g|k'\rangle\ , \tag{14.86a}$$

or

$$T(k,k';E^+) = g(k)\,\tau(E^+)\,g(k')\ . \tag{14.86b}$$

To establish the relation between the wave function for bound and scattering states and the T-matrix, let us now consider the Schrödinger equation for the bound state problem using this separable potential. In operator form, the Schrödinger equation can be written as

$$(E - H_0)|\psi\rangle = V|\psi\rangle\ . \tag{14.87}$$

Here again, we are considering the equation for a given angular momentum ℓ. For bound states we have that $E < 0$. Since the eigenvalues of H_0, the kinetic energy in the center of mass, is positive, we have no problems dividing by the operator $(E - H_0)$. Also, the homogenous equation $(E - H_0)|\phi\rangle = 0$ has no solution. All this allows us to write the solution of Eq. (14.87) for $E < 0$ as

$$|\psi\rangle = \frac{1}{E - H_0} V|\psi\rangle = G_0(E)\, V|\psi\rangle \; . \tag{14.88}$$

Since $E < 0$ and $(E - H_0)^{-1}$ is not singular, we need not introduce the $+i\epsilon$ required in Sec. 14.1. In Eq. (14.88), we have a homogeneous integral equation, and if we replace the integral by a sum, the integral equation will be converted to a set of homogeneous algebraic equations that have a solution only for certain values of E, i.e., we have an eigenvalue problem.

For the separable potential considered above, Eq. (14.88) reduces to

$$|\psi\rangle = G_0(E)|g\rangle\, \lambda\, \langle g|\psi\rangle \quad \text{for} \quad E < 0 \; . \tag{14.89}$$

We note here, that $\langle g|\psi\rangle$ is a matrix element of the unit operator in the sense of Eq. (14.78), and therefore a constant that can be incorporated into the normalization of the wave function. In momentum representation, this wave function takes the simple form

$$\begin{aligned} \langle k|\psi\rangle &= \langle k|(E - H_0)^{-1}|g\rangle\, \lambda\, \langle g|\psi\rangle \\ &= \frac{1}{E - k^2} \langle k|g\rangle\, \lambda\, \langle g|\psi\rangle \; , \end{aligned} \tag{14.90a}$$

or in functional form

$$\psi(k) = \mathcal{N} \frac{g(k)}{E - k^2} \; . \tag{14.90b}$$

Here, $\mathcal{N} = \lambda\, \langle g|\psi\rangle$ is the normalization of the wave function. To determine this normalization we need to determine $\langle g|\psi\rangle$. This we get by multiplying Eq. (14.89) from the left by $\langle g|$. This gives us the result that

$$\langle g|\psi\rangle = \langle g|G_0(E)|g\rangle\, \lambda\, \langle g|\psi\rangle \; ,$$

or

$$\left\{\lambda^{-1} - \langle g|G_0(E)|g\rangle\right\} \lambda\, \langle g|\psi\rangle = 0 \; . \tag{14.91}$$

For this equation to be satisfied, we either have

$$\langle g|\psi\rangle = 0 \; ,$$

which corresponds to the normalization being zero, i.e., a zero wave function, or we have that

$$\left\{\lambda^{-1} - \langle g|G_0(E)|g\rangle\right\} = 0 \; . \tag{14.92}$$

This condition will determine the energy ($E < 0$) at which we have a bound state. If we now compare this condition with Eq. (14.85), we find that the energy at which the system has a bound state, the T-matrix has a singularity. To determine the form of this singularity, let us take the energy of the bound state to be $E = B < 0$. Then

$$\left\{\lambda^{-1} - \langle g|G_0(B)|g\rangle\right\} = 0 \ . \tag{14.93}$$

But we have that

$$\begin{aligned} G_0(E) = \frac{1}{E - H_0} &= \frac{1}{B - H_0}\,(B - H_0)\,\frac{1}{E - H_0} \\ &= \frac{1}{B - H_0} - \frac{1}{B - H_0}\,(E - B)\,\frac{1}{E - H_0} \ . \end{aligned}$$

This equation can be written as

$$G_0(E) = G_0(B) - (E - B)\,G_0(B)\,G_0(E) \ . \tag{14.94}$$

With this result in hand we can write

$$\langle g|G_0(E)|g\rangle = \langle g|G_0(B)|g\rangle - (E - B)\,\langle g|G_0(B)\,G_0(E)|g\rangle \ .$$

Making use of this result and Eq. (14.85), we can write $\tau(E)$ as

$$\tau(E) = \frac{1}{(E - B)\,\langle g|G_0(B)\,G_0(E)|g\rangle} \ . \tag{14.95}$$

Therefore at $E \approx B$, the T-matrix has a simple pole in the complex energy plane along the negative real axis. Having established the energy at which we have a bound state, we now can write the wave function given in Eq. (14.90) as

$$|\psi_B\rangle = \frac{\mathcal{N}}{B - H_0}\,|g\rangle \ , \tag{14.96}$$

which in momentum space takes the simple form

$$\psi_B(\vec{k}) = \mathcal{N}\,\frac{g(\vec{k})}{B - k^2} \ . \tag{14.97}$$

The normalization $\mathcal{N}$ is now determined by the condition that

$$1 = \langle\psi_B|\psi_B\rangle = \langle g|G_0(B)\,G_0(B)|g\rangle\,\mathcal{N}^2$$

and therefore

$$\mathcal{N} = \{\,\langle g|G_0(B)\,G_0(B)|g\rangle\,\}^{-\frac{1}{2}} \ . \tag{14.98}$$

Comparing this result for the normalization with Eq. (14.95), we observe that $\mathcal{N}^2$ is the residue of $\tau(E)$ at $E = B$, and the residue of the T-matrix is given by

$$|g\rangle\,\mathcal{N}^2\,\langle g| \ ,$$

which is related to the square of the bound state wave function.

We now turn to the scattering wave function which is given in Eq. (14.6). In momentum space, this wave function takes the form

$$\begin{aligned} \langle \vec{k}\,|\psi^{(+)}_{\vec{k}_i}\rangle &= \langle \vec{k}\,|\vec{k}_i\rangle + \langle \vec{k}\,|G_0(E^+)\,V|\psi^{(+)}_{\vec{k}_i}\rangle \\ &= \delta(\vec{k}-\vec{k}_i) + \frac{1}{E^+ - k^2}\,\langle \vec{k}\,|V|\psi^{(+)}_{\vec{k}_i}\rangle\ . \end{aligned} \tag{14.99}$$

But we have from the definition of the T-matrix in Eq. (14.18) that

$$\begin{aligned} \langle \vec{k}\,|V|\psi^{(+)}_{\vec{k}_i}\rangle &= \langle \vec{k}\,|T(E^+)|\vec{k}_i\rangle \\ &= \langle \vec{k}\,|g\rangle\,\tau(E^+)\,\langle g|\vec{k}_i\rangle\ . \end{aligned} \tag{14.100}$$

With this result we can write the scattering wave function in momentum space as

$$\langle \vec{k}\,|\psi^{(+)}_{\vec{k}_i}\rangle = \delta(\vec{k}-\vec{k}_i) + \frac{\langle \vec{k}\,|g\rangle}{E^+ - k^2}\,\tau(E^+)\,\langle g|\vec{k}_i\rangle\ , \tag{14.101a}$$

or in functional form this wave function is given by

$$\psi^{(+)}_{\vec{k}_i}(\vec{k}\,) = \delta(\vec{k}-\vec{k}_i) + \frac{g(\vec{k})}{E^+ - k^2}\,\tau(E^+)\,g(\vec{k}_i)\ . \tag{14.101b}$$

Here we observe that the scattering wave function is related to the T-matrix. However, this T-matrix has the initial and final momentum as independent variables since $\vec{k}_i$ is the initial momentum of the incident beam and is related to the energy $E = |\vec{k}_i|^2$, while the momentum $\vec{k}$ is a variable whose presence is the result of the fact that we are considering the scattering wave function in momentum representation. This T-matrix that is a function of two variables $\vec{k}_i$ and $\vec{k}$ is known as the *half-off-shell* T-matrix.

To get the scattering wave function in coordinate space, we need to transform the wave function, i.e.,

$$\begin{aligned} \langle \vec{r}\,|\psi^{(+)}_{\vec{k}_i}\rangle &= \int d^3k\,\langle \vec{r}\,|\vec{k}\rangle\,\langle \vec{k}\,|\psi_{\vec{k}_i}\rangle \\ &= \langle \vec{r}\,|\vec{k}_i\rangle + \int d^3k\,\frac{\langle \vec{r}\,|\vec{k}\rangle\,\langle \vec{k}\,|g\rangle}{E^+ - k^2}\,\tau(E^+)\,\langle g|\vec{k}_i\rangle\ . \end{aligned} \tag{14.102a}$$

This equation can be written in functional form as

$$\psi^{(+)}_{\vec{k}_i}(\vec{r}\,) = \frac{1}{(2\pi)^{3/2}}\left\{e^{i\vec{k}_i\cdot\vec{r}} + \int d^3k\,e^{i\vec{k}\cdot\vec{r}}\,\frac{g(\vec{k})}{E^+ - k^2}\,\tau(E^+)\,g(\vec{k}_i)\right\}\ , \tag{14.102b}$$

where the form factors $g(\vec{k})$ are given by

$$\langle \vec{k}\,|g\rangle = g(\vec{k}) = \frac{1}{\sqrt{4\pi}}\,g(k) = \frac{1}{\sqrt{4\pi}}\,\langle k|g\rangle\ . \tag{14.103}$$

The factor of $\frac{1}{\sqrt{4\pi}}$ in the above equation is due to the fact that we have restricted our analysis to the S-wave, i.e., $\ell = 0$, and $Y_{00}(\hat{k}) = \frac{1}{\sqrt{4\pi}}$.

All of the above results are valid in general for any potential even though we have only derived the equation for a separable potential. Also the extension to all angular momentum proceeds in the same identical steps with the additional feature that the potential and T-matrix have to be expanded in partial wave form. Finally we should emphasize that the determination of the T-matrix, the bound state wave function, and the scattering wave function, have all been reduced to the evaluation of the integral

$$\langle g|G_0(E^+)|g\rangle = \int_0^\infty dk\, k^2\, \frac{[g(k)]^2}{E - k^2}\,, \tag{14.104}$$

which is required in determining the function $\tau(E)$.

To get explicit expressions for the wave function and T-matrix, we need to specify the form factor $g(k)$. For S-waves, the simplest form factor we can have is [10]

$$g(k) = \frac{1}{k^2 + \beta^2}\,. \tag{14.105}$$

where β is a parameter that is referred to as the range of the interaction. For this form factor, the integral in Eq. (14.104) is given by

$$\langle g|G_0(E^+)|g\rangle = \frac{\pi}{2}\frac{1}{(k_0^2 + \beta^2)^2}\left(\frac{k_0^2 - \beta^2}{2\beta} - ik_0\right)\,, \tag{14.106}$$

where $E = k_0^2$. Making use of this result in Eq. (14.85) we can determine $\tau(E^+)$ which is used in Eq. (14.86) to give us the on-shell T-matrix as

$$\begin{aligned} T(k_0) &= g(k_0)\,\tau(E^+)\,g(k_0) \\ &= \left\{\lambda^{-1}\left(k_0^2 + \beta^2\right)^2 - \frac{\pi}{2}\left(\frac{k_0^2 - \beta^2}{2\beta} - ik_0\right)\right\}^{-1}\,. \end{aligned} \tag{14.107}$$

This is the inverse of a second order polynomial whose zeros could give rise to bound states if the corresponding energy was negative. To determine the phase shifts for this potential, we make use of Eq. (14.34) to write the partial wave amplitude as [11]

$$f(k_0) = e^{i\delta}\sin\delta = -\frac{\pi k_0}{2}\,T(k_0)\,. \tag{14.108}$$

We now can write the phase shifts as

$$\cot\delta = \frac{\mathcal{R}e\,T(k_0)}{\mathcal{I}m\,T(k_0)}\,,$$

[10] This form factor was first suggested by Yamaguchi [51].

[11] We have modified Eq. (14.34) to take into account the fact that in this section we have taken $\frac{2\mu}{\hbar^2} = 1$.

and therefore

$$k_0 \cot\delta = -\frac{\beta}{2}\left(1+\frac{4\beta^3}{\pi\lambda}\right)+\frac{k_0^2}{2\beta}\left(1-\frac{8\beta^3}{\pi\lambda}\right)-\frac{2}{\pi\lambda}k_0^4\,. \tag{14.109}$$

To determine the scattering wave function in coordinate space, we need to calculate the integral

$$\begin{aligned}\int d^3k\, e^{i\vec{k}\cdot\vec{r}}\,\frac{g(\vec{k})}{E^+-k^2} &= \frac{\sqrt{\pi}}{ir}\int_0^\infty dk\,k\,\frac{\left(e^{ikr}-e^{-ikr}\right)}{(k_0^2+i\epsilon-k^2)\,(k^2+\beta^2)}\\ &= \frac{\sqrt{\pi}}{ir}\int_{-\infty}^{\infty} dk\,\frac{k\,e^{ikr}}{(k_0^2+i\epsilon-k^2)\,(k^2+\beta^2)}\,. \end{aligned}\tag{14.110}$$

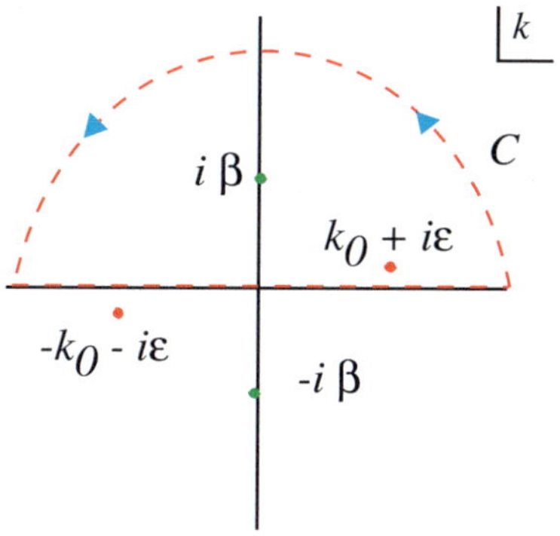

Figure 14.3: The contour of integration used to calculate the scattering wave function in coordinate representation.

This integral can be evaluated by closing the contour of integration in the upper half of the complex k-plane, and then by using Cauchy's theorem (see Appendix B Sec. B.3) to write the integral in terms of the residues at the poles inside the contour. In this case the integrand has four poles in the complex k-plane (see Figure 14.3), and only two of the poles are inside the contour. These poles are at $k = i\beta$ and $k = k_0 + i\epsilon$. The pole at $k = i\beta$ comes from the form factor and its position depends on the parameters of the potential and in particular the range of the interaction. On the other hand, the pole at $k = k_0 + i\epsilon$ is from the free-particle Green's function, and its position depends only on the energy of the incident beam in the center of mass. In other words, this second pole is independent of the form of the potential. Using the residue theorem to calculate the integral in Eq. (14.110), we get

$$\begin{aligned}\int d^3k\, e^{i\vec{k}\cdot\vec{r}}\,\frac{g(\vec{k})}{E^+-k^2} &= \frac{2\pi^{3/2}}{r}\left\{\left.\frac{k\,e^{ikr}}{(k_0^2+i\epsilon-k^2)\,(k+i\beta)}\right|_{k=i\beta}\right.\\ &\qquad\left.-\left.\frac{k\,e^{ikr}}{(k_0+i\epsilon+k)\,(k^2+\beta^2)}\right|_{k=k_0+i\epsilon}\right\}\end{aligned}$$

$$= \pi^{3/2} \left(\frac{e^{-\beta r} - e^{ik_0 r}}{r} \right) \frac{1}{k_0^2 + \beta^2} . \tag{14.111}$$

We now can write the scattering wave function as

$$\begin{aligned} \langle \vec{r} | \psi^{(+)}_{\vec{k}_i} \rangle &= \langle \vec{r} | \vec{k}_i \rangle - \sqrt{\frac{\pi}{2}} \left(\frac{e^{ik_f r} - e^{-\beta r}}{r} \right) \langle \vec{k}_f | g \rangle \, \tau(E^+) \, \langle g | \vec{k}_i \rangle \\ &= \frac{1}{(2\pi)^{3/2}} \left\{ e^{i\vec{k}_i \cdot \vec{r}} - 2\pi^2 \left(\frac{e^{ik_f r} - e^{-\beta r}}{r} \right) \langle \vec{k}_f | T(E^+) | \vec{k}_i \rangle \right\} , \end{aligned} \tag{14.112a}$$

where we have taken $k_0 = |\vec{k}_i| = |\vec{k}_f|$. This equation can be written in a functional form as

$$\psi^{(+)}_{\vec{k}_i}(\vec{r}) = \frac{1}{(2\pi)^{3/2}} \left\{ e^{i\vec{k}_i \cdot \vec{r}} - 2\pi^2 \left(\frac{e^{ik_f r} - e^{-\beta r}}{r} \right) T(\vec{k}_f, \vec{k}_i; E^+) \right\} . \tag{14.112b}$$

This scattering wave function, which is valid for all $\vec{r}$, is the sum of three parts:

1. The first part is the incident plane wave with momentum $\vec{k}_i$ and given by

   $$e^{i\vec{k}_i \cdot \vec{r}} .$$

   The amplitude of this wave is unity.

2. The second term is a spherical outgoing wave with momentum $\vec{k}_f$ pointing in the radial direction. This part is given by

   $$-4\pi^2 \, T(\vec{k}_f, \vec{k}_i, E^+) \, \frac{e^{ik_f r}}{r} .$$

   The amplitude of this wave is the scattering amplitude given by [12]

   $$f(k_0, \theta) = -4\pi^2 \, T(\vec{k}_f, \vec{k}_i; E^+) ,$$

   with $\cos\theta = \hat{k}_f \cdot \hat{k}_i$.

3. Finally, the scattering wave function has a component that is decaying because it is proportional to

   $$\frac{e^{-\beta r}}{r} .$$

   The rate of decay of this component depends on the parameter of the potential β. This is why we referred to this parameter as the range of the potential. Here we note that the longer the range of the potential, the further we have to go from the scattering center before we get the asymptotic solution that has the scattering amplitude and therefore the phase shifts.

[12] Recall that we have taken $\frac{\hbar^2}{2\mu} = 1$ when comparing this result with Eq. (14.19).

We next turn to the bound state wave function for the form factor given in Eq. (14.105). In momentum space this wave function is of the form

$$
\begin{aligned}
\langle \vec{k}\,|\psi_B\rangle &= \mathcal{N}\frac{\langle \vec{k}\,|g\rangle}{B-k^2} \\
&= -\frac{\mathcal{N}}{\sqrt{4\pi}}\frac{g(k)}{k^2+\alpha^2} \\
&= -\frac{\mathcal{N}}{\sqrt{4\pi}}\frac{1}{(k^2+\alpha^2)(k^2+\beta^2)} \ ,
\end{aligned}
\tag{14.113}
$$

where $\alpha^2 = -B$. In coordinate space this wave function takes the form

$$
\begin{aligned}
\langle \vec{r}\,|\psi_B\rangle &= \int d^3k\, \langle \vec{r}\,|\vec{k}\,\rangle\, \langle \vec{k}\,|\psi_B\rangle \\
&= \frac{1}{(2\pi)^{3/2}}\int d^3k\, e^{i\vec{k}\cdot\vec{r}}\, \langle \vec{k}\,|\psi_B\rangle \ .
\end{aligned}
\tag{14.114}
$$

Making use of Eq. (14.113), we have after performing the angular integration

$$
\langle \vec{r}\,|\psi_B\rangle = -\frac{\mathcal{N}}{\sqrt{2}}\frac{1}{2\pi i}\frac{1}{r}\int_{-\infty}^{+\infty} dk\, \frac{k\, e^{ikr}}{(k^2+\alpha^2)\,(k^2+\beta^2)} \ .
\tag{14.115}
$$

This integral can be converted, as was the case for the scattering wave function, to a contour integral by closing the contour in the upper half k-plane. In this case there are two poles inside the contour. One pole at $k = i\beta$ comes from the form factor and its position is the same as in the scattering wave function case. The second pole at $k = i\alpha$ comes from the Green's function. This pole was on the positive real axis for the scattering case, and has in a continuous manner moved to the positive imaginary axis as we changed our energy from a positive value for the scattering case to a negative value for the bound state. Using the residue theorem we can perform the integration to get

$$
\langle \vec{r}\,|\psi_B\rangle = -\frac{\mathcal{N}}{\sqrt{2}}\frac{1}{2(\beta^2-\alpha^2)}\left(\frac{e^{-\alpha r}-e^{-\beta r}}{r}\right) \ .
\tag{14.116}
$$

This bound state wave function, up to a normalization, is an analytic continuation of the scattered part of the scattering wave function.

14.6 Spin Dependent Scattering

So far we have considered the scattering of particles with no spin. In nature, most scattering experiments will entail either the target the beam or both having spin. To illustrate how spin is included in the calculation of cross section with minimal additional complexity, we consider the problem of nucleon-nucleon scattering as an example in which both the beam and target have spin.

Since the two nucleons have spin one half, the total spin can be zero (singlet) or one (triplet). As a result, the total scattering amplitude is the sum of an the amplitude for scattering in spin singlet and the amplitude for scattering in spin triplet, i.e., we can write our total scattering amplitude as

$$f(k,\theta) = P_s\, f_s(k,\theta) + P_t\, f_t(k,\theta) \ , \tag{14.117}$$

where P_s and P_t are the spin singlet and spin triplet projection operators. Making use of the facts that the total spin $\vec{S} = \vec{s}_1 + \vec{s}_2$ and that the nucleon spin can be written in terms of the Pauli spin matrices as $\vec{s}_i = \frac{1}{2}\hbar\vec{\sigma}(i)$, we can determine the matrix elements of $\vec{\sigma}(1)\cdot\vec{\sigma}(2)$ between states of total spin to be

$$\langle\vec{\sigma}(1)\cdot\vec{\sigma}(2)\rangle = \left\{ \begin{array}{ll} -3 & \text{for} \quad S = 0 \\ +1 & \text{for} \quad S = 1 \end{array} \right. . \tag{14.118}$$

With this result in hand, we can write the spin single and spin triplet projection operators as

$$P_s = \frac{1}{4}\left[1 - \vec{\sigma}(1)\cdot\vec{\sigma}(2)\right] \tag{14.119a}$$

$$P_t = \frac{1}{4}\left[3 + \vec{\sigma}(1)\cdot\vec{\sigma}(2)\right] \ . \tag{14.119b}$$

As we would expect, the projection operators satisfy the condition $P_s + P_t = 1$. Thus if the singlet amplitude was the same as the triplet amplitude, i.e., $f_s = f_t$, then we would not have any spin dependence in the interaction.

The cross section for scattering in this case will depend on our measurement of the direction of the spin in the initial and final state. If we carry out no measurement on the orientation of the spin in the initial and final state, then we need to average over all possible orientations of the spin of the two particles in the initial state, and sum over all possible orientations of the spin of the final two particles, i.e., the differential cross is given by section

$$\frac{d\sigma}{d\Omega} = \frac{1}{(2s_1+1)(2s_2+1)} \sum_{m_1\,m_2} \sum_{m_3\,m_4} \left|\langle m_1, m_2|f(k,\theta)|m_3, m_4\rangle\right|^2 \ , \tag{14.120}$$

where m_1 and m_2 are the projection of the spin s_1 and s_2 of the two particles in the initial state, and m_3 and m_4 are the projection of the spin of the particles in the final states. This result is valid for the scattering of any two particles of spin s_1 and s_2 which produces two particles with spin s_3 and s_4. For the case of two nucleon scattering, this equation reduces to

$$\frac{d\sigma}{d\Omega} = \frac{1}{4} \sum_{m_1\,m_2} \sum_{m_3\,m_4} \left|\langle m_1, m_2|f(k,\theta)|m_3, m_4\rangle\right|^2 \ . \tag{14.121}$$

If we couple the spin in the initial and final states to states of total spin, and take into consideration the fact that spin is a good quantum number so that the matrix elements of the amplitude are diagonal in the total spin, we can write

$$\begin{aligned}\langle m_1, m_2|f(k,\theta)|m_3, m_4\rangle &= \sum_{S\,M_S} (\tfrac{1}{2}m_1, \tfrac{1}{2}m_2|\tfrac{1}{2}\tfrac{1}{2}SM_S)\,(\tfrac{1}{2}m_3, \tfrac{1}{2}m_4|\tfrac{1}{2}\tfrac{1}{2}SM_S) \\ &\qquad\qquad \times\langle S|f(k,\theta)|S\rangle\ . \qquad (14.122)\end{aligned}$$

Using the orthogonality of the Clebsch-Gordan coefficients Eqs. (12.66) and (12.67), we can write the cross section as

$$\begin{aligned}\frac{d\sigma}{d\Omega} &= \frac{1}{4}\sum_{S\,M_S} |\langle S|f(k,\theta)|S\rangle|^2 \\ &= \frac{3}{4}|f_t(k.\theta)|^2 + \frac{1}{4}|f_s(k,\theta)|^2\ . \qquad (14.123)\end{aligned}$$

We note again that if $f_t = f_s$, then our expression for the differential cross section reduces to the one for spinless particles, which is what we expect.

14.7 Effective Range Parameters

In Sec. 14.2 we considered the Born approximation which can have good convergence at high energies since it can be thought of as a power series in E^{-1}. We now turn to an approximation that is valid at low energies. For low energies, the scattering amplitude is dominated by S-wave scattering, i.e., $\ell = 0$, and the scattering amplitude can be written (see Eqs. (10.45) and (10.46)) as

$$\begin{aligned}f(k,\theta) &= \frac{1}{k}\sum_{\ell}(2\ell+1)\,e^{i\delta_\ell}\sin\delta_\ell\,P_\ell(cos\theta) \\ &\approx \frac{1}{k}\,e^{i\delta_0}\sin\delta_0 \\ &= \frac{1}{k\cot\delta - ik}\ , \qquad (14.124)\end{aligned}$$

where in the last line we dropped the subscript 0 on the δ with the understanding that we will be considering S-wave scattering only. Since we are considering low energy scattering, we should be able to expand $k\cot\delta$ as a power series in the energy or k^2, i.e.,

$$k\cot\delta = -\frac{1}{a} + \frac{1}{2}r_e k^2 + \cdots\ . \qquad (14.125)$$

This series is known as the effective range expansion, with a being the *scattering length* and r_e the *effective range*.

To develop this expansion, we consider the radial Schrödinger equation at two energies E_1 and E_2, i.e.,

$$\frac{d^2u_1}{dr^2} + \frac{2\mu}{\hbar^2}\left[E_1 - V(r)\right]u_1 = 0\ , \tag{14.126a}$$

and

$$\frac{d^2u_2}{dr^2} + \frac{2\mu}{\hbar^2}\left[E_2 - V(r)\right]u_2 = 0\ , \tag{14.126b}$$

where $V(r)$ is the potential. If we multiply Eq. (14.126a) by u_2 and Eq. (14.126b) by u_1 and subtract, we get

$$u_2\frac{d^2u_1}{dr^2} - u_1\frac{d^2u_2}{dr^2} = \left(k_2^2 - k_1^2\right)u_2(r)\,u_1(r)\ . \tag{14.127}$$

We now integrate this equation to get

$$\int_0^{r_0} dr\left\{u_2\frac{d^2u_1}{dr^2} - u_1\frac{d^2u_2}{dr^2}\right\} = \left(k_2^2 - k_1^2\right)\int_0^{r_0} dr\,u_2(r)\,u_1(r)\ .$$

If we now integrate the left hand side of the equation by parts, i.e.,

$$\int_0^{r_0} dr\left\{u_2\frac{d^2u_1}{dr^2} - u_1\frac{d^2u_2}{dr^2}\right\} = \left(u_2(r)\frac{du_1}{dr} - u_1(r)\frac{du_2}{dr}\right)\Bigg|_0^{r_0}$$

we get

$$\left(u_2(r)\frac{du_1}{dr} - u_1(r)\frac{du_2}{dr}\right)\Bigg|_0^{r_0} = \left(k_2^2 - k_1^2\right)\int_0^{r_0} dr\,u_2(r)\,u_1(r)\ . \tag{14.128}$$

If r_0 is taken to be the range of the potential, then for $r > r_0$ the radial wave function takes the form

$$u_i(r) \propto \sin(k_i r + \delta_i) \quad \text{for} \quad i = 1, 2\ . \tag{14.129}$$

We now define the function $v_i(r)$ as

$$v_i(r) \equiv \frac{\sin(k_i r + \delta_i)}{\sin\delta_i} \qquad \text{for} \quad i = 1, 2\ . \tag{14.130}$$

Since the function $v_i(r)$ satisfies the Schrödinger equation, it also satisfies the equation

$$\left(v_2(r)\frac{dv_1}{dr} - v_1(r)\frac{dv_2}{dr}\right)\Bigg|_0^{r_0} = \left(k_2^2 - k_1^2\right)\int_0^{r_0} dr\,v_1(r)\,v_2(r)\ . \tag{14.131}$$

If $u(r)$ is chosen such that $u(r) = v(r)$ for $r \geq r_0$, and since $u(0) = 0$ from the boundary condition at the origin and $v_i(0) = 1$, then subtracting Eq. (14.128) from Eq. (14.131) we get

$$\left(\frac{dv_2}{dr} - \frac{dv_1}{dr}\right)\bigg|_{r=0} = \left(k_2^2 - k_1^2\right) \int_0^\infty dr\, [v_1(r)\, v_2(r) - u_1(r)\, u_2(r)] \ . \tag{14.132}$$

But we have that

$$\frac{dv_i}{dr}\bigg|_{r=0} = k \cot \delta_i \ , \tag{14.133}$$

and therefore

$$k_2 \cot \delta_2 = k_1 \cot \delta_1 + \left(k_2^2 - k_1^2\right) \int_0^\infty dr\, [v_1(r)\, v_2(r) - u_1(r)\, u_2(r)] \ . \tag{14.134}$$

Taking $k_1 \to 0$ and $k_2 \to k$, we get

$$k \cot \delta = -\frac{1}{a} + \frac{1}{2} k^2 \rho(0, E) \ , \tag{14.135}$$

where

$$\begin{aligned} \frac{1}{2}\rho(E_1, E_2) &= \int_0^\infty dr\, [v_1(r)\, v_2(r) - u_1(r)\, u_2(r)] \\ &= \int_0^{r_0} dr\, [v_1(r)\, v_2(r) - u_1(r)\, u_2(r)] \ . \end{aligned} \tag{14.136}$$

For low energies, the wave function $u(r)$ is not a very sensitive function of E. Thus we can define the effective range as

$$r_e = \rho(0, 0) \ , \tag{14.137}$$

and we have

$$k \cot \delta = -\frac{1}{a} + \frac{1}{2} r_e k^2 + \cdots \ . \tag{14.138}$$

If we now go through the above procedure with $E_2 = -B$, where B is the binding energy of the two-body system, then

$$v_2(r) = e^{-\alpha r} \qquad \text{where} \quad \alpha^2 = \frac{2\mu B}{\hbar^2} \ , \tag{14.139}$$

and then

$$\frac{dv_2}{dr}\bigg|_{r=0} = -\alpha \ . \tag{14.140}$$

In this way we can write the binding energy in terms of the effective range parameters, i.e.,

$$\begin{aligned} -\alpha &= -\frac{1}{a} - \frac{1}{2}\alpha^2\rho(0,B) \\ &\approx -\frac{1}{a} - \frac{1}{2}\alpha^2\rho(0,0) = -\frac{1}{a} - \frac{1}{2}\alpha^2 r_e \ . \end{aligned} \tag{14.141}$$

In other words we can use this effective range expansion to determine the behavior of the system in the neighborhood of the origin in the energy variable. We will next apply this effective range expansion to neutron-proton and proton-proton scattering.

With the above results, we are in a position to write the phase shifts in terms of the cross section measured experimentally. Since we are considering low energy scattering, i.e., $E < 10$ MeV and therefore $k < \frac{1}{2}$, the only important partial wave is the $\ell = 0$, and we can use effective range theory as a parameterization of the S-wave phase shifts or $k \cot\delta$, i.e.,

$$k\cot\delta = -\frac{1}{a} + \frac{1}{2}k^2 r_e - Pk^4 r_e^3 + \cdots \ , \tag{14.142}$$

where

$$\begin{aligned} a &= \text{Scattering length} \\ r_e &= \text{Effective range} \\ P &= \text{Shape parameter} \ . \end{aligned}$$

Since we have that

$$\lim_{k\to 0} f(k,\theta) = -a$$

then the differential cross section at zero energy is given by

$$\frac{d\sigma}{d\Omega} = a^2$$

and the total cross section is

$$\lim_{E\to 0} \sigma_T = 4\pi a^2 \ . \tag{14.143}$$

Thus the scattering length is a measure of the cross section at zero energy.

We now turn to the determination of the scattering length for the singlet and triplet. First let us consider the scattering of neutrons from molecular hydrogen. Molecular hydrogen is a mixture of

ortho-hydrogen	parallel spin for the two protons. $\uparrow\uparrow$
para-hydrogen	antiparallel spin for the two protons. $\uparrow\downarrow$.

The para-hydrogen is a lower energy state than the ortho-hydrogen, and therefore as we lower the temperature of molecular hydrogen more of the molecules drop into the lower energy state of para-hydrogen until at a temperature of $\approx 20^\circ$ K, molecular hydrogen is predominantly para-hydrogen. We can take advantage of this property of molecular hydrogen to carry out experiments in which neutrons are scattered at low energy from the two protons in the molecule coherently and in this way determine the singlet and triplet scattering length.

The scattering amplitude in Eq. (14.117) can be written with the help of Eq. (14.119) as

$$f(k,\theta) = \frac{1}{4}\left(3f_t + f_s\right) + \frac{1}{4}\left(f_t - f_s\right)\vec{\sigma}(1)\cdot\vec{\sigma}(2) \ . \tag{14.144}$$

At very low energies we have that $f \approx -a$ and we can write an effective scattering length in terms of the singlet and triplet scattering length, i.e.,

$$\begin{aligned} a &= \frac{1}{4}\left(3a_s + a_t\right) + \frac{1}{4}\left(a_t - a_s\right)\vec{\sigma}_n\cdot\vec{\sigma}_p \\ &= \frac{1}{4}\left(3a_s + a_t\right) + \left(a_t - a_s\right)\vec{s}_n\cdot\vec{s}_p \ , \end{aligned} \tag{14.145}$$

where $\vec{s}_n$ and $\vec{s}_p$ are the spin operators for the neutron and proton respectively. For the low energy scattering of neutrons from molecular hydrogen, the scattering off the two protons in the molecule is coherent, and in that case we can write the effective scattering length as

$$\begin{aligned} a_{H_2} &= \frac{1}{2}\left(3a_t + a_s\right) + \left(a_t - a_s\right)\vec{s}_n\cdot\left(\vec{s}_{p1} + \vec{s}_{p2}\right) \\ &= \frac{1}{2}\left(3a_t + a_s\right) + \left(a_t - a_s\right)\vec{s}_n\cdot\vec{S}_{H_2} \ . \end{aligned} \tag{14.146}$$

The cross section involves squaring this and averaging over spin to get [13]

$$a^2 = \frac{1}{4}\left(3a_t + a_s\right)^2 + \frac{1}{4}\left(a_t - a_s\right)^2 S_{H_2}(S_{H_2}+1)$$

[13] The squaring of the effective scattering length a gives an expression that includes the operator $\vec{s}_n\cdot\vec{S}_{H_2}$ and its square. The average over spin of this operator is zero, i.e.,

$$\langle\vec{s}_n\cdot\vec{S}_{H_2}\rangle = 0 \ .$$

On the other hand, we have that

$$\begin{aligned} \left(\vec{s}_n\cdot\vec{S}_{H_2}\right)^2 &= \frac{1}{4}\left(\vec{\sigma}_n\cdot\vec{S}_{H_2}\right)\left(\vec{\sigma}_n\cdot\vec{S}_{H_2}\right) \\ &= \frac{1}{4}\left[\vec{S}_{H_2}\cdot\vec{S}_{H_2} + i\vec{\sigma}_n\cdot\left(\vec{S}_{H_2}\times\vec{S}_{H_2}\right)\right] \\ &= \frac{1}{4}\left[\vec{S}_{H_2}\cdot\vec{S}_{H_2} - \vec{\sigma}_n\cdot\vec{S}_{H_2}\right] \ . \end{aligned}$$

Averaging this over spin, we get zero from the second term on the right hand side, while the first term gives $S_H(S_H+1)/4$.

$$= \begin{cases} \frac{1}{4}\left(3a_t + a_s\right)^2 + \frac{1}{2}\left(a_t - a_s\right)^2 & \text{for ortho-hydrogen } S_H = 1 \\ \frac{1}{4}\left(3a_t + a_s\right)^2 & \text{for para-hydrogen } S_H = 0 \end{cases} . \quad (14.147)$$

From experiment, we have that

$$\sigma_{\text{para}} \approx 4 \text{ barns} \quad (14.148a)$$

$$\sigma_{\text{ortho}} \approx 125 \text{ barns} \quad (14.148b)$$

where 1 barn $= 10^{-24}$ cm^2. Comparing these cross sections with the expression given in Eq. (14.147), we may conclude that $a_s < 0$ if $a_t > 0$ and vice versa. We now can determine the sign of a_t from the knowledge that the triplet $\ell = 0$ channel has a bound state, the deuteron, with a binding energy of 2.2246 MeV. Since this is a small binding on the nuclear scale, we could use the effective range expansion given in Eq. (14.141), i.e.,

$$\alpha = \frac{1}{a} + \frac{1}{2}\alpha^2 r_e \quad (14.149)$$

to determine the scattering length. For a binding energy of 2.2246 MeV, we have that $\alpha = 0.23$, and therefore taking the effective range to be of the order of magnitude of the range of the nucleon-nucleon interaction, i.e., $r_e \approx 1.4$ fm, we get the triplet scattering length to be

$$a_t \approx 5.15 \text{ fm} .$$

Since the triplet scattering length is positive, then the singlet scattering length must be negative. To determine the magnitude of singlet scattering length, we make use of the fact that according to our estimate the triplet total cross section is given by

$$\sigma_t \approx 4\pi a_t^2 \approx 3.0 \text{ barnes} .$$

But the experimental total cross section is given by

$$\sigma_T = \frac{1}{4}\sigma_s + \frac{3}{4}\sigma_t \approx 20. \text{ barns} ,$$

and therefore

$$\sigma_s = 4\sigma_T - 3\sigma_t \approx 71 \text{ barns.}$$

This corresponds to a singlet scattering length of

$$a_s \approx -24 \text{ fm.}$$

A more accurate analysis of the data gives us the triplet and singlet neutron-proton scattering length at [52]

$$a_t = 5.419 \pm 0.007 \text{ fm} \qquad a_s = -23.720 \pm 0.02 \text{ fm} . \quad (14.150a)$$

To extract the effective range, we need to get measurements of the cross section at low energy and analyze the data using the effective range expansion. This gives, for the effective ranges for the triplet and singlet channel,

$$r_{et} = 1.753 \pm 0.008 \text{ fm} \qquad r_{es} = 2.77 \pm 0.05 \text{ fm} . \tag{14.150b}$$

When comparing the zero energy cross section for the singlet (71 barns) and triplet (3 barns), we find that the singlet cross section is much larger than the triplet cross section. To see if this is due to a bound state or "resonance" near the zero of the energy, we use the scattering length and effective range in Eq. (14.133) to determine the binding energy in that state, i.e., we solve the equation

$$\alpha_s = \frac{1}{a_s} + \frac{1}{2}\alpha_s^2 r_{es} .$$

This equation is a quadratic in α_s with two solutions corresponding to

$$\alpha_s = 0.04 \text{ fm} \quad \text{and} \quad 1.39 \text{ fm} .$$

Clearly, $\alpha_s = 1.39$ fm is not a solution, since there is no such nucleus with spin zero. However the large cross section at zero energy suggests that there is a *virtual* or *anti-bound* singlet state at

$$\alpha_s = -0.05 \text{ fm} .$$

The wrong sign for α_s in our analysis is due to the fact that the singlet cross section σ_s does not determine the sign of the scattering length.

14.8 Problems

1. Show that the Green's function with the boundary condition corresponding to incoming spherical wave, i.e. $E \to E - i\epsilon$ is given by

$$G(\vec{r}, \vec{r}\,'; E^-) = \frac{1}{4\pi}\frac{2\mu}{\hbar^2}\frac{e^{-ik_0|\vec{r}-\vec{r}\,'|}}{|\vec{r}-\vec{r}\,'|} ,$$

   where $E = \frac{\hbar^2 k_0^2}{2\mu}$. Show explicitly that the corresponding wave function for large r has a spherical incoming wave.

2. The potential between a proton and a neutron is due to the exchange of a pion. This potential is given by

$$V(r) = V_0 \frac{e^{-m_\pi r}}{r} ,$$

   where m_π is the mass of the pion and V_0 is a constant.

(a) Calculate the Born amplitude for proton-neutron scattering.

(b) Write the differential cross section for proton-neutron scattering, in the Born approximation, as a function of the scattering angle.

3. The potential for electron-hydrogen scattering can be written as

$$V(r) = -\frac{e^2}{r} + \int d^3r' |\phi_{100}(r')|^2 \, \frac{e^2}{|\vec{r} - \vec{r}'|}$$

(a) Using the expression

$$\frac{1}{|\vec{r} - \vec{r}'|} = 4\pi \sum_{\ell m} \frac{1}{2\ell + 1} \frac{r_<^\ell}{r_>^{\ell+1}} Y_{\ell m}(\hat{r}) \, Y^*_{\ell m}(\hat{r}')$$

simplify the potential $V(r)$ taking for the wave function $\phi_{100}(r)$ the ground state of the hydrogen atom.

(b) Calculate the differential cross section in the Born Approximation as a function of the angle at intervals of $10°$ for $0 < \theta < 60°$, with incident electrons of energy of 1 KeV.

4. Consider the S-wave separable potential

$$V(k, k') = g(k) \, \lambda \, g(k') \ ,$$

with $g(k) = (k^2 + \beta^2)^{-1}$.

(a) Calculate the T-matrix for S-wave scattering, and show that this T-matrix satisfies unitarity.

(b) Write an expression for $k \cot \delta$ as a polynomial in k. If the scattering length a, and the effective range r_e are defined by the expansion

$$k \cot \delta = -\frac{1}{a} + \frac{1}{2} r_e \, k^2 + O(k^4) \ ,$$

determine a and r_e in terms of the parameters of the potential λ and β.

(c) For the neutron-proton system in the triplet state, i.e. $S = 1$, there is a bound state, the deuteron. Determine the binding energy for the parameters

$$\beta = 1.4384 \text{ fm}^{-1} \ , \qquad C = -1.07339 \, \hbar c \text{ fm}^{-2} \ .$$

Hint: The units of the potential strength C are chosen such that

$$\frac{2\mu c^2}{\hbar c} = \frac{939 \text{ MeV}}{\hbar c} = 4.76 \text{ fm}^{-1}$$

(d) Calculate the phase shifts for this potential for energies between zero and 200 MeV.

Appendix A

Atomic Units

In most fields of physics the units are chosen such that all observables are of the order one. This facilitates most computational problems by avoiding underflows and overflows on modern computers. In Atomic physics the most common choice of units are atomic units in which $\hbar = e = m_e = 1$, where e is the magnitude of the charge on the electron, and m_e is the mass of the electron. To see how these units come about, let us consider the Schrödinger equation for the Coulomb problem as a model of the Hydrogen atom. In units where length is measured in nm ($1\ nm = 10^{-9}\,m$) or fm ($1\ fm = 10^{-15}\,m$), and energy is measured in eV or MeV, we have

$$\left[-\frac{\hbar^2}{2m_e}\nabla^2 - \frac{e^2}{r}\right]\psi = E\,\psi\ . \tag{A.1}$$

In these units $\hbar c = 197.3\ nm\ eV = 197.3\,fm\ MeV$. The Bohr radius is then given by

$$a_B = \frac{\hbar^2}{m_e e^2} = 0.0529\ nm\ , \tag{A.2}$$

while the Rydberg, R_∞, is given by

$$R_\infty = \frac{m_e e^4}{2\hbar^2} = 13.6058\ eV\ . \tag{A.3}$$

We now can write the Schrödinger equation for the Coulomb potential as

$$\left[-\frac{1}{2}\nabla^2 - \frac{m_e\,e^2}{\hbar^2}\,\frac{1}{r}\right]\psi = \frac{m_e\,E}{\hbar^2}\,\psi\ , \tag{A.4}$$

or

$$\left[-\frac{1}{2}\nabla^2 - -\frac{1}{a_B\,r}\right]\psi = \frac{m_e\,E}{\hbar^2}\,\psi\ , \tag{A.5}$$

and multiply this equation by a_B^2 to get

$$\left[-\frac{a_B^2}{2}\nabla^2 - \frac{a_B}{r}\right]\psi = \left(\frac{\hbar^2}{m_e e^2}\right)^2\frac{m_e E}{\hbar^2}\,\psi = \frac{E}{2R_\infty}\,\psi\ . \tag{A.6}$$

Iraj R. Afnan

If we now measure lengths in units of a_B and energies in units of $2R_\infty$, our equation reduces to

$$\left[-\frac{1}{2}\nabla^2 - \frac{1}{r}\right]\psi = E\,\psi\ . \tag{A.7}$$

This is equivalent to taking

$$\hbar = e = m_e = 1 \tag{A.8}$$

in the Schrödinger Eq. (A.1). In these units the ground state energy of the Hydrogen atom is $-\frac{1}{2}$.

Appendix B

Complex Analysis in a Nutshell

In this appendix we would like to very briefly summarize those features of complex analysis often encountered in a course on quantum mechanics:

1. The Fourier transforms of the wave function from coordinate to momentum space, and vice versa, often requires the use of Cauchy's theorem and the analytic structure of integrand in the complex plane.

2. The determination of the Green's function for the differential equation, and in particular the free Green's function, requires the evaluation of integrals that can be determined with the help of Cauchy's theorem. This in turn allows the inclusion of the boundary condition of the differential equation into the Green's function.

3. In scattering theory the amplitude for scattering is a complex function of the energy. The structure of this function often reflects the physics of the scattering process.

In all three cases a sound understanding of complex analysis, and in particular the use of Cauchy's theorem plays a central role in understanding the mathematical framework, as well as the physical phenomena.

B.1 Functions of a Complex Variable

We can represent a complex variable either in rectangular coordinates as $z = x + iy$, or in polar coordinates as $z = re^{i\theta}$. These two representations of a complex number are related by

$$r = \sqrt{x^2 + y^2} \qquad \tan\theta = \frac{y}{x} \; . \tag{B.1}$$

We now can define a function of a complex variable as $f(z)$, e.g.,

$$\begin{aligned} f(z) &= z^2 = x^2 - y^2 + 2ixy = r^2\, e^{2i\theta} \\ &= f_R(x, y) + i f_I(x, y) \end{aligned} \tag{B.2}$$

Iraj R. Afnan

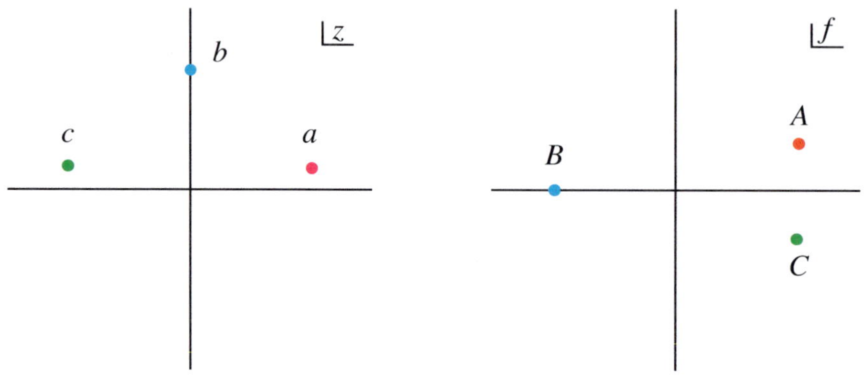

Figure B.1: The complex z-plane and the corresponding f-plane, where $f(z) = z^2$. The points a, b and c in the z-plane are mapped into the points A, B and C in the f-plane.

where

$$f_R(x,y) = x^2 - y^2 = r^2 \cos 2\theta \quad f_I(x,y) = 2xy = r^2 \sin 2\theta \ . \tag{B.3}$$

In other words for every point (x, y) in the complex z-plane we have a point (f_R, f_I) in the complex f-plane, see Figure B.1. However, we observe in this example that the points in the upper half of the z-plane are mapped onto the full f-plane. Similarly, the points in the lower half of the z-plane are mapped onto the full f-plane. In fact, the points $+z$ and $-z$ are mapped onto the same point in the f-plane.

The inverse transformation

$$f(z) = z^{1/2} = r^{1/2}\, e^{i\theta/2} \tag{B.4}$$

has the problem that for a given z we have two possible values of $f(z)$, i.e., the function is multi-valued. For every value of (x, y) we have the two values of $f(z) = (f_R, f_I)$ corresponding to $\theta/2$ and $(\theta + 2\pi)/2$. In this case the z-plane is mapped onto the upper half of the f-plane **OR** the lower half of the f-plane, see Figure B.2.

There are many other mappings that we could consider, e.g.,

$$f(z) = \log(z) = \log(re^{i\theta}) = \log r + i\theta \ . \tag{B.5}$$

In this case, replacing $\theta \to \theta + 2n\pi$ does not change z but changes $f(z)$. In fact we should write

$$f_n(z) = \log(z) = \log r + i(\theta + 2n\pi) \ . \tag{B.6}$$

In this case for a given value of z we can generate an infinite number of possible values of $f(z) = \log z$ by changing the integer n.

To overcome this problem of the multi-valued nature of mappings, to allow for the function $f(z)$ to change in a continuous manner when z is varied continuously, Riemann suggested the following procedure, whereby he assigned the same values of $f(z)$ corresponding to different values of z to different complex planes. To illustrate this, consider

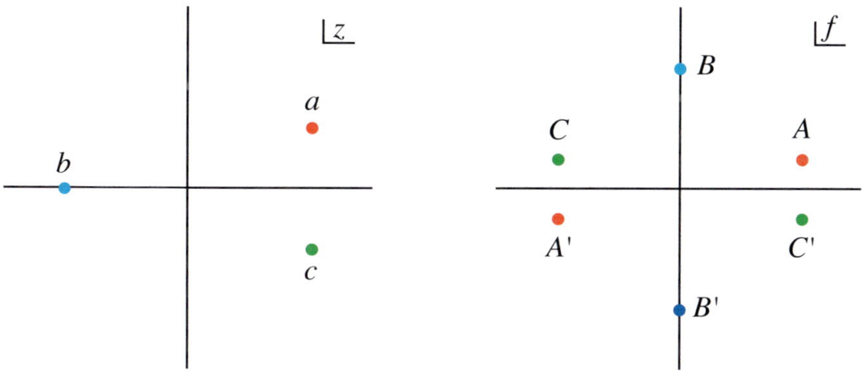

Figure B.2: The complex z-plane and the corresponding f-plane, where $f(z) = z^{1/2}$. The points a, b and c in the z-plane are mapped into the points A, B and C, or the points A', B' and C'.

the function

$$f(z) = z^2 .$$

We could assign the first complex f-plane to those values of z corresponding to the upper half of the z-plane, i.e., $Im(z) > 0$, while the second complex f-plane could correspond to those values of z with $Im(z) < 0$. If we now write z in polar coordinates, i.e., $z = re^{i\theta}$, then for $r > 0$ as we increase θ from 0 to $\pi/4$ to $\pi/2$ to $3\pi/4$ to π, the point $f(z)$ in the f-plane proceeds from the first quadrant to the second, third and then fourth quadrant of that plane. Now as we continue to vary the point z by changing θ from $\theta = \pi - \epsilon$ to $\theta = \pi + \epsilon$ with $\epsilon \ll 1$, we move from the fourth quadrant of the first f-plane, to the first quadrant of the second f-plane. In this way we get a continuous change in $f(z)$ as we change z. To get back onto the first f-plane we need change θ from π to $3\pi/2$ to 2π. This change in θ from π to $3\pi/2$ to 2π will allow us to scan the second f-plane. As we proceed from $\theta = 2\pi - \epsilon$ to $\theta = 2\pi + \epsilon$, which is the same as $\theta = \epsilon$, we return from the second f-plane to the first f-plane. Thus we have two planes that are continuously joined. These planes are referred to as *Riemann sheets* and we say that the function $f(z) = z^2$ has a two Riemann sheet structure. The point $r = 0$ is a problem, in that it either belongs to both sheets, or the function $f(z)$ is not continuous at this point. This point is known as a *branch point.* The lines along which the two planes join is known as a *branch cut.* In the above analysis and Figure B.3 this branch cut is taken to be along the positive real axis. However, the branch cut could be from the point $z = 0$ in any radial direction to the infinity circle.

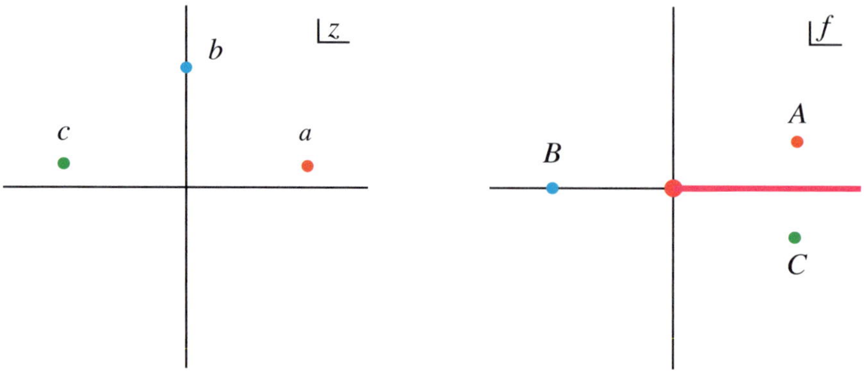

Figure B.3: The complex z-plane and the corresponding f-plane, where $f(z) = z^2$. In the f-plane we have a branch point at the origin and a branch cut along the positive real axis.

B.2 Analytic Functions

In the last section we considered the function of a complex variable $f(z)$ that can be written as

$$f(z) = f_R(x, y) + i f_I(x, y) \ , \tag{B.7}$$

where both $f_R(x, y)$ and $f_I(x, y)$ are real functions of (x, y). We now define the derivative of this function of a complex variable as

$$\frac{df}{dz} = \lim_{h \to 0} \frac{f(z + h) - f(z)}{h} \ , \tag{B.8}$$

where h is a complex number. Unlike a function of a real variable, there are an infinite number of ways we can approach the point z in the complex z-plane. The derivative of the function $f(z)$ along each direction could be different, in which case the concept of a derivative is not meaningful. Thus for the derivative to be meaningful, we have to make sure that the above definition of a derivative does not depend on the direction. This can be achieved by requiring that the derivative of $f(z)$ be the same with respect to the two directions, x and iy, in the complex z-plane. We have that the derivative with respect to the $Re(z) = x$ is

$$\frac{\partial f}{\partial x} = \frac{\partial f_R}{\partial x} + i \frac{\partial f_I}{\partial x} \ , \tag{B.9}$$

while the derivative with respect to the $Im(z) = iy$ is

$$\frac{1}{i} \frac{\partial f}{\partial y} = \frac{1}{i} \left(\frac{\partial f_R}{\partial y} + i \frac{\partial f_I}{\partial y} \right) = \frac{\partial f_I}{\partial y} - i \frac{\partial f_R}{\partial y} \ . \tag{B.10}$$

For these two derivatives to be identical we require that,

$$\frac{\partial f_R}{\partial x} = \frac{\partial f_I}{\partial y} \qquad \frac{\partial f_I}{\partial x} = -\frac{\partial f_R}{\partial y} \ . \tag{B.11}$$

These two conditions are known as the *Cauchy-Riemann condition.* If these conditions are satisfied, we can say that the function $f(z)$ is differentiable, and if the function $f(z)$ is differentiable at z it is said to be an *analytic function* at z. If the function $f(z)$ is not differentiable at z, then the function has a *singularity* at z.

If the function $f(z)$ is analytic for all z except for $z = a$, then the point $z = a$ is an *isolated singularity.* If the function $f(z)$ now behaves like $(z - a)^{-n}$ with the integer $n > 0$, for z in the neighborhood of a, then the singularity of $f(z)$ is an nth-*order pole*, and for $n = 1$ we have a *simple pole.*

B.3 Cauchy's Theorem

We now turn to the integral of a complex function $f(z)$. For functions of a real variable, the integral was defined on a segment of a line, e.g., the integral

$$\int_a^b dx\, f(x) \tag{B.12}$$

is over the interval $a < x < b$. On the other hand, for functions of a complex variable z, the integral is from one point in the z-plane (x_1, y_1) to another point in the z-plane (x_2, y_2) along a specified path. Thus the integral of a function of a complex variable can be written as

$$\int_C dz\, f(z) = \int_C (dx + idy)\, [f_R(x, y) + if_I(x, y)] \ , \tag{B.13}$$

where C specifies the path of integration in the complex z-plane. If the path is a closed path specified by C, we write the integral as

$$\oint_C dz\, f(z) \ . \tag{B.14}$$

If the function $f(z)$ is analytic in a domain D, and the contour C is in this domain and is piecewise continuous, see Figure B.4, then

$$\oint_C dz\, f(z) = 0 \ . \tag{B.15}$$

This is known as the *Cauchy Integral Theorem.* This theorem has strong implications to the extent that we can deform the integration path C without changing the value of the integral, **provided** the function $f(z)$ is analytic over the domain in which the contour of integration resides. However, we can not deform the contour of integration to include new poles or singularities of the function $f(z)$ without changing the value of the integral. The simplest case of an integral over a contour which includes a singularity is the case where the singularity is a simple pole, i.e.,

$$\oint_C dz \frac{1}{z - a} \ ,$$

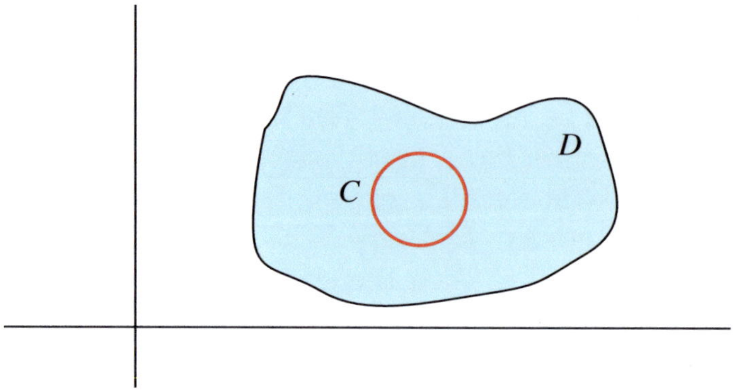

Figure B.4: The function $f(z)$ is analytic over the domain D, the circle labeled C is the contour of integration. This contour is in the domain D.

where the point $z = a$ is inside the contour C. We now can change the contour C to be a circle of radius ϵ centered about the pole at $z = a$. The integral in this case can be written as an integral over an angle θ. This can be achieved by defining $z - a = \epsilon e^{i\theta}$, with $dz = i\epsilon e^{i\theta}\, d\theta$, so that the integral over the contour C can be written as

$$\oint_C dz \frac{1}{z-a} = i \int_0^{2\pi} d\theta = 2\pi i \ . \tag{B.16}$$

In the event that we have a higher order pole we have

$$\begin{aligned} \oint_C dz \frac{1}{(z-a)^n} &= i \int_0^{2\pi} d\theta \, (\epsilon e^{i\theta})^{1-n} \\ &= i\epsilon^{1-n} \int_0^{2\pi} [\cos(n-1)\theta - i \sin(n-1)\theta] = 0 \quad \text{for} \quad n \neq 1 \ . \end{aligned} \tag{B.17}$$

We now consider the more general case where we have a function $f(z)$ that is analytic over the domain D, with the contour C also in this domain. Then we have that

$$\oint_C dz \frac{f(z)}{z-a} = 2\pi i f(a) \ . \tag{B.18}$$

To prove this result we collapse the contour to a circle of radius $\epsilon \ll 1$. In this way the function $f(z)$ can be taken out of the integral and given the value $f(a)$. Now with the help of Eq. (B.16) we have established the result in Eq. (B.18). Eq. (B.18) is a special case of a more general result where we can write the value of a function $f(z)$ in terms of a contour integral that includes the point z, i.e.,

$$f(z) = \frac{1}{2\pi i} \oint_C dz' \frac{f(z')}{z'-z} \tag{B.19}$$

This result is often used in physics to determine the value of a function at a point z in the complex plane given the measured values of the function along the contour C.

B.4 Taylor and Laurent Series

For functions of a real variable we can expand a function $f(x)$ about a point $x = a$ as a Taylor series of the form

$$f(x) = f(a) + (x-a)\frac{df}{dx}\bigg|_{x=a} + \cdots + \frac{(x-a)^n}{n!}\frac{d^n}{dx^n}\bigg|_{x=a} + \cdots \ . \tag{B.20}$$

This series converges for $|x - a| < 1$. We can carry over this Taylor expansion to functions of a complex variable z by writing

$$f(z) = \sum_{n=0}^{\infty} c_n (z-a)^n \ . \tag{B.21}$$

This series has a radius of convergence that is given by $|z - a| < 1$. Since the right hand side of this expansion is analytic, we are restricted to the application of this expansion over domains where the function $f(z)$ is analytic. To extend this expansion to include functions that have singularities at a point a, with this singularity being either an nth order pole or essential singularity, we introduce the Laurent series, i.e.,

$$f(z) = \sum_{n=-\infty}^{+\infty} c_n \, (z-a)^n \ . \tag{B.22}$$

Then the coefficient c_m can be determined by multiplying Eq. (B.22) by $(z-a)^{-(m+1)}$ and integrating the resultant expression over a closed contour C that includes the point $z = a$, i.e.,

$$\begin{aligned}
\oint_C dz \frac{f(z)}{(z-a)^{m+1}} &= \sum_{n=-\infty}^{+\infty} c_n \oint_C dz\,(z-a)^{n-m-1} \\
&= \sum_{n=-\infty}^{m-1} c_n \oint_C (z-a)^{n-m-1} + c_m \oint_C \frac{dz}{z-a} \\
&\quad + \sum_{n=m+1}^{+\infty} c_n \oint_C dz\,(z-a)^{n-m-1} \ .
\end{aligned} \tag{B.23}$$

The first integral on the right hand side is zero according to Eq. (B.17), while the last integral on the right hand side is zero because the integrand is analytic in the domain that includes the contour C. Using Eq. (B.16) for the remaining integrals on the right hand side we get

$$c_m = \frac{1}{2\pi i} \oint_C dz \, \frac{f(z)}{(z-a)^{m+1}} \ . \tag{B.24}$$

This allows us to determine the coefficients of the Laurent series.

B.5 Residue Theorem

Let us now consider the Laurent series for the function $f(z)$. The integral of this function over a closed contour is given by

$$\begin{aligned} \oint_C dz\, f(z) &= \sum_{n=-\infty}^{+\infty} c_n \oint_C dz\, (z-z_0)^n \\ &= c_{-1} \oint_C \frac{dz}{z-z_0} = 2\pi i c_{-1} \ . \end{aligned} \tag{B.25}$$

We now can define the residue of the function $f(z)$ as

$$Res[f(z_0)] = c_{-1} = \frac{1}{2\pi i} \oint_C dz\, f(z) \ . \tag{B.26}$$

This assumes that $f(z)$ has only one simple pole inside the contour C. In the event that we have several poles inside the contour C, we can write the *Residue Theorem* as

$$\oint_C dz\, f(z) = 2\pi i \sum_{i=1}^{n} Res[f(z_i)] \ , \tag{B.27}$$

where $z_i, i = 1, \ldots, n$ are the positions of poles of $f(z)$ inside the contour C. This Residue Theorem is very useful in calculating Fourier transform and the determination of the Green's function for the Schrödinger equations and the Laplace's equation in Electromagnetic theory.

For further reading on the subject and examples of the application of complex analysis to physical problems you can consider any book on Mathematical Physics or Complex Analysis, e.g., [47], [48], or [49].

B.6 Problems

1. Use the Residue Theorem to calculate the following integrals;

   (a)

   $$\int_{-\infty}^{+\infty} dx\, \frac{1}{(x^2+b^2)^2}$$

   (b)

   $$\int_{-\infty}^{+\infty} dx\, \frac{e^{ikx}}{(x^2+b^2)^2}$$

(c)

$$\int_{-\infty}^{+\infty} dx \frac{\cos kx}{x^2 + a^2}$$

Appendix C

Numerical Solution of the Schrödinger Equation

To calculate the phase shifts for a general central potential $V(r)$, we need to numerically solve the Schrödinger equation, Eq. (10.33), for the radial wave function $\psi_\ell(r)$. This is achieved by integrating the differential equation from the origin up to a point r_0 beyond the range of the potential $V(r)$, and then by matching this numerical solution, and its first derivative, to the asymptotic solution, which is given in Eq. (10.39). In this appendix we present the numerical procedure for integrating the Schrödinger equation for the radial wave function.

If we write the radial wave function, $\psi_\ell(r)$, as

$$\psi_\ell(r) = \frac{u_\ell(r)}{r} \tag{C.1}$$

then the Schrödinger equation for $u_\ell(r)$ can be written as a second order differential equation without a first order derivative, and of the form

$$\frac{d^2 u_\ell}{dr^2} + w(r)\, u_\ell(r) = 0 \ , \tag{C.2}$$

or

$$u''_\ell(r) + w(r)\, u_\ell(r) = 0 \ , \tag{C.3}$$

where

$$w(r) = \frac{2\mu}{\hbar^2}\,[E - V(r)] - \frac{\ell(\ell+1)}{r^2} \ . \tag{C.4}$$

Here μ is the reduced mass, E is the energy, and ℓ the angular momentum.

To solve the above differential equation numerically, we are going to take advantage of the fact that the equation does not include terms proportional to the first derivative, $u'_\ell(r)$. This will improve the numerical accuracy of the procedure. In general, to solve a

Iraj R. Afnan

differential equation numerically, we convert the equation to a difference equation, and then solve the difference equation. This involves writing the first derivative as

$$\begin{aligned}\frac{du}{dr} &= \lim_{h\to 0}\frac{u(r+h)-u(r-h)}{2h}\\ &\approx \frac{u(r+h)-u(r-h)}{2h}\,,\end{aligned} \tag{C.5}$$

where h is taken to be small. Here we note that on most computers the value of the functions $u(r+h)$ and $u(r-h)$ are known to a finite number of significant figures, e.g., we have seven significant figures in single precision on most 32 bit computers. This means that for small h, when we take the difference $\Delta u = u(r+h) - u(r-h)$ in Eq. (C.5), we lose accuracy as a result of the reduction in the number of significant figures to which the difference Δu is known. On the other hand, for h large, the above approximation to the derivative has an error that depends on h. To determine this error let us write a Taylor series expansion for $u(r \pm h)$ about the point r, i.e.,

$$u(r+h) = u + h\,u^{(1)} + \frac{h^2}{2!}u^{(2)} + \cdots = \sum_{n=0}^{\infty}\frac{h^n}{n!}u^{(n)} \tag{C.6}$$

$$u(r-h) = u - h\,u^{(1)} + \frac{h^2}{2!}u^{(2)} + \cdots = \sum_{n=0}^{\infty}(-1)^n\frac{h^n}{n!}u^{(n)}\,, \tag{C.7}$$

where $u^{(n)}$ is the nth derivative of the function u calculated at r. In particular $u^{(0)} = u(r)$. We now can write, the first derivative by using Eq. (C.5), as

$$\frac{1}{2}\left[\,u(r+h) - u(r-h)\,\right] = h\,u^{(1)} + \frac{h^3}{3!}u^{(3)} + \cdots\,. \tag{C.8}$$

This gives us an error of the order of $\frac{h^2}{3!}u^{(3)}$ in the first derivative. In a similar manner we can write the second derivative as

$$\begin{aligned}\frac{d^2u}{dr^2} &= \frac{\frac{1}{h}\left[u(r+h)-u(r)\right] - \frac{1}{h}\left[u(r)-u(r-h)\right]}{h}\\ &= \frac{1}{h^2}\left[\,u(r+h)+u(r-h)-2u(r)\,\right]\\ &= u^{(2)} + \frac{h^2}{12}u^{(4)} + \cdots\,.\end{aligned} \tag{C.9}$$

Here again, we have an error of the order of h^2. To improve our accuracy when replacing the derivative by a difference, without reducing the value of h, we have to define the derivative so that the error is proportional to a higher power of h than Eqs. (C.8) and (C.9) provide us with. This is possible for the Schrödinger equation because we have a second order differential equation with no first order derivative. To see how this can be achieved, let us consider the relation

$$\frac{1}{2}\left[u(r+h)+u(r-h)\right]=u+\frac{h^2}{2!}u^{(2)}+\frac{h^4}{4!}u^{(4)}+\frac{h^6}{6!}u^{(6)}+\cdots\ . \tag{C.10}$$

If we differentiate this equation twice, or take an expansion similar to Eqs. (C.6) and (C.7) for $u''(r)$, we get

$$\frac{1}{2}\left[u''(r+h)+u''(r-h)\right]=u^{(2)}+\frac{h^2}{2!}u^{(4)}+\frac{h^4}{4!}u^{(6)}+\cdots\ . \tag{C.11}$$

We now can eliminate the error of the order of h^4 by multiplying Eq. (C.11) by $\frac{h^2}{12}$ and subtracting the result from Eq. (C.10) to get

$$\begin{aligned}\frac{1}{2}\left[\left(u(r+h)-\frac{h^2}{12}u''(r+h)\right)\right. &+ \left.\left(u(r-h)-\frac{h^2}{12}u''(r-h)\right)\right]\\ &= u(r)+\frac{5h^2}{12}u^{(2)}-\frac{h^6}{480}u^{(6)}+\cdots\end{aligned} \tag{C.12}$$

Using the differential equation, Eq. (C.2), we can write

$$\frac{h^2}{12}u''(x)=-\frac{h^2}{12}w(x)\,u(x)\equiv -t(x)\,u(x)\ . \tag{C.13}$$

This in turn allows us to write Eq. (C.12) as

$$\begin{aligned}(1+t(r+h))\,u(r+h) &= -(1+t(r-h))\,u(r-h)\\ &\quad +(2-10\,t(r))\,u(r)-\frac{h^6}{240}u^{(6)}+\cdots\ .\end{aligned} \tag{C.14}$$

The error in this case is of the order of h^6, rather than h^2. This allows us to consider a much larger value of h yet maintain the accuracy needed for the calculation. Finally we can solve Eq. (C.14) for $u(r+h)$ in terms of $u(r)$ and $u(r-h)$ to get

$$u(r+h)=\frac{(2-10\,t(r))\,u(r)-(1+t(r-h))\,u(r-h)}{1+t(r+h)}\ . \tag{C.15}$$

With this result in hand we can calculate the wave function at $(r+h)$, given the wave function at $(r-h)$ and r. In other words, we need to know the wave function at two points a distance h apart to determine the wave function for all values of r. From Eq. (C.1) we require that $u_\ell(r)$ at $r=0$ has to be zero for the radial wave function $\psi_\ell(r)$ to be finite at the origin. Thus by taking

$$u(0)=0 \qquad \text{and} \qquad u(h)=\text{const.}\ , \tag{C.16}$$

we can calculate the wave function at $u(2h)$. We then can use our knowledge of the wave function at h and $2h$ to calculate the wave function at $r = 3h$. In this way we can integrate the Schrödinger equation to determine the wave function for all r. This procedure gives us the wave function up to the constant used for $u(h)$. For the bound state problem, the value of this constant is determined by the fact that the wave function has to be normalized. For the case when $u(r)$ is the scattering wave function, in order to determine the phase shifts, we need to calculate the logarithmic derivative at some large value of r. In this case the logarithmic derivative is independent of the constant used for $u(h)$.

This leads us to the problem of how to calculate the first derivative of the wave function at large r to the same degree of accuracy used to integrate the differential equation. Here again we resort to the same procedure used above by calculating the difference

$$\frac{1}{2}\left[u(r+h) - u(r-h)\right] = h\,u^{(1)} + \frac{h^3}{3!}u^{(3)} + \frac{h^5}{5!}u^{(5)} + \cdots \,. \tag{C.17}$$

To eliminate the error term proportional to $u^{(3)}$, we write the similar expansion for the second derivative, i.e.,

$$\frac{1}{2}\left[u''(r+h) - u''(r-h)\right] = h\,u^{(3)} + \frac{h^3}{3!}u^{(5)} + \frac{h^5}{5!}u^{(7)} + \cdots \,. \tag{C.18}$$

We now multiply Eq. (C.18) by $\frac{h^2}{3!}$ and subtract the result from Eq. (C.17) to get

$$\begin{aligned}\frac{1}{2}\left[\left(u(r+h) - \frac{h^2}{3!}u''(r+h)\right) \right. & - \left.\left(u(r-h) - \frac{h^2}{3!}u''(r-h)\right)\right] \\ & = h\,u^{(1)} - \frac{7\,h^5}{360}u^{(5)} + \cdots \,.\end{aligned} \tag{C.19}$$

In this way the error in our first derivative is of the order of h^4,[1] rather than h^2. Making use of the differential equation, Eq. (C.2), to replace $u''(r \pm h)$ by $-w(r \pm h)\,u(r \pm h)$, we find that the first derivative at r is

$$u^{(1)} = \frac{1}{2h}\left[\left(1 + \frac{h^2}{3!}w(r+h)\right)u(r+h) - \left(1 + \frac{h^2}{3!}w(r-h)\right)u(r-h)\right] \,. \tag{C.20}$$

The error in this derivative is of the order if $h^4\,u^{(5)}$ which is comparable to the error we had in integrating the differential equation. We now can calculate the logarithmic derivative at $r = r_0$ from the wave function at $r = r_0 \pm h$, which we have from integrating the differential equation starting at the origin and finishing at $r = r_0 + h$.

[1]This can be further improved so that the error in the first derivative is of the order of h^9 [53].

C.1 Problems

1. Calculate the S-wave phase shifts as a function of energy for the potential

$$V(r) = V_0\, e^{-\beta r^2} ,$$

using the numerical procedure described above. For $\alpha - \alpha$ scattering the parameters of the potential are

$$V_0 = -0.62178 \text{ fm}^{-1} \qquad \text{and} \qquad \beta = 0.22 \text{ fm}^{-2} .$$

All masses and energies are in units of fm^{-1}. To convert units of MeV to fm^{-1}, we divide energies and mass by $\hbar c = 197.327$ Mev fm.

Show that in the above problem the phase shifts are not sensitive to the value of r_0, provided r_0 is greater than the range of the potential.

References

[1] W. Heisenberg, *Z. Physik*, **43**, 127 (1927).

[2] W. Heisenberg, *Physics and Beyond: Encounters and conversations*, Harper & Row (1971).

[3] James Gleick, *CHAOS: Making a New Science*, Viking (1987).

[4] J. C. Maxwell, *Treatise on Electricity and Magnetism*, 3rd ed., 2 vols., reprinted by Dover, New York (1954).

[5] M. Planck, *Ann. Physik*, **4**, 553 (1901).

[6] A. Einstein, *Ann. Physik*, **17**, 132 (1905).

[7] A. H. Compton, *Phys. Rev.*, **21**, 715 (1923); **22**, 409 (1923).

[8] Louis de Broglie, *Phil. Mag.*, **47**,446 (1924); Ann. Phys. **3**, 22 (1925).

[9] C. J. Davisson and L. H. Germer, *Phys. Rev.*, **30**, 705 (1927).

[10] J. J. Balmer, *Ann. Physik*, **25**, 80 (1885).

[11] E. Rutherford, *Phil. Mag.*, **21**, 669 (1911).

[12] N. Bohr, *Phil. Mag.*, **26**, 1 (1913).

[13] R.P. Feynman, *Rev. Mod. Phys.*, **20**, 267 (1948).

[14] J. B. J. Fourier *Theory Analytique de la Chaleur* (1822); English translation in *Analyitc Theory of Heat*, Dover, New York (1955).

[15] P. A. M. Dirac, *The Principles of Quantum Mechanics*, 4th Edition, Oxford (1958).

[16] E. Schrödinger, Ann. der Phys. **79**, 361 (1926); English translation in *Collected papers on wave mechanics by E. Schrodinger*, trans. J. Shearer and W. Deans, Blackie, Glasgow (1928).

Iraj R. Afnan

[17] Leonard I. Schiff, *Quantum Mechanics*, 3rd Edition, McGraw-Hill, New York (1968).

[18] O. Klein, *Z. Physik*, **41**, 407 (1927).

[19] W. Gordon, *Z. Physik*, **40**, 117 (1926).

[20] E. Schrödinger, *Proc. Cambridge Phil. Soc.*, **31**, 555 (1935).

[21] A. Einstein, B. Podolsky and N. Rosen, *Phys. Rev.*, **47**, 777 (1935).

[22] George Greenstein and Arther G Zajonc, *The Quantum Challenge: Modern Research on the Foundation of Quantum Mechanics*, Jones and Bartlett, 2nd Ed. (2005).

[23] W. Pauli and V. F. Weisskopf, *Helv. Phys. Acta*, **7**, 709 (1934).

[24] Silvan S. Shweber, *An Introduction to Relativistic Quantum Field Theory*, Harper & Row (1962).

[25] Henri Becquerel, *Comptes Rendus*, **122**, 420 (1896).

[26] G. Gamow, *Z. Physik*, **51**, 204 (1928).

[27] E. U. Condon and R. W. Gurney, *Nature*, **122**, 439 (1928) and *Phys. Rev.*, **33**, 127 (1929).

[28] Felix Bloch, *Z. Physik*, **52**, 555 (1928).

[29] R. de L. Kronig and W. G. Penney, *Proc. Roy. Soc. (London)*, **A130**, 499 (1931).

[30] J.E. Lennard-Jones, *Proc. Roy. Soc. (London)*, **A106**, 463 (1924).

[31] Milton Abramowitz and Irene A Stegun *Handbook of Mathematical Functions*, Dover, New York (1965).

[32] M. Moshinsky, *The harmonic oscillator in modern physics:from atoms to quarks*, Gordon and Breach (1969).

[33] W. Pauli, *Z. Physik*, **36**, 336 (1926).

[34] B. Lohmann and E. Weigold, *Phys. Lett.*, **86A**, 139 (1981).

[35] Marvin L. Goldberger and Kenneth M. Watson, *Collision Theory*, John Wiley & Sons, New York (1964).

[36] A. S. Davydov, *Quantum Mechanics*, Pergamon Press (1965).

[37] P. A. M. Dirac, *Proc. Roy. Soc. (London)*, **A117**, 610 (1928).

[38] A. R. Edmonds, *Angular Momentum in Quantum Mechanics*, Princeton University Press (1996).

[39] Stephen Walfram, *The Mathematica Book*, Cambridge University Press (1999); www.wolfram.com/products/mathematica/index.html

[40] T.D. Lee and C.N. Yang, *Phys. Rev.*, **104**, 254 (1956).

[41] W. Heisenberg, *Z. Physik*, **77**, 1 (1932).

[42] C. Amsler *et al.*, *Phys. Lett.*, **B 667**, 1 (2008).

[43] John David Jackson, *Classical Electrodynamics*, 3dr Edition, Joh Wiley & Sons, New York (1999).

[44] P. Zeeman, *Phil. Mag.*, **43**, 226 (1897).

[45] B. A. Lippmann and Julian Schwinger, *Phys. Rev.*, **79**, 469 (1950).

[46] M. Born, *Z. Physik*, **37**, 863 (1926); **38**, 803 (1926).

[47] C. W. Wong, *Introduction to Mathematical Physics: Methods & Concepts*, Oxford University Press (1991).

[48] G. Arfken, *Mathematical Methods for Physicists*, Academic Press (1985).

[49] M. L. Boas, *Mathematical Method for Physical Sciences*, Wiley (1983).

[50] I. M. Gel'fand and G. E. Shilov, *Generalized Functions*, Vol. 1, Sec. 2.4, Academic Press (1964).

[51] Yoshio Yamaguchi, *Phys. Rev.*, **95**, 1628 (1954).

[52] R. Machleid, *Phys. Rev. C*, **63**, 024001 (2001).

[53] J. M. Blatt, *J. Comp. Phys.*, **1**, 382 (1967).

Index

Iraj R. Afnan

Made in the USA
Las Vegas, NV
03 November 2026

57423713R00214